ENVIRONMENTAL POLLUTION, BIODIVERSITY, AND SUSTAINABLE DEVELOPMENT

Issues and Remediation

ENVIRONMENTAL POLLUTION, BIODIVERSITY, AND SUSTAINABLE DEVELOPMENT

Issues and Remediation

Edited by

Hasnain Nangyal

Muhammad Saleem Khan

APPLE ACADEMIC PRESS

Apple Academic Press Inc.
4164 Lakeshore Road
Burlington ON L7L 1A4
Canada

Apple Academic Press Inc.
1265 Goldenrod Circle NE
Palm Bay, Florida 32905
USA

Exclusive worldwide distribution by CRC Press, a member of Taylor & Francis Group

No claim to original U.S. Government works

International Standard Book Number-13: 978-1-77188-805-9 (Hardcover)
International Standard Book Number-13: 978-0-42926-501-3 (eBook)

Library and Archives Canada Cataloguing in Publication

Title: Environmental pollution, biodiversity, and sustainable development : issues and remediation / edited by Hasnain Nangyal, Muhammad Saleem Khan.

Names: Nangyal, Hasnain, 1987- editor. | Khan, Muhammad Saleem, 1985- editor.

Description: Includes bibliographical references and index.

Identifiers: Canadiana (print) 20190180463 | Canadiana (ebook) 20190180595 | ISBN 9781771888059 (hardcover) | ISBN 9780429265013 (ebook)

Subjects: LCSH: Biodiversity.

Classification: LCC QH541.15.B56 E58 2020 | DDC 578.7—dc23

Library of Congress Cataloging-in-Publication Data

Names: Nangyal, Hasnain, 1987- editor. | Khan, Muhammad Saleem, 1985-editor.

Title: Environmental pollution, biodiversity, and sustainable development : issues and remediation / edited by Hasnain Nangyal, Muhammad Saleem Khan.

Description: Oakville, ON ; Palm Bay, Florida : Apple Academic Press, [2020] | Includes bibliographical references and index. | Summary: "Environmental Pollution, Biodiversity, and Sustainable Development: Issues and Remediation provides an extensive summary of biodiversity. It is the result of the assistance of environmentalists, researchers, policy experts, and academicians from across the globe sharing their research and knowledge on biodiversity and ways to mitigate the threat from climate change, over-utilization of natural resources, pollution, and more. The volume considers that biodiversity encompasses a wide range of biological processes, ranging from genetic diversity, species, populations, communities and ecosystems, landscapes and regions. This book, written by a panel of international experts in biodiversity, conservation biology, and evolution from different countries, including Iran, Pakistan, India, Bangladesh, Tajikistan, Russia, and others, highlights the human impact on biodiversity hotspots on a global scale. Topics include: The potential of phytoremediation Community-based conservation efforts Environmental pollution and the increasing risk of neurodegenerative diseases Ways to determine the most appropriate diversity indices The volume provides an abundance of valuable research for faculty, students, and researchers in environmental sciences; government agencies, and many others"-- Provided by publisher.

Identifiers: LCCN 2019040168 (print) | LCCN 2019040169 (ebook) | ISBN 9781771888059 (hardcover) | ISBN 9780429265013 (ebook)

Subjects: LCSH: Biodiversity conservation. | Pollution. | Bioremediation.

Classification: LCC QH75 .E69 2020 (print) | LCC QH75 (ebook) | DDC 333.95/16--dc23

LC record available at https://lccn.loc.gov/2019040168

LC ebook record available at https://lccn.loc.gov/2019040169

Apple Academic Press also publishes its books in a variety of electronic formats. Some content that appears in print may not be available in electronic format. For information about Apple Academic Press products, visit our website at **www.appleacademicpress.com** and the CRC Press website at **www.crcpress.com**

Dedicated to

My Father and Advocate Syed Bahader
of whose authentic gray hairs
not a few were caused by myself

And

Professor Arman Loni (1983–2019)

About the Editors

Hasnain Nangyal

Hasnain Nangyal is a research fellow in the Department of Botany, Faculty of Sciences, Hazara University, Mansehra, Khyber Pakhtunkhwa, Pakistan. He has published several review articles, research papers, and books in reputed scientific journals from publishers such as Bentham Scientific Publishers, Netherlands, and others. He is an active member of many national and international research organizations, including the Pakistan Medical Microbiology Association; Pakistan Medicinal Plants Association; North Carolina Herb Association (NCHA), USA; Pakistan Society of Psycophysiology; and Arab Society for Fungal Conservation. He is also an editorial board member of scientific magazine *Microbiology World,* a bimonthly magazine (ISSN 2350-8744) published from Nepal; *Advances in Biomedicine and Pharmacy* (an international journal of molecular medicine and pharmacy) published from UK; *International Journal of Endorsing Health Science Research* (impact factor 3.6), indexed by Eastern Mediterranean Index Medicus, WHO; *Journal of Biosciences and Agricultural Research* (ISSN 2312-7945), published from Bioscience Information Network, Bangladesh, *World Journal of Clinical Pharmacology, Microbiology and Toxicology,* published from Mexico; and *Paradise of Science,* a bimonthly multidisciplinary magazine published from Cairo University, Cairo, Egypt. He is also an invited reviewer of *FUUAST Journal of Biology,* a Joint Secretary of the Pakistan Herbs Society, a sister organization of NCHA (North Carolina Herbs Association) North Carolina, USA; a member of the working group "By Science It Will be a Paradise" (Scientific Group) Cairo, Egypt; and lifetime member of Biocognizance.

He is a member of the publishing team of the book series Biodiversity Distribution & Conservation, published from India, and has published three chapters in an international book entitled *Biodiversity Distribution*

& Conservation. He has attended an international scientific conference of young scientists entitled "The role of multidisciplinary approach in solution of actual problems of fundamental and applied sciences" (Earth, Chemical, Technical) in October 2014, organized by the National Academy of Sciences, Azerbaijan. He was invited as a keynote speaker by the Academy of Sciences, Azerbaijan. He was also invited as a keynote speaker by the National Science Congress held at Bombay University in January 10 2015, and the international symposium entitled "Biotechnology, Recent Advances," organized by the Department of Biotechnology, University of Malakand, Khyber Pakhtunkhwa.

Muhammad Saleem Khan

Muhammad Saleem Khan is a research fellow in the Department of Zoology, Government College University, Faisalabad, Pakistan, and is currently working as Assistant Professor of Zoology (Guest) in the Department of Biological Sciences, University of Okara, Pakistan. Dr. Khan recently completed his PhD in Toxicology (Zoology) from Government College University, Faisalabad, Pakistan. He remained part of the HEC project on sustainable biocontrol of *Helicoverpa armigera* (Hub) through Araneid fauna in the Central Punjab from 2009 to 2011. He also worked as an internship fellow in the Bioassay Lab of the Nuclear Institute of Agriculture and Biology (NIAB) in 2009 and the Bioassay Lab of Ayuab Agricultural Institute Faisalabad in 2010. He has published more than 40 research and review articles in reputed journals. He is also working as reviewer for many high-impact factor journals. Currently, he is reviewing articles for *Ecological Indicators* (impact factor: 3.983) and *Applied Biological Chemistry* (impact factor: 1.362). His current area of reach is biodiversity and pollution, aquaculture and fishery, toxicology (genotoxicity and cytotoxicity), synthesis of nanomaterials, biodiversity and ecology, histopathology, and physiology.

Contents

Contributors

Heba S. Abbas
Microbiology Department, National Organization for Drug Control and Research (NODCAR), Giza, Egypt

A. M. Abdel-Azeem
Department of Sciences, University of Suez Canal, Ismailia 41522, Egypt

Ali Akbar
Department of Microbiology, Faculty of Life Sciences, University of Balochistan, Quetta 87300, Balochistan, Pakistan

Imran Ali
Institute of Biochemistry, Faculty of Life Sciences, University of Balochistan, Quetta, Pakistan

N. Akhtar
Department of Botany, G. C. Women University, Madina Town, Faisalabad 38000, Pakistan

Muhammad Anwar
Institute of Biochemistry, Faculty of Life Sciences, University of Balochistan, Quetta, Pakistan

H. S. El-Tahawe
Plant Protection Research Institute, Dokki (Sharkia branch), Agricultural Research Center, Giza, Egypt

R. M. Fikry
Department of Chemistry, Faculty of Science, Zagazig University, Zagazig, Egypt

Nasyrova Firuza
Institute of Botany, Plant Physiology and Genetics, Tajikistan Academy of Sciences, Dushanbe, Tajikistan

Soma Ghosh
School of Environmental Studies, Jadavpur University, Kolkata, West Bengal, India

H. Ikram
Department of Biochemistry, University of Karachi, Karachi 75270, Pakistan

N. A. Ismail
Department of Chemistry, Faculty of Science, Zagazig University, Zagazig, Egypt

F. A. Khan
Dr. A. Q. Khan Institute of Biotechnology and Genetic Engineering (KIBGE), University of Karachi, Karachi, Pakistan

M. S. Khan
Department of Zoology, Government College University, Faisalabad, Pakistan

S. Kulshreshtha
Department of Zoology, Government Shakambhar P. G. College, Sambhar Lake, Jaipur, Rajasthan, India

Fatima Abdul Majeed
Department of Microbiology, University of Balochistan, Quetta 87300, Balochistan, Pakistan

N. Malik
Department of Botany and Microbiology, S.G.R.R. (P.G.) College,
Constituent of S.G.R.R Education Mission, Dehradun 248001, Uttarakhand, India

Meenakshi Mukherjee
School of Environmental Studies, Jadavpur University, Kolkata, West Bengal, India

K. Melikov
Department of Botany, Hazara University, Mansehra, Khyber Pakhtunkhwa, Pakistan

A. Naimov
Institute of Botany, Plant Physiology and Genetics, Academy of Sciences, Republic of Tajikistan, Tajikistan

H. Nangyal
Department of Botany, Faculty of Sciences, Hazara University, Mansehra, Khyber Pakhtunkhwa, Pakistan

A. Nawaz
Department of Zoology, University of Lahore, Sargodha Campus, Sargodha, Pakistan

J. Parchizadeh
Tehran, Tehran Province, Iran

K. Partoev
Institute of Botany, Plant Physiology and Genetics, Academy of Sciences, Republic of Tajikistan, Tajikistan

R. Qureshi
Department of Botany, Pir Mehr Ali Shah Arid Agriculture University, Rawalpindi, Pakistan

S. A. Raslan
Plant Protection Research Institute, Dokki, (Sharkia Branch), Agricultural Research Center, Giza, Egypt

G. V. Reddy
Department of Forests, Government of Rajasthan, Jaipur, India

F. M. Salem
Department of Science, University of Suez Canal, Ismailia 41522, Egypt

Pinaki Sar
Department of Biotechnology, Indian Institute of Technology Kharagpur, Kharagpur, West Bengal, India

S. Saxena
Department of Botany and Microbiology, S.G.R.R. (P.G.) College,
Constituent of S.G.R.R Education Mission, Dehradun 248001, Uttarakhand, India

M. M. Shahnaz
Government Postgraduate College of Science, Faisalabad, Pakistan

Uzma Sitara
Food Quality Storage Research Institute, Southern-Zone Agricultural Research Centre, Pakistan Agricultural Research Council, Karachi, Pakistan

T. Zohra
Department of Botany, PMAS - Arid Agriculture University, Rawalpindi, Pakistan
Department of Biotechnology, Quaid-i-Azam University, Islamabad, Pakistan

Abbreviations

AD	Alzheimer's disease
ALS	amyotrophic lateral sclerosis
AMDs	acid mine drainages
ANOVA	analysis of variance
ARMs	As-resistant microorganisms
BBB	blood–brain barrier
BMCs	biodiversity management committees
CAOs	chemoautotrophic As(III) oxidizers
CBD	convention on biological diversity
CRD	completely randomized design
DARPs	dissimilatory As(V)-reducing prokaryotes
FeRBs	Fe reducing bacteria
HAOs	heterotrophic As(III) oxidizers
OS	oxidative stress
PAOs	phototrophic As(III) oxidizers
PBRs	people's biodiversity registers
PCB	phycocyanobilin
PCR	polymerase chain reaction
RAPD	randomly amplified polymorphic DNA
RFLP	restriction fragment length polymorphism
ROS	reactive oxygen species
SRBs	sulfate-reducing bacteria
TCE	trichloroethylene
TG	triglyceride
UPGMA	unweighted paired group arithmetic mean analysis

Preface

Biodiversity encompasses a wide range of biological processes ranging from genetic diversity, species, populations, communities and ecosystems, to landscapes and regions, all of which provide innumerable benefits to humankind. Indeed, it is not just an accretion of species, but in fact it is the existence of species *in situ* with all their multifaceted and ecological and evolutionary communications. Regardless of being in the public eye from decades, still the conception as well as quintessence of biodiversity is a poorly implicit but precious natural resource that is vanishing and in many regions is at the stage of extermination.

This current book is one among the many honest efforts to breach this cavernous gap by presenting an extensive summary of biodiversity. This overview is just not a record but also an elucidation of inherent remuneration that needs to be apparent, which is far more imperative than the economic reinforcement that the human mind leans to comprehend.

It is obvious that biodiversity in all of its types is under menace as a consequence of human dealings and other associated anthropogenic hassles. To comprehend better the many and often restrained behaviors in which human associations depend on their biological resources, to isolate why those possessions are endangered, and to execute proficient and steadfast procedures to safeguard these resources for prospect cohorts will necessitate an amalgamation of pioneering economic assessment with a sound ecological foundation.

In this manner, this book is the result of the assistance of a class of environmentalists, researchers, policy experts, and academicians all around the world coming jointly to present the affluence of biodiversity and, in the same stratum, refreshing their apprehensions beneath it for the very information that human variety can assemble their own objectives of well-being. In addition, the book highlights the need for mutual sharing and collaboration among world societies, which would portend well for explicit preservation of all species, whether they are ecologically comparable or nonequivalent. The trouncing of biodiversity somewhere is a menace to affluence ubiquitously. We earnestly anticipate that readers would be inspired by the exhilarating assistance in this volume that encourages initiating that course. On the other hand, we are only just providing instigation, and many important contests lie ahead. In this association, it will be valuable for the reader to be introduced to the central message of each chapter in this book.

Chapter 1 gives a general overview of ecologically different areas of Punjab that are metal- and salt-polluted and are under study and the different ecotypes of *Typha domingensis* Pers. that have been collected from such areas. The ecotypes from these metal-polluted and saline areas were exposed to metal (nickel) stress to assess their degree of tolerance and adaptive response under measured environmental conditions. Four levels of nickel were maintained (0, 50, 100, and 150 mg kg^{-1}) in the growth medium. Different data were recorded for morpho-anatomical, physiological, and biochemical characteristics. The treatments were placed in a completely randomized design (CRD) with six replications.

Chapter 2 discusses the genetic resources of Tajikistan as an inexhaustible source in biodiversity. We all know that the human population is increasing day by day. According to research there is a need to increase food potential by 2% annually. To protect the world from the shortage of food, it is necessary to protect world genetic resources. For this we have to protect wild species. Wild species are genetically stable. We should put emphasis on local varieties.

Chapter 3 provides detailed information on the existing flora, threats, and conservation issues of the Nara Desert, Pakistan. It highlights the importance of the desert ecosystem that supports human and livestock populations for fulfilling their basic needs, such as fuel, food, fodder, sheltering, ethnomedicine, etc. Furthermore, certain anthropogenic activities leading to the desertification are discussed along with some possible remedial measures.

Chapter 4 investigates different trials of chemicals on larvae of spiny bollworm, corresponding to the acute toxicity of different chemicals that showed high significant decrease in larval weight; the sex ratio of emerged adult spiny bollworm was insignificantly affected. All treatments decreased the number of deposited eggs, which proved significant with A, B, and C chemicals. Incubation periods were unaffected. Significantly B, A, and C compounds caused reduction in the hatchability of deposited eggs by spiny bollworm females that survived from treated newly hatched larvae.

Chapter 5 determines the most appropriate diversity indices in a study area that is one of the necessary things to preserve animal species.

Therefore, the simplest way to do it seems to be quite important in ecological communities. Here, with the help of a case study in Iran, I explained the simplest and easiest way to determine the most sensitive biological indices in a study area. My survey showed that Menhinick's species richness, Shannon–Weaver diversity, and Simpson's measure of evenness were the most appropriate indices to calculate the richness, the diversity, and the evenness in the study area.

Chapter 6 gives a detailed overview on the study of pollen grain fertility and hybridization of potato varieties (*Solanum tuberosum* L.) that have been carried out for the first time in the mountainous area of Tajikistan. In this study, the degree of pollen grain fertility and great variability has been identified. It has been stated that some potato clones have great variability according to flower and anther color and flower stamen filament shape. The clones of potato can be a good source for selection and genetic work aimed at the raising of new potato hybrids and grades. The result of the study reveals that the cardinal grade has the least percentage of fertile pollen grains among cultivated potatoes, whereas dusti grade has the greatest percentage of fertile pollen grains.

Chapter 7 gives a detailed overview of some threatened and endangered plants, which are recognized and classified as RET medicinal plants depending on their "threat status" RET medicinal plants are Rare, Endangered, and Threatened forest species growing and multiplying naturally in restricted areas. These plant species play a predominant role as folk remedies against many diseases. In the present study, 33 RET plant species are explored, identified, and documented from Talakona, Seshachalam Hill Ranges of Eastern Ghats, Andhra Pradesh. The specific objective of the study is to generate the database of RET plants of the Talakona region.

Chapter 8 highlights the importance of *Hippophae* L. (sea buckthorn), a multipurpose wonder plant in biodiversity conservation. This berry crop has a wide ecological adaptation, fast growth, strong coppicing, and suckering habit, which makes the plant suitable for planting in degraded soils. The authors document the importance of sea buckthorn as biofertilizer, protection against wind and for preventing sand from drifting. This important bioresource can also control soil erosion and water loss and effectively serves as a soil-binder species in the cold desert areas of the Indian Himalayan Region. It is also utilized in farmstead protection, habitat restoration of wildlife, riparian protection, soil improvement, and marginal land reclamation via its ability for nitrogen fixation. It also promotes growth of other species as well. The authors stress the urgent need to launch a worldwide awareness campaign about sustainable utilization of this novel plant for the purpose of biodiversity conservation.

Chapter 9 highlights several different factors that contribute to the predisposition and precipitation of neurodegenerative disorders. This may also involve interaction between genetic predisposition and environmental factors. As a result of industrialization and bad waste management strategies, people are becoming more exposed to the toxic effects of environmental pollutants. Globally, neurodegenerative disorders are becoming a major threat to mental health with more cases reported each year. Environmental pollutants (serving as environmental cues) affect mental and physical wellness in the elderly as epigenetics suggests that environmental factors affect genetic imprinting. Pre/postnatal exposure to these environmental cues predisposes the individual to neurodegenerative disorders due to defective growth and development of brain.

Chapter 10 highlights saline lakes of arid zones that are of considerable economic, ecological, and scientific values besides being a fragile ecosystem. Sambhar Lake, Rajasthan, a "Wetland of International Importance" and a "Ramsar Site" are struggling with an array of conservation threats like siltation and disturbances in the lake bed, excessive resource-utilization, and unsustainable development, causing a massive shrinkage of the watersheet.

Chapter 11 focuses on analyzing the risk and benefits of genetically modified (GM) crops to the environment and biodiversity. Additionally, socioeconomic impacts and benefits of GM crops are also reviewed comprehensively. The main advantage of GM crops is their potential promise for future food security for small agriculture in developing countries. However, the critics of genetic engineering of food have concerns, not only for safety, toxicity, allergincity, carcinogenicity, and altered nutritional quality of food, but also for human health and environment. "Transgenic Revolution" might not appear to have been particularly special, but until the fears of the public are assuaged, it will continue to be ethically contentious.

Chapter 12 presents the real situation of fungal (Orphans of Rio) biodiversity and conservation in Northern Africa countries. It sheds light on what Abdel-Azeem and coworkers have done in the last decade in Egypt (as a case study) toward the fungal conservation. The authors stress how they are obtaining the attention of Egyptian politicians and decision makers to conserve fungi and promote best practice and to solve such problems through the Arab Society for Fungal Conservation, Egypt's Mycologists Network, Egypt's National Fungus Day, collaboration between mycologists, amateur fungal groups, fungal conservation societies, regional natural parks, and environmental agencies.

—**Hasnain Nangyal**
Department of Botany, Faculty of Sciences
Hazara University, Khyber Pakhtunkhwa
(Independent Researcher)

Introduction

Biodiversity refers to the variety of life on the earth, from the genetic level to terrestrial, freshwater, and marine habitats along with their ecosystems. It strengthens our global economy as well as human well-being. Biodiversity offers essential benefits to people and contributes to society as a whole by providing knowledge, protection, medicine, and community identity. Ecosystems in their turn provide a range of vital services, including regulation of nutrient and carbon cycles. It supplies several indispensable services to society and strengthening global economy. The variety of services consists of goods (i.e., food, fiber, timber, and medicines), ecosystem functioning (flood control, climate regulation, nutrient cycling) and amusement that strengthen agriculture by employing them for the development of new varieties, pollination and pest control, carbon storage and sequestration, and improvement of human health.

The genetic diversity is referred to as variability amongst the functional units of inheritance in plants, animals, and microbes. Diversity in species portrays the variety of species in all taxonomic units of geographically marked areas. It has been estimated that approximately 2–100 million species exist across the world, and of them less than 1.5 million have been explained. The ecosystem diversity encompasses the association and interaction between plants, animals, micro-organisms, and ecological processes that function together. Thus, biodiversity is the structural unit that acclimatizes in changing environmental conditions. The variability in native germ plasm facilitates scientists to develop better varieties of animals and plants for human requirements. The genetic diversity is the foundation for the development of new varieties, and its loss may imperil crop productivity. These issues are very important from various perspectives, such as historical, economic, social, ethical, esthetic, and ecological points of view.

The species richness signifies the biodiversity of any area that shows the overall complexity of an ecosystem. Every species exhibits genetic diversity between individuals. This disparity conveys natural selection and adaptability in response to climate that eventually guarantees the survival of species. Biodiversity supplies free of cost various services such as clean water, pure air, pollination, soil development and conservation, pest control, food, fuel, fiber, and medicine, all of which are vital to the well-being of human society. Many of the wild plants are important components of medicinal resources either used in traditional healthcare systems of medicines or in synthetic drugs. In addition these are indemnities for food security issues. The sustainability of ecosystems depends on rich and healthy genetic diversity, species, and habitats, and loss of biodiversity eventually affects these services.

Due to the indispensable role of biodiversity in human development, there is escalating concern about the species loss and the declining of genetic diversity of cultivated crops and wild species. This apprehension takes place because the natural lands are very rapidly being deteriorated and transformed by human interference in the world, especially in the tropics. Scientists also believe that the loss of species diversity may result in instability and functioning of the ecosystem. Therefore, conserving biodiversity is of utmost importance for the survival of human beings on the earth. For instance, the air for breathing, the water for drinking, and the soil for crop production resulted due to multifaceted interactions that occur between living beings. Loss of any keystone species would harm the services upheld by the species and degrade ecosystem.

Mankind has altered the ecosystems and affected the composition of biodiversity by undertaking a range of activities that augmented species loss and invasions at local and global levels. This kind of modification in biodiversity composition amends ecosystem chattels and services offered to mankind. The scientific studies have illustrated that properties of an ecosystem largely depend on biodiversity at the extant of efficient distinctiveness of prevailing organisms in the ecosystem and their abundance and distribution in space and time.

Conservation of biodiversity is crucial to the sustainability of sectors as diverse as energy, agriculture, forestry, fisheries, wildlife, industry, health, tourism, commerce, irrigation, and power. Pakistan's development in the future will continue to depend on the foundation provided by living resources, and conserving biodiversity will ensure this foundation is strong.

Biodiversity is a very broad topic, and despite the ample literature available in the market, there is a need to enhance the understanding of this notion. This present endeavor is the heartfelt effort that portrays various researches on biodiversity. It is a matter of great concern that biodiversity is under threat due to the anthropogenic activities by humankind. For better understanding the way of utilizing biological resources by human societies, why those are threatened, and to implement the conservation methods for future generations will require a range of novel cost-effective protocols. Keeping in view, this book is a dedicated work contributed by renowned environmentalists, researchers, and academicians around the world, highlighting the biodiversity, ecosystem services and their utilization, threats, and conservation issues coupled with their possible remedial measures. Additionally, the book highlights the need for the participatory approach and collaboration in this interdisciplinary science. We sincerely anticipate that readers will be motivated by the contributions of this volume. In addition this book will be valuable for the readers to comprehend the central message of each given chapter.

Foreword

I am honored to have been asked to write the foreword for this edition. In this edition, we aimed at finding productive knowledge to provide integrated information on different aspects of application in one text. This ambitious goal was set with the purpose to provide readers with the most comprehensive and current information in this field. Reflecting the global relevance of the topic, the authors come from a number of countries. It is hoped that this volume will capture the current state of this vital and dynamic science from an international perspective.

The realization of the message delivered in the book would possibly make certain students of all disciplines of plant sciences fulfill their needs according to need of the day with reference to previous literature. This book is designed for those students and researchers who wish to become familiar with basic and advanced features of biodiversity, while the output of these efforts can be implemented to ensure sustainability and preservation of biodiversity.

This book, edited by Hasnain Nangyal and Muhammad Saleem Khan, provides a broad overview of different promising aspects of biodiversity including its different diverse aspects. Various chapters written by international experts in their respective specialties not only allow the reader to understand better the wide range of medical problems linked with environmental pollution, biodiversity, and sustainable development, but also propose strategies to reduce the threat to biodiversity and sustainable development. The book should be useful to all those working on biodiversity, pollution eradication, and sustainable development

—Prof. Dr. Andjelka Mihajlov
Professor at University Educons and University of Novi Sad
Ambassador for Sustainable Development, Serbia
Chair Environmental Sciences Rensselaer
Polytechnic Institute University of Belgrade

CHAPTER 1

Phytoremediation Potential of *Typha*: An Approach to Biodiversity Conservation

H. NANGYAL[1*], N. AKHTAR[2], M. M. SHAHNAZ[3], A. NAWAZ[4], and M. S. KHAN[5]

[1]*Department of Botany, Hazara University, Mansehra, Khyber Pakhtunkhwa, Pakistan*

[2]*Department of Botany, Government College Women University, Faisalabad 38000, Pakistan*

[3]*Department of Botany, Government Postgraduate College of Science, Faisalabad, Pakistan*

[4]*Department of Zoology, University of Lahore, Sargodha Campus, Sargodha, Pakistan*

[5]*Department of Zoology, Government College University, Faisalabad, Pakistan*

[*]*Corresponding author. E-mail: hasnain308@gmail.com*

ABSTRACT

Natural biodiversity of plants faces different abiotic problems, most noticeable are drought, salinity, and metals toxicity. Physiological and biological functions of various plant biodiversity are harshly affected by different metal (Ni, Cd, Pb) stress. Different areas of Punjab are metal-polluted due to various activities, like unsatisfactory irrigation system, industrial pollution, etc. It is hypothesized that plants grown in such habitats are able to tolerate metal contamination. Ecologically different areas of Punjab that are metal and salt polluted are under study and different ecotypes of *Typhadomingensis* Pers. have been collected from such areas. The ecotypes from these metals polluted and saline areas were exposed to metal (nickel) stress to assess their degree of tolerance and adaptive response under measured environmental conditions. Four levels of nickel were maintained (0, 50, 100, and 150 mg kg^{-1}) in the growth medium. Different data were recorded for morpho-anatomical, physiological, and biochemical characteristics. The treatments were placed in a completely randomized design (CRD) with six replications.

1.1 INTRODUCTION

The rapid increase of industrialization results in metal-polluted environment. Metal pollution is most lethal contaminants that are severely hazardous to the biodiversity of aquatic habitats and drinking filthy water results in many health difficulties in humans and aquatic animals. The economic features and negative effects of conservative management skills in aquatic ecosystems covered the technique of phytoremediation. In phytoremediation, different plant species are used to amend the environment from numerous dangerous contaminants. It is a low-cost technique for environmental cleaning. The importance of biodiversity is gradually more painstaking for the cleaning of the metal-polluted ecosystems. Discharge of industrial and domestic wastewater, increased use of manures and conveyance activities are key aspects that pollute drinking water and disturb the excellence of ground and aquatic habitats (Ezzat et al., 2002).

Among metals, nickel is the 22nd most ample element on the biosphere (Sunderman and Oskarsson, 1991). Nickel is naturally present everywhere, and its level in the soil is less than 100 ppm (McIlveen and Negusanti, 1994). It occurs in air, water, and soil, which is carried away into the atmosphere from various sources (Barrie, 1981) and is dumped into cultivated lands. The chief cause of nickel in the air is the burning of fuels to produce energy. Additional sources include steel, cement manufacturing, and nickel withdrawal.

Many plants need a trace quantity of nickel for physiological processes (Welch, 1981) and encourage plant growth, but above 5 μg g^{-1} Ni is toxic to plants (Allen, 1989). A higher concentration of nickel affects plant growth and decrease photosynthesis by decreasing Hill activity and disturbing photosystem-II in plants. Nickel causes toxicity in two means: (1) meddling with other metallic ions and (2) introduction of oxidative stress. Anatomical structure modifications are also an adaptive strategy for different plants species (Grigore and Toma, 2007). The root system of plant negatively altered by many environmental (Malamy, 2005) and modifications in root and shoot anatomical features during growth have been observed at early growth phases (Maggio et al., 2007).

Water ecosystems are polluted mainly by contaminated metals, which are unsafe to lives in water habitats. These water ecosystems are metal reservoirs and are exposed to heavy metals (Kanoun-Boule et al., 2009) are vulnerable to waste product due to the occurrence of discreetly minor biomass in a variety of trophic levels. Consequently, water plant species are often directly affected by the metals contaminations in the aquatic habitats.

Many researchers have described the proficiency of wetland plants to absorb metals from aquatic ecosystems (Tremp and Kohler, 1995) and thus can be utilized in the removal of pollutants from the contaminated water. Exclusion of metal toxins from the water habitats brings about by many plants that have the capacity to gather metals in many organs. This ability of some plants to sequester metals has given rise to new techniques of remediation called phytoremediation (Black, 1995). Many plant species store metals, which are essential for their development, but the unnecessary quantity of these toxins become lethal to the plants. Metal in large amount (including nickel) poisons aquatic habitats but aquatic flora plays a most important role in remediating polluted water of industry and daily use (Wang et al., 1997).

Water plants have a pronounced potential to store metal pollutants, plants developing in contaminated water provide a model for studying metal mechanisms of accumulation and tolerance. *Typhadomingensis* is believed to remove heavy metals from water as an efficient metal accumulator so it is considered as bio absorbent (Abdel-Ghani et al., 2009). It, therefore, has great prospective as phytoremediation (Hegazy et al., 2011).

1.2 MATERIALS AND METHODS

Six ecotypes of *Typhadomingensis* Pers. under study were taken from six different sites of the Punjab, Pakistan, which were hypothesized to adopt various abiotic stresses. Normal biodiversity of *Typha* was existing in freshwater spring of Knotti Garden (Salt Range) and canals of Gatwala (Faisalabad). Salinity adapted ecotypes were collected from Jahlar and Sahianwala areas. Saline water with a low level of salinity was collected from Treemu region, Ecotype from Sheikhupura was collected with industrial wastewater that contain different metals (Ni, Pb, and Cd) was severely contaminated the water.

PLATE 1 Ecotypes of *Typhadomingensis* from river, Canal banks, and industrial-polluted waters.

Hyper-saline Jahlar Lake, the Salt Range, Chakwal.

Hyper-saline wetland Knotti Garden, the Salt Range, Chakwal

Saline-waterlogged wetland Sahianwala, Faisalabad

PLATE 2 Ecotypes of *Typhadomingensis*from salt-affected wetland.

Six *T. domingensis* ecotypes were established in the Botanic Garden Area, University of Agriculture, Faisalabad for 6 months. Equal size vegetative buds were removed from each plant and transferred in plastic tubs (capacity 20 L). Plastic tubs contained a mixture of sand and clayey-loam soils in 1:1 ratio and 10 L water. The present investigation was accompanied during the month May–July 2011; the average temperatures of the experiment were 38–41°C and 24–26°C, respectively, and 13–16 h was photoperiod length.

In this study, four levels of nickel, namely 0, 50, 100, and 150 mg kg^{-1}, were applied to the *Typha* plant and different morpho-anatomical, biochemical, and physiological parameters were determined. The experiment with six replications was planned in 2-factorial CRD. The final data was analyzed to the analysis of variance (ANOVA) and means were compared by least significant range (LSD) test.

1.3 RESULTS

1.3.1 MORPHOLOGICAL PARAMETER

Ni at a moderate level (50 mg L^{-1}) in external growth medium showed nonsignificant result related to fresh weight of shoot, but as the level of nickel was increased it brings about a progressive and significant reduction in this variable in five ecotypes of *Typha* with the exception of Sheikhupura. The ecotype from Sheikhupura revealed no variation in shoot fresh weight by the introduction of nickel in the external medium. In comparison to other ecotypes, the ecotype from Knotti was fairly least affected due to nickel stress. The significant and gradual reduction was noticed in root fresh weight in five ecotypes of *Typhadomingensis* under nickel stress with the exception of Sheikhupura ecotype. The ecotype from Sheikhupura showed stable response under nickel stress. Treemu ecotype was relatively least affected by nickel stress. Similar observations were recorded for Knotti ecotype.

Similarly, dry weight of shoot variation takes place in accordance with fresh weight of shoot. Nickel at 50 mg kg^{-1} level had not altered the value of this parameter. Except for Sheikhupura ecotype, all ecotypes showed a gradual and significant decrease in response to an increased level of the medium. The Sheikhupura ecotype revealed stability at all levels of nickel stress. Nickel at moderate level showed stability to dry weight of root. Except for Sheikhupura ecotype, the other five ecotypes revealed a progressive and gradual decrease in this variable under increasing levels of nickel stress. Ecotype from Sheikhupura remained stable and did not show any variation under nickel stress.

Under nickel stress, plant's height was probably the most affected variable. *Typha* ecotypes, except ecotypes from Sheikhupura and Knotti, no remarkable enhancement or reduction at a moderate level of nickel stress (100 mg kg^{-1} Ni), but thereafter, a consistent decrease in plant height was recorded at 100 mg kg^{-1} Ni and 150 mg kg^{-1}of nickel stress. The Sheikhupura and Knotti ecotypes showed different pattern of variation, in these ecotypes, there was no any significant increase or decrease in this variable in response to nickel toxicity (Fig. 1.1 and Table 1.1).

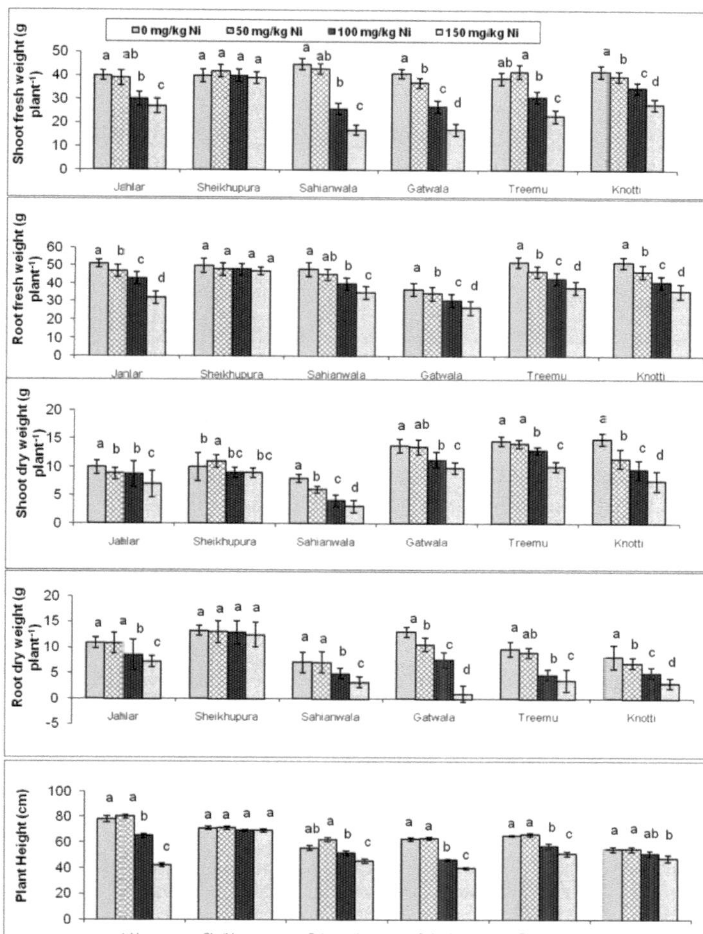

FIGURE 1.1 Morphological parameters of six ecotypes of *Typhadomungensis* Pers. under different levels of nickel stress.

TABLE 1.1 Analysis of Variance (ANOVA) for Morphological Characteristics of *Typhadomingensis* Pers. Ecotypes Under Different Levels of Ni.

S.O.V	df	Shoot fresh wt.	Root fresh wt.	Shoot dry wt.	Root dry wt.	Plant height
Ecotype (E)	5	1626.78***	1591.51***	650.87***	217.23***	7.82***
Treatment (T)	3	764.37***	1331.81***	286.15***	342.17***	30.25***
E x T	15	773.78***	402.87***	258.03***	75.12***	10.01***
Error	48	0.59	0.42	0.12	0.17	37.16

*, **, *** = significance levels at 0.05, 0.01, and 0.001, respectively, ns = nonsignificant.

1.3.2 PHYSIOLOGICAL PARAMETER

A progressive and gradual reduction in Chl "*a*" value was observed in Gatwala, Treemu, and Knott ecotypes of *Typha* as nickel concentrations was increased. The Jahlar and Sahianwala region plants revealed a severe decline at 150 mg kg^{-1} level of nickel. The Sheikhupuraeco type did not reveal any significant variation and showed stability under different levels of nickel stress.

The Jahlar, Sheikhupura, Sahianwala, and Knotti areas plants were least affected for Chl "*b*" value at moderate nickel levels and then decreased consistently and gradually under the higher concentration of nickel toxicity. A clear and significant reduction was observed in Gatwala and Treemu site inhabiting plants in response to nickel contamination.

Under nickel stress, total chlorophyll value in Sahianwala and Knotti remained stable by the induction of nickel toxicity. A sharp reduction in this variable was noted in ecotypes from Jahlar and Sheikhupuraat 150 mg kg^{-1} of nickel. A prominent, significant, and gradual decrease was recorded in Gatwala and Treemu region plants under different levels of nickel.

Under increased levels of nickel, the carotenoids value was progressively and significantly enhanced in five ecotypes of *Typha* with the exception of Jahlar. At 50 mg kg^{-1} level of nickel no significant variation was recorded but as the level of nickel was increase the value related to this variable was reduced significantly (Fig. 1.2 and Table 1.2).

1.3.3 IONIC CONTENTS

Sodium (Na^+) content of root revealed no distinct pattern of variation under nickel contamination of different levels. Conversely, a slight reduction was recorded in *Typha* plants of Treemu and Knotti habitats. However, in Gatwala ecotype, this variable enhanced with increase in nickel levels.

Potassium (K^+) content of root was increased consistently up to 100 $mgkg^{-1}$ level of nickel, but afterward, at higher nickel concentration this variable reduced severely in Jahlar, Sheikhupura, and Knotti region plants. The ecotype from Sheikhupura showed relatively less decrease in comparison with other ecotypes under different levels of nickel. A gradual and progressive decrease was observed in three ecotypes that were from Sahianwala, Gatwala, and Treemu areas under metal toxicity.

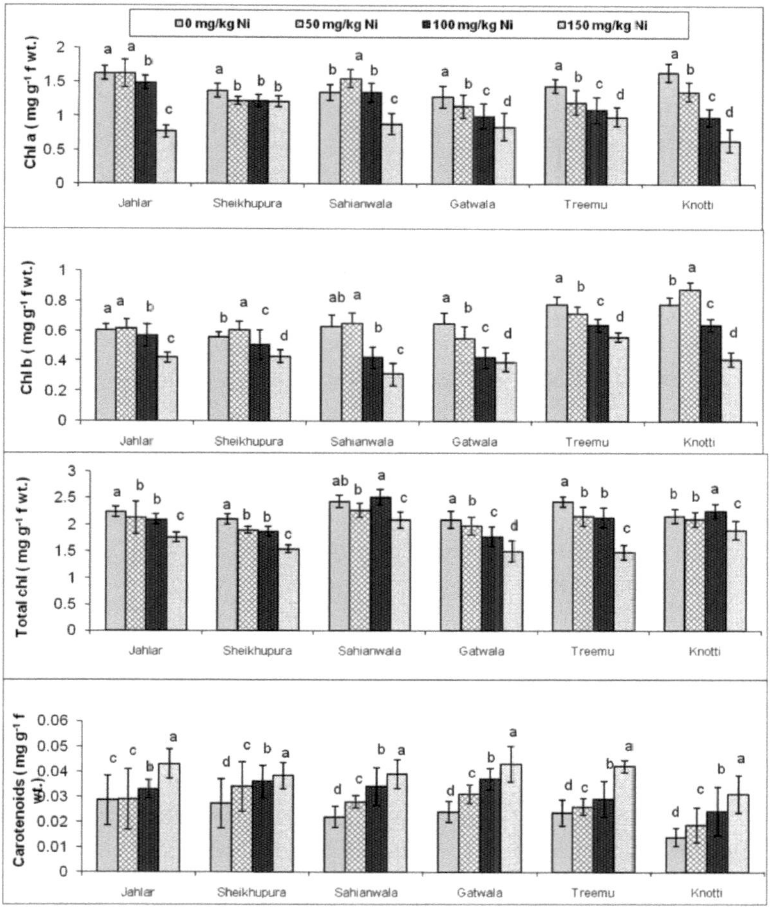

FIGURE 1.2 Photosynthetic pigments of six ecotypes of *Typhadomingensis* Pers. under different levels of nickel stress.

TABLE 1.2 Analysis of Variance (ANOVA) for Photosynthetic Pigments of *Typhadomingensis* Pers. Ecotypes Under Different Levels of Ni.

S.O.V	d.f	Chl. A	Chl. b	Total Chl.	Carotenoids
Ecotype (E)	5	976.32^{***}	587.03^{***}	3459.2^{***}	28.87^{***}
Treatment (T)	3	742.88^{***}	730.97^{***}	3219.16^{***}	74.61^{***}
E x T	15	521.86^{***}	270.63^{z}	1763.76^{***}	50.75^{***}
Error	48	0.0009	0.003	0.005	0.0009

$^{*, **, ***}$ = significance levels at 0.05, 0.01, and 0.001, respectively, ns = nonsignificant.

Freshwater Gatwala and Knotti ecotypes and moderately saline Treemu ecotype showed a consistent and significant decrease in nitrogen (N) content of root by the induction of nickel of different levels in the external medium. The metal-polluted Sheikhupura and Saline area Sahianwala ecotype revealed no remarkable change in value related to root N under nickel contamination. The Jahlat ecotype that was from the hypersaline area showed an irregular pattern of variation under moderate and high levels of nickel treatment.

The Cl^- value of root was least affected parameter under high and low levels of nickel stress in all five ecotypes of *Typha* with the exception of Sahianwala site plants. A significant increasing trend was recorded in Sahianwala ecotype by the introduction of nickel of culture medium.

The metal-contaminated Sheikhupura ecotype showed large accumulation of nickel in root as compared with other five ecotypes of *T. domingensis* Pers. All the ecotypes showed a similar pattern of variation under nickel stress. Six ecotypes revealed consistent enhancement at all levels of nickel in growth culture. (Fig. 1.3 and Table 1.3).

1.3.4 ROOT ANATOMY

The root cross-sectional area was increased significantly at a moderate level of nickel (50 mg kg^{-1} and 100 mg kg^{-1} Ni), but afterward, at a higher level of 150 mg kg^{-1} of Ni this variable decrease noticeably incomparable with control. In contrast, ecotype from Sheikhupura showing a distinct increase in root cross-sectional area when compared with control. The ecotype from Sahianwala remained stable and did not give any variation in response to the introduction of nickel in the culture medium. A severe reduction was noticed in Treemu ecotype by the induction of different levels of nickel in the medium.

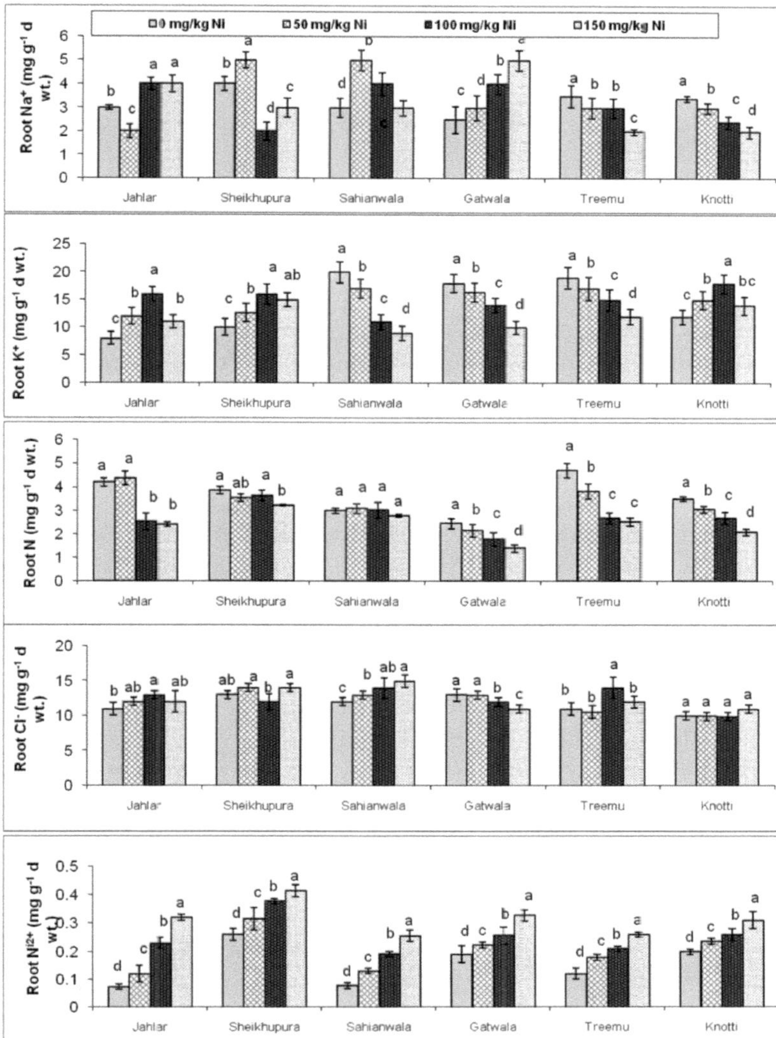

FIGURE 1.3 Root ionic contents of six ecotypes of *Typhadomingensis* Pers. under different levels of nickel stress.

TABLE 1.3 Analysis of variance (ANOVA) for root ionic contents of *Typhadomingensis* Pers. ecotypes under different levels of Ni.

S.O.V	df	Root Na^+	Root K^+	Root N	Root Cl^-	Root Ni
Ecotype (E)	5	0.26 ns	2.13 ns	0.63 ns	12.98***	115.82***
Treatment (T)	3	0.048 ns	3.92*	0.45 ns	14.78***	211.56***
E x T	15	0.74 ns	3.27***	0.006 ns	3.05***	4.58***
Error	48	8.15	10.6	3.99	14.37	4.2

*, **, *** = significance levels at 0.05, 0.01, and 0.001, respectively, ns = nonsignificant.

With the exception of Sheikhupura, all other five ecotypes revealed a distinct reduction in root epidermal cell area in reaction to different levels of nickel contamination. The Sheikhupura ecotype showed a reduction in this variable at 50 and 100 mg kg^{-1} level of Ni, but at 150 mg kg^{-1} Ni it remains stable and gives no differences.

An enhancement behavior in root epidermal thickness was observed in Sheikhupura, Sahianwala, Treemu, and Knotti ecotypes of *Typha*at 50 mg kg^{-1} Ni, but afterward, a significant decrease was observed at 100 and 150 mg kg^{-1} Ni.

Root epidermal thickness showed an increasing behavior up to 50 mg kg^{-1} Ni, thereafter this variable reduced as the level of nickel was increase. These findings were recorded in ecotypes that were from Sheikhupura, Sahianwala, Treemu, and Knotti. The enormous difference in this variable was not detected in Jahlar and Gatwala region ecotypes that showed stability in regard to this parameter under nickel toxicity.

Root aerenchyma area exhibited a minor but nonsignificant proliferation in different *Typha* ecotypes accompanied by higher nickel levels. Ecotype of Gatwala site presented a large aerenchyma area in their root as compared with all other ecotypes.

Endodermal cell area of root revealed an increase up to 50 mg kg^{-1} Ni but after this, a progressive but significant lessening was documented at succeeding higher nickel levels of the medium in all ecotypes of *Typha*. A harsh decrease was seen in ecotype from Knotti at 150 mg kg^{-1} Ni (Fig. 1.4 and Plate 3).

The four ecotypes of *Typha* that were from Jahlar, Sheikhupura, Gatwala, and Treemu, the vascular bundle area increased progressively at a moderate level (50 mg kg^{-1} Ni) of nickel treatment, but further increase in nickel level ultimately decrease the value of this variable. The two other ecotypes that were from Sahianwala and Knotti revealed consistent increase by the stimulation of metal in the medium. A severe decline Meta-xylem cell area of the root was noted in Gatwala, Jahlar, Treemu, and Sheikhupura inhabiting plants with the increase of nickel level. *Typha* ecotypes of Knotti and Sahianwala areas were not considerably affected by different levels of nickel in the culture medium.

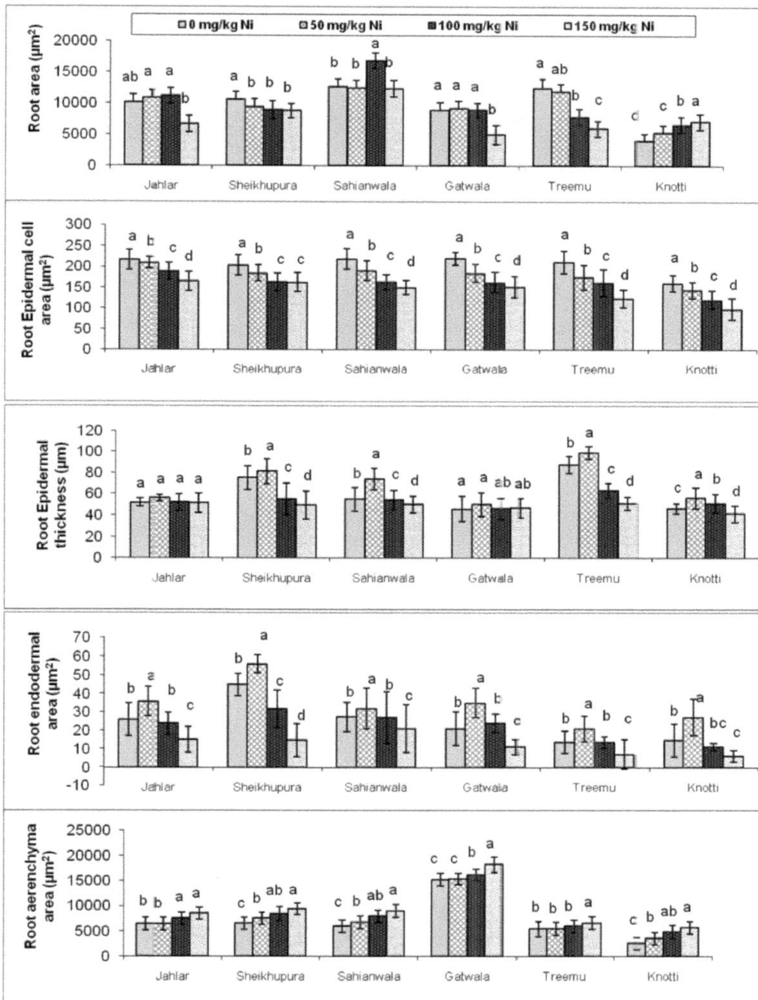

FIGURE 1.4 Root anatomical characteristics of six ecotypes of *Typhadomingensis* Pers. under different levels of nickel stress.

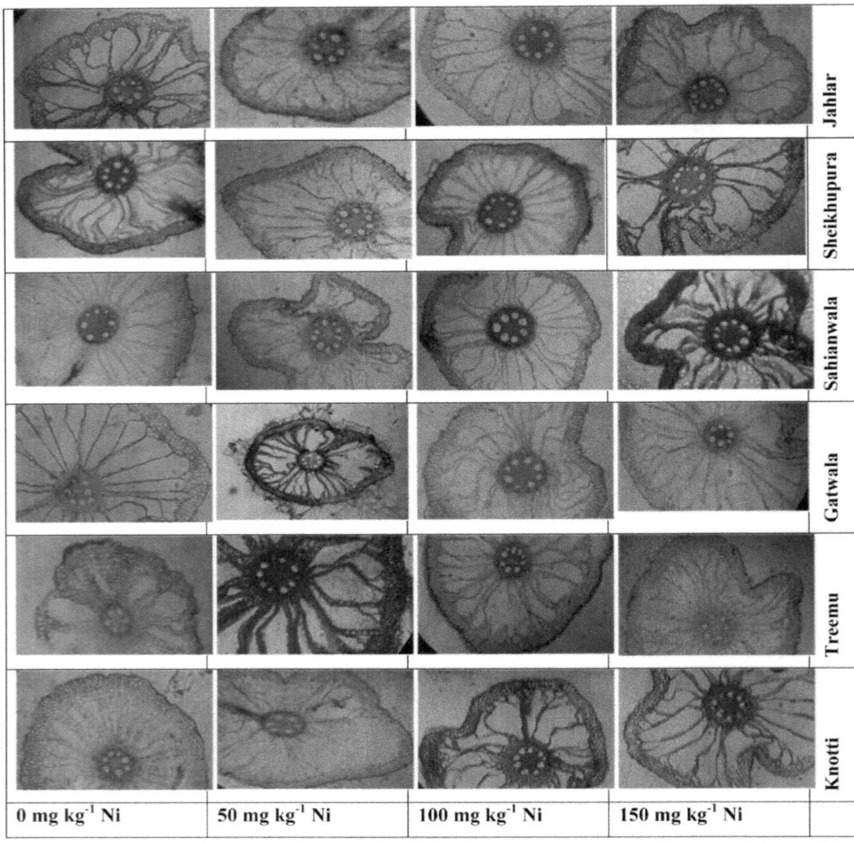

PLATE 3 Root transverse sections of six ecotypes of *Typhadomingensis* Pers. under different levels of nickel stress.

Cortical cell area of the root was consistently reduced in Treemu, Gatwala, and Sahianwala eco types of *T. domingensis* Pers. under different treatment levels of nickel. The ecotype from Sheikhupura was least affected at a low level of nickel but, subsequently, at higher nickel concentration considerably lessened this variable. Jahlar and Knotti ecotypes were not as much affected under nickel stress in comparison with other ecotypes. Six ecotypes exhibited alike form of differences in pith area of root under metal toxicity. Six ecotypes were designated a consistent enhancement in pith area of root up to 100 mg kg^{-1} Ni and at higher level the response in inverse (it reduced sharply at 150 mg kg^{-1}Ni). In metal contamination, phloem area of root presented an improvement up to 50 mg kg^{-1} Ni but later increase in nickel treatment levels decreased this variable in ecotypes of Jahlar, Knotti, and Sheikhupura habitats. Treemu and Gatwala revealed a significant decline at higher nickel levels of the medium. Ecotype from Sahianwala exhibited an increasing tendency up to 100 mg kg^{-1} Ni but at 150 mg kg^{-1} Ni reduced this variable (Fig. 1.5 and Plate 3).

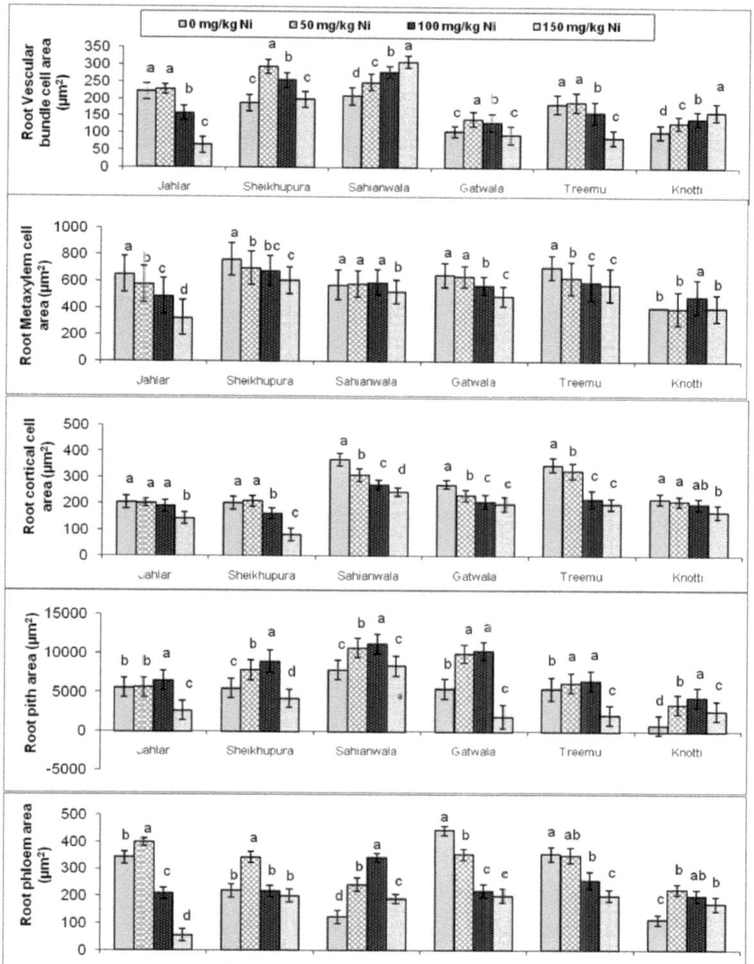

FIGURE 1.5 Root anatomical characteristics of six ecotypes of *Typhadomingensis* Pers. under different levels of nickel stress.

1.4 DISCUSSION

By considering the value of root and shoot fresh weights, the ecotype from Sheikhupura can be regarded as the best performer under nickel stress. The capability of differentially adapted six ecotypes of *Typha* for nickel stress tolerance from best tolerant to most effective are as follows, Shei khupura>Jahlar>Knotti>Treemu>Gatwala>Sahianwala. The production of dry biomass and fresh weight of shoot revealed no alteration under different concentrations of nickel contamination in the metal-polluted area of Sheikhupura ecotype, which markedly states its better performance under the metal-polluted environmental conditions. The nickel at a low or moderate level has no impact on fresh weight of shoot in nearly all ecotypes of *Typhadomingensis*. This finding indicates the application of nickel in various functioning of the plants at its low concentration (Hagemeyer, 1999). Nevertheless, nickel at higher concentration distinctly declines the height of the plant and other growth parameters in virtually all ecotypes with the exception of Sheikhupurain habiting plants.

A prominent variance of nickel toxicity on photosynthetic pigments was noticed in five ecotypes of *Typha* excluding the ecotype from Sheikhupura, which showed the clear reduction in the chlorophyll *a* and *b* contents. A comprehensive reduction in photosynthetic pigments has been documented by many researchers in the past (Bajguz, 2011; Perveen et al., 2012), whereas the enhancement in carotenoids value is an unusual phenomenon. Approaching many scientists through analysis proved the decline in carotenoids value in response to nickel toxicity. Conversely, Campenni et al. (2013) proved an amplified value in carotenoids.

Nickel at higher concentration reported a severe negative impact on the absorbance, accumulation, and transmission of ions, like K^+ and nitrogen. The root ionic content reduced as the concentration of nickel of culture medium was improved. Same observations were documented by a number of authors (Cuypers et al., 2011; Wang et al., 2013). Nitrogen contents though increase at a low level of nickel in most of the experiments but its higher concentration imparted a very negative impact on plants. Cl^- content of root, however, indicate no clear pattern of variation and showed irregular differences under nickel stress, where enhancement was noted in some ecotypes and decline in others ecotypes of *Typha* regardless of nickel level in the culture medium.

The thickness of root may contribute to the survival of plant species under salinity environment. But, in case of survival of plants under metal toxicity, the role of root thickness was not clearly documented. Almost all six ecotypes of *Typha* showed different behavior of variation under metal contamination. The ecotype from Sheikhupura region indicates no distinctive performance in relation to anatomical characteristics of the root. That is why, it was considered that this ecotype modifies itself to physiological attributed rather than anatomical features.

The Sahianwala ecotype, which performs poorly under nickel stress, revealed some discrete structural variations including, enhancement of root area, cortical cell area, and vascular bundle area. The second sensitive Gatwala eco type exhibited increase root aerenchyma formation and phloem area, this specifies that sensitive ecotypes were most receptive of nickel toxicity. When compared with tolerant ecotype, which generally depends on the physiological functioning for their survival and tolerate metal-polluted environment.

KEYWORDS

- ***Typhadomingensis***
- **nickel stress**
- **anatomical modifications**
- **plant biodiversity**
- **metal-polluted environment**

REFERENCES

Abdel-Ghani, N. T.; Hegaz AK, El-Chaghaby, G. A.; *Typhadomingensis* Leaf Powder for Decontamination of Aluminium, Iron, Zinc, and Lead Biosorption Kinetics and Equilibrium Modeling. *Int. J. Environ. Sci. Tech.* **2009,** *6* (2), 243–248.

Allen, S. E.; Grimshaw, H. M.; Parkinson, J. A.; Quarmby, C. Chemical Analysis of Ecological Materials. In *Chemical Analysis of Ecological Materials;* 2nd ed.; pp 565, Halsted Press, 1974.

Bajguz A. Suppression of Chlorella vulgaris Growth by Cadmium, Lead, and Copper Stress and its Restoration by Endogenous Brassinolide. *Arch. Environ. Contam. Toxicol.* **2011,** *60*, 406–416.

Barrie, L. A. Atmospheric Nickel in Canada. In *Effect of Nickel in the Canadian Environment;* National Research Council of Canada: *Ottawa*, 1981; pp 55–76.
Black, H. Absorbing Possibilities. Phytoremediation. *Environ. Health. Pres.* **1995,** *103*, 1106–1108.
Campenni, L., Nobre, B. P.; Santos, C. A. Carotenoid and Lipid Production by the Autotrophic Microalga *Chlorella protothecoides* Under Nutritional, Salinity, and Luminosity Stress Conditions. *Appl. Microbiol. Biotechnol.* **2013,** *97*, 1383–1393.
Cuypers, A.; Karen, S.; Jos, R. The Cellular Redox State as a Modulator in Cadmium and Copper Responses in *Arabidopsis thaliana* seedlings. *J. Plant Physiol.* **2011,** *168*, 309–316.
Ezzat, M. N.; Shehab, H.; Hassan, A. A. Survey of Nile System Pollution Sources. Report No. 64. APRP Water Policy Program, 2002.
Grigore, M. N.; Toma, C. Histo-anatomical Strategies of Chenopodiaceae halophytes: Adaptive, Ecological and Evolutionary Implications. *Trans. Biol. Biomed.* **2007,** 12, 204–218.
Hagemeyer, J. Ecophysiology of Plant Growth Under Heavy Metal Stress. In *Heavy Metal Stress in Plants;* Springer: Berlin, Heidelberg, 1999; pp 157–181.
Hegazy, A. K.; Abdel-Ghani, N. T.; El-Chaghaby, G. A. Phytoremediation of Industrial Wastewater Potentiality by *Typhadomingensis. Int. J. Environ. Sci. Tech.* **2001,** *8,* 639–648.
Kanoun-Boule, M.; Vicente, J. A. F.; Nabais, C.; Prasad, M. N. V.; Freitas, H. Ecophysiological Tolerance of Duckweeds Exposed to Copper. *Aquat. Toxicol.* **2009,** *91*, 1–9.
Maggio, A.; Raimondi, G.; Martino, A.; De Pascale, S. Salt Stress Response in Tomato Beyond the Salinity Tolerance Threshold. *Environ. Exp. Bot.* **2007,** *59*, 276–282.
Malamy, J. Intrinsic and Environmental Response Pathways that Regulate Root System Architecture. *Plant Cell Environ.* **2005,** 28, 67–77.
McIlveen, W. D.; Negusanti, J. J. Nickel in the Terrestrial Environment. *Sci. Environ.* **1994,** *148*, 109–138.
Perveen, S.; Anis, M.; Aref, I. M. In Vitro Morphogenic Response and Metal Accumulation in *Albizialebbeck* (L.) Cultures Grown Under Metal Stress. *Eur. J. Forest. Res.* **2012,** *131*, 669–681.
Sunderman, F. W.; Oskarsson, A. Metals and Their Compounds in the Environment. In Merian, E., Ed; V. C. H: Weinheim, 1991; pp 1101–1126.
Tremp, H.; Kohler, A. The Usefulness of Macrophyte-monitoring Systems, Exemplified on Eutrophication and Acidification of Running Waters. *Acta. Bot. Gallica.* **1995,** *142*, 541–550.
Wang, Q.; Liang, X.; Dong, Y. Effects of Exogenous Salicylic Acid and Nitric Oxide on Physiological Characteristics of Perennial Ryegrass Under Cadmium Stress. *J. Plant Growth Regul.* 2013, *32*, 721–731.
Wang, W. J.; Gorsuch, W.; Hughes, J. S. *Plants for Environmental Studies;* CRC Lewis Publishers: New York, 1997.
Welch, R. M. The Biological Significance of Nickel. *J. Plant. Nutr.* **1981,** *3*, 345–356.

CHAPTER 2

The Genetic Resources as an Inexhaustible Source of Biodiversity in Tajikistan

N. FIRUZA[1*], H. NANGYAL[2], A. NAWAZ[3], and M. S. KHAN[4]

[1]*Institute of Botany, Plant Physiology and Genetics, Tajikistan Academy of Sciences, Dushanbe, Tajikistan*

[2]*Department of Botany, Hazara University, Khyber, Pakhtunkhwa, Pakistan*

[3]*Department of Zoology, University of Lahore, Sargodha Campus, Sargodha, Pakistan*

[4]*Department of Zoology, Government College University, Faisalabad, Pakistan*

[*]*Corresponding author. E-mail: firuza_nasyrova@mail.ru*

ABSTRACT

Given the population projections for the growing human population, we need to increase food potential annual average up to 2%. Since time immemorial, to ensure the development of agriculture plant genetic resources are an important source for meeting the food requirement. Therefore, the primary task is to protect the world's genetic resources while preserving them for management.

In the current context of climate change, the reserve of arable land is near the end of the limit on the yield of important crops almost achieved as a result of intensive breeding, held in the last 100 years; the remaining reserves are very small and do not in any way ensure population growth. Hence, we need new approaches to solve the food problem that can only come from the analysis of the fundamental achievements of modern science. Naturally, in order to double the volume in the future of food production, it is necessary to create a fundamentally new form with the reconstructed genomes which are more productive, qualitative, and resistance to abiotic and biotic stress. For this purpose, wild relatives are extensively used to search for and transfer to new modern commercial varieties stress resistance gene. The work begun by Vavilov and colleagues has shown the exceptionally high efficiency of selective breeding to create high-yielding crop varieties.

2.1 INTRODUCTION

In view of the demographic projections for a growing human population, we need to increase the capacity of food each year by an average of up to 2%. It is estimated that in order to ensure food grain for the population of our planet, which will reach by 2020 to 8 billion people, its production should be increased by 1.3% per year in order to reach the average level of productivity 4 ton/ha. In the current context of climate change, when the reserve of arable land is almost exhausted, a limit on the yield of important crops almost achieved as a result of intensive breeding, conducted in the last 100 years, the remaining reserves are very small and in no way ensure population growth. Hence, we need new approaches to solve the food problem, which can only occur on the basis of the analysis of the achievements of modern fundamental science (Karimov, 2008).

2.2 TO MEET THE REQUIREMENT OF FOOD

Naturally, in order to double the volume of food production in future, it is necessary to create a fundamentally new form—a reconstructed genome and more productive, qualitative, and resistance to abiotic and biotic stress. For this purpose, the wild relatives are used extensively for search and transfer to the modern commercial varieties of new genes for resistance to stress. Works started by Vavilov and colleagues have shown exceptionally high-efficiency selective breeding to create high-yielding varieties of crops. Vavilov said repeatedly that plant collections are created for use in the national economy, particularly in the selection. His favorite expression was "Selection—an evolution, guided by the human will."

Since the dawn of the century, for the development of agriculture plant genetic resources are an important source to meet food needs. In recent decades, the problems of conservation and sustainable use of plant genetic resources that provide the basis for the development of breeding, sustainable agriculture, and food security, received great attention worldwide. The problem of food security in the world is directly related to the need to improve the productivity of crop plants. The changing climate of the planet caused by human activity influence confronts the problem by breeders to create new forms of cultivated plants that are resistant to extreme environmental factors (drought, salinity, low temperatures). In addition, increasing demand for food and land loss due to salinization due to irrigation demand the creation of highly productive and resistant varieties of important crops. To meet the food requirement of the growing population, it is necessary to create new crop varieties with valuable properties, is resistant to adverse environmental factors, diseases. New varieties, of course, are far superior to their wild relatives on productivity, however, in pursuit of the creation of new varieties with improved properties can lead to varietal homogeneity, the so-called genetic erosion. Vavilov emphasized that "among the plants protection against a variety of diseases the most radical way to combat is the introduction to the culture of immune varieties or the creation of such by crossing" (Vavilov, 1935).

2.3 THE WILD SPECIES: THE MAIN GENETIC STABILITY OF RESISTANCE OF CULTIVATED PLANTS

The main genetic stability reserves are the wild species, relatives of cultivated plants. Therefore, the collection, study, preservation, and use of wild species in breeding work are extremely important in creating new productive varieties. It is necessary to collect local varieties, create or gene banks gene pools of wild species for the purposes of future crop breeding. Any genetic seed bank is of particular value because it contains substantial margin allele combinations of genes that determine the fitness of plants to different soil and climatic conditions and resistance to biotic factors. Formation of the gene pool of samples of plant collections is due to the introduction and bringing samples of their distribution on the nature of origin; the study on the basis of complex samples (methods, classifications); selection on the basis of the reference samples and the level of their manifestation; systematization of samples and their distribution by types of collections.

Each gene bank have a passport underlying collection that includes information about culture, categories collections, number of samples, the number of plant species, subspecies and varieties, the number of countries of origin of the samples, the country of origin of the samples and their quantity, distribution of samples on the nature of origin (country grade, the variety of foreign countries, old landraces varieties, breeding lines of the country, genetic lines, and wild and related lines), the degree of genetic diversity in the samples on the grounds and the level of expression, the level of knowledge of the collection, the number of registered designs with the issuance of the certificate, and the number of samples whose seeds laid down for storage in a national repository (for crops and species that reproduce by seeds).

Formed genetic collection—a source of identified genes, genomes and gene complexes genomic complexes, substitution and translocation chromosome aneuploidy, nuclear-cytoplasmic connections, etc. In addition, there is training in gene banks collections for educational programs that covers issues of specific and intraspecific diversity, direction of use, breeding history, genetic diversity on the basic characteristics (reference samples). Gene bank transmits samples users for the implementation of breeding, scientific, educational and other programs. Formation and long-term preservation of basic, attributive, genetic, study collection allows users to stably secure the necessary samples. Exchange of samples of the gene pool of plants is based on the International Agreement on Plant Genetic Resources for Food and Agriculture and bilateral agreements.

The notion of a perfect grade was put into use in 1895 and then updated by N. I. Vavilov. Kumak (1985) defines a model of varieties as a scientific outlook, describes the combination of signs of the plant required to provide a given level of productivity, sustainability, and other productive properties. Altergot et al. (1974) mentioned that the intensive exploitation of the variety of natural resources is only possible with high resistance to local complex of unfavorable factors. Currently, under the model implied grade technical specification for the creation of varieties are mentioned, that is, detailed description of the business, the morphological and physiological characteristics, as well as the ways (mating combinations and methods of selection patterns) by which the set parameters are achieved.

For the selection of the plants on the grounds, which are difficult to test, using easily recognizable phenotype genes in the process of selection-markers is helpful. Genetic markers—monogenic inherited features, analysis of the relationship that certain important features to solve a number of theoretical and practical issues (Bezrukov, 1995). The main idea of the markers of the concept lies in the fact that the marker locus reflects the expression of either the gene that determines what that economically valuable feature, or used as a label for the chromosome region near this marker that lets you find and learn this segment, as well as to spend with them various manipulations (Hawking, 1997).

In the last two decades, major discoveries in the field of the structural and functional organization of plant genome were made; acquired clusters and repetitive DNA sequence transposable elements and structural genes encoding various biochemical and morphological characters. Displaying mosaic structure of the structural genes; studied the genetic mechanisms of resistance to abiotic and biotic stresses. New methods for isolating and studying proteins, nucleic acids significantly improve existing and established (Sozinov, 1989). Recent advances in modern biology have allowed developing a new "marker technology". What is considered the most promising molecular labeling, based on the polymorphism of proteins and nucleic acids? According to modern concepts, any substance that claims to be the marker must meet certain requirements: a high level of polymorphism, codominant nature of inheritance, neutral behavior in relation to the phenotype, the optimal level of frequency of occurrence in the genome for specific tasks, a uniform distribution in the chromosomes in the genome, easy evaluation of the marker parameters, the ability to automate the marker assessment parameters, and high reproducibility marker assessment (Konarév, 1998).

2.4 PROTEIN MARKERS

Theoretically, any marker polypeptide can serve its encoding gene (Sozinov, 1985). Modern methods of separation and identification of proteins (electrophoresis, isoelectric focusing, gel filtration, immune electrophoresis, and others) allowed obtaining reliable information about the variety of protein molecules and, consequently, the difference of genotypes. Markers for different solutions genetics and breeding problems commonly use two main groups of proteins: enzyme systems and seed storage proteins (tubers). *Seed, a fixed phase of ontogeny*: Seed storage proteins often remain unchanged for many years. It was possible to obtain satisfactory spectra prolamin storage protein for wheat and barley seeds were stored at room temperature for more than 110 years (Konarév, 1998). It should be noted the high reproducibility of the results of electrophoresis and chromatography storage proteins. An advantage of these markers is also a fairly complete study of the genetics and biochemistry of storage proteins of seeds and tubers at the most widely cultivated species of plants; simplicity and low cost of operation of these facilities, the possibility of standardization of protein markers assessment methods (Konarév, 1983).

Storage protein from most studied and widely used varieties are used as markers of cereal prolamins, specifically wheat gliadins (Sozinov, 1979; Konarev, 1983; Sozinov, 1985). Soft spring wheat (*Triticum aestivum*) and barley (*Hordeum vulgare*) are the most studied in the field of polymorphism of storage proteins gliadin and hordein. Cereal seeds contain a complex group of proteins, differing in composition, properties, and functions. Gliadins and glutenins were called gluten proteins, as they play a key role in the formation of gluten, the unique properties of which determine the technological, food, and the taste of white bread. Spare wheat proteins actively synthesized during seed maturation and mainly accumulated in the electron-dense protein bodies, the value of which ranges from 1 to 15 μm (Graham et al., 1963).

Gliadin is characterized by a very high content of the amino acids glutamine and proline, which provide nitrogen and energy germinating embryo. They are an important part of many metabolic processes, where the biosynthesis of amino acids is carried out. By physicochemical properties of gliadin, it may be divided into two groups. The first group consists of α-, β-, and γ-gliadins, which include molecules of sulfur amino acids and form the tertiary structure of the protein through the formation of intermolecular bonds. The second group of ω-gliadins is practically devoid of sulfur-containing amino acids. An important feature of gliadin is the complete absence of intermolecular bonds. The molecular weight of α-, β-, and γ-gliadin ranges from 32 to 42 kD (Koval, 1999).

2.5 MOLECULAR MARKERS

Development of methods of molecular genetics over the past decade has led to the emergence of a new class of molecular markers, DNA fragments corresponding to nucleotide sequences that are included directly in the structure of genes of agronomic traits or linked with this gene. The number of DNA markers is many times greater than the number of loci isoenzymes and storage proteins. Furthermore, expression of these molecular markers in relation to a neutral phenotype is not tissue-specific and can be detected at any stage of plant development. The advent of DNA markers has radically changed methods for assessing the genetic diversity of plants, certification, and classification of grades, mapping and determining the physical nature of the gene introgression of new genes and genetic monitoring in breeding and seed production (Hawking, 1997). Currently, there are many modern methods of using the DNA fragments as genetic markers, among which two main ones are: The analysis of restriction fragment length polymorphism (restriction fragment length polymorphism—RFLP) and analysis of polymorphism fragments obtained through polymerase chain reaction (PCR).

RFLP: For the analysis of length polymorphisms of restriction fragments of DNA isolated from plant tissues, "cut" specific bacterial restriction enzymes. Each of them produces a dissection (restriction) site in the DNA molecule docking certain combinations of nucleotides. Application on a single preparation of several DNA restriction endonucleases allows us to obtain small fragments whose size in each case depend on the genetic characteristics of the organism. By selecting the number and combination of restriction enzymes can get a great (and specific to this group of organisms), the spectrum of the DNA fragments.

The restriction fragments were separated by gel electrophoresis, transferred to a nylon membrane, "show" by hybridization with radiolabeled probes (DNA fragments of the size of bases 500–2000) and subsequently analyzed by autoradiography to position bands. As the probes can be random products of genomic DNA with restriction, however, often use specially selected clones are unique and rarely repeated DNA sequences. Like protein markers, RFLP-markers are codominant, which allows distinguishing between homo- and heterozygotes (Hawking, 1997). This is an important difference from their morphological markers.

Main advantages of RFLP-analysis are its versatility and high reproducibility. However, the main advantage of RFLP-markers is the ability to determine the position of marked genes of agronomic traits at the molecular maps. Such localization of genes involves preliminary preparation of molecular maps for specific plant populations and is associated with the necessary application to map several hundred-RFLP probes, which requires a lot of labor and resources. This work is a preparatory stage for the cloning of genes and their subsequent transfer to other crops.

Chain reaction polymerization PCR amplify allows (amplified) by means of specific enzymes polymerases genome regions, which correspond to randomly selected seed or pre-DNA sequences (primers). Markers PCR-amplified sequences differ in length. In comparison with RFLP-analysis, this method requires the allocation of significantly lower (sometimes three orders of magnitude) amounts of DNA from plant tissue. However, the latter must be free from contamination of the DNA of fungi and bacteria. Due to the simplicity and accessibility for any molecular genetics laboratory of the PCR-technology produced in the last 5 years a real revolution in the application of molecular markers in genetics and plant breeding (Hawking, 1997).

Depending on the nature of the primers and method for identifying the amplification products are several types of PCR-analysis. First, there was a method of randomly selected DNA sequences [randomly amplified polymorphic DNA (RAPD)], which uses a standard set of primers, amplification products are separated by electrophoresis and stained. Embodiments RAPD primers differ in length, the ratio of primer/template DNA and gel separation of the amplification products. RAPD analysis requires a minimum amount of DNA (10 ng per reaction) and do not need to know in advance exactly what the DNA sequence to be amplified. The method is very simple to perform, although it requires strict standardization of amplification conditions.

Molecular genetic markers used in genetics, breeding, and seed production for the identification of varieties and genotypes, in the determination of varietal purity, that is of great importance in the creation of classes and their reproduction, as the differences between the varieties on morphological characters are not always sufficiently clear. On this is based the idea of using molecular genetic markers to identify varieties (Sozinov, 1985). The most convenient markers varietal identification cereals were storage proteins. When identifying the recognized varieties of wheat, it is generally sufficient to investigate the spectrum of storage protein gliadin (Sozinov, 1972).

Protein and DNA markers are now used to assess the genetic differences between samples of different geographical origin, cultivated and wild forms. And as for the analysis of kinship species phylogeny audit schemes and the origin of genomes (Konarév, 1998). Back in the 1970s in VIR using immunochemical methods of data were received, which forced a fresh look at the origins of the genomes of soft and durum wheat. Using

RAPD-analysis genetic differentiation of 316 samples of cultivated and wild barley of different origin was conducted. The data allowed dividing the studied varieties and populations into three main groups, which obviously reflect the main trends in the evolution and geographical spread of cultural forms of barley. And also set the membership of these groups the world's crop diversity centers and modern ecological and geographical classification of barley (Konarév, 1998).

One of the objectives of applied genetics is the study of the nature of complex traits: adaptability, resistance to adverse biotic and abiotic factors, crop quality and productivity. Significant among these methods are the molecular-genetic markers (Konarév, 1994). Now we firmly established the link between the allelic state of gliadin loci and quantitative traits, such as the technical quality of the wheat grain, winter hardiness, drought tolerance, and productivity.

Great expectations in marking and mapping of agronomic traits are assigned to the DNA-markers. The greatest advances in this field were achieved by RFLP-markers. In many laboratories of the world are being mapped loci such important agronomic traits as the structural features of plant development rate (earliness), productivity and quality of the crop. The use of molecular genetic markers significantly increased the possibility of evaluating genes of resistance to unfavorable factors. Among the many examples of the successful use of DNA probes in the selection of cereals, fruit and vegetable crops for resistance to diseases and pests, most prominent considered are mapping resistance genes of rice blast, late blight of potato, gray leaf spot of corn, wheat brown rust, bacterial wilt of tomatoes, nematode soybean (Havkin, 1997).

In recent years, in the world are actively developing research in the field of salt tolerance gene search, research is underway to create salt-resistant forms of crops. This work is extremely relevant for the countries located in the arid and semi-arid areas of agriculture, such as Saudi Arabia, Australia, Central Asia, and others. The selection of salt-tolerant specimens held in the world collection of seeds. Research carried out on the salt tolerance of localization of genes on chromosomes, which are interspecific hybrids and substituted line. The aim of this work is to increase the productivity of plants for food, raw materials, and energy in areas with a shortage of moisture. Using the genotypes resistant to salinity allow to engage in agricultural practices, new land and reduce crop losses from adverse climatic conditions.

Wheat is one of the most important crops in the world, accounting for almost one-quarter of the world's food. The genetic potential of wheat productivity is now almost exhausted. Mainly cultivated varieties of wheat are (especially strong) not resistant to salinity. Therefore, the work in the study of mechanisms for increasing the salt tolerance of wheat is extremely relevant, since it has practical significance. One way to solve such problems can be an expansion of the arsenal of methods for increasing genetic diversity through the use of cultured plant cells. At the moment of the cell selection techniques have created new salt-resistant forms of alfalfa, rice, tomatoes, and other plants. Increasing resistance of wheat plants to abiotic stresses will reduce crop losses from adverse environmental factors.

Salt tolerance in hexaploid wheat *Triticum aestivum* (genome AA BB DD) is controlled by the set of genes that manifest as at the cellular level and at the level of the whole plant. The work of Gorham et al. (1987) showed that for control of the whole organism of the transport of sodium ions in the above-ground parts of the shoot in hexaploid wheat are responsible genes located in the long arm of chromosome 4D. Tetraploid wheat (*Triticum durum* L.) has a genome AABB, deprived of organismic mechanism confrontation salt stress due to lack of D genome. In this case, at the forefront are the cell-resistance mechanisms. Therefore, the study of the effect of salt stress on cell growth of hard and soft wheat and selection of salinity resistant cell lines followed by regeneration of the modified plant acquires not only theoretical but practical significance. DNA-markers—DNA fragments corresponding nucleotide sequences are included in the structure or economically important gene linked to it. The use of DNA technology can significantly extend the capabilities of conventional plant breeding. The manifestation of molecular markers in relation to a neutral phenotype is not tissue-specific and can be detected at any stage of plant development. DNA genotyping methods and selection using molecular markers [marker-assisted selection (MAS)] can speed up the transfer of commercially valuable genes in the process of selection and ensure the creation of new varieties with a complex set of properties (Hawking, 2003). DNA technology in breeding, protection of copyright, and protection of breeders of plant products from possible tampering are being increasingly used. Only DNA analysis that directly characterizes the gene, not the phenotype, can provide stable characteristics of the plant, almost suitable for the identification of genotypes, registration and labeling varieties of commercially valuable genes and traits.

Modern farming practices that lead to increased concentrations of greenhouse gases in the atmosphere, desertification and soil degradation, reduction of forest area, and accelerate the loss of biodiversity is largely a negative impact on the environment and cause irreversible environmental damage. Plant genetic resources are the basis for improving its quality, ensure food security of the state and bioresource. However, it was their disappearance occurs at an alarming rate. Genetic erosion in some places reaches 20–35% of the known plant diversity. According to Food and Agriculture Organization (FAO) experts, about 75% of the global genetic diversity of agricultural crops has been lost in the course of the twentieth century. For example, in the USA 97% of the varieties registered in the State Registry of the old US Department of Agriculture, have already lost; in Germany lost 90% of the historical diversity of cultivated plants, and in Italy 70% old local varieties disappeared forever. Over the last 50 years, genetically uniform modern varieties have replaced thousands of landraces adapted to the environmental conditions of its cultivation (World Report, 1998). The world has a variety of vegetation types according to different estimates of 250,000–270,000 species of higher plants and only 7000 of them are used by the world community for their livelihoods. A total of 15 crops provide 90% of food energy needs of humanity. Chief among them are: rice (26%), wheat (23%), and corn (7%) (Fowler, 2005). Most of the plant wealth of the planet has not been studied, and according to the theory of Vavilov centers of origin of cultivated plants growing in the developing countries. Now is exactly the time, which in the late twenties of the last century, wrote N. I. Vavilov (1987): "Only through the establishment of international communication, by an international organization of research will be to get close to the study of the most interesting and most important centers of concentration of genes (crops)."

Since the mid-eighties of 20th century, under the auspices of the UN has developed and adopted a number of conventions and agreements of the intergovernmental level in order to minimize the threat of global ecological crisis, food and bioresource security of countries and entire regions, as well as the conservation and sustainable use of genetic resources for the benefit of mankind. The most important of these are the Convention on Biological Diversity (CBD, 1992.; signed by 180 countries); The Global Plan of Action for the Conservation and Sustainable Use of Plant Genetic Resources (GPA, 1996; adopted by 146 countries); The International Treaty on Plant Genetic Resources for Food and Agriculture (MD, 2004;

ratified by 150 countries and the EU). They are of particular importance for the countries of the world, uniting their efforts to effectively address the conservation of agro-biodiversity and use its components for sustainable agriculture, reducing poverty, improving the welfare of the world's population, and ensuring its quality food. These international agreements establish the legal responsibility of States for the preservation of biodiversity, growing on their territory and providing them with a legal basis for cooperation on the basis of mechanisms of interaction worked. At the same time, the level of demand from the germplasm collections has increased dramatically. Primarily, rapid development of biotechnology has caused the establishment of rights of intellectual property in the bioproduction, the acceleration of the privatization of breeding organizations, and tightening competition in the global food market. A key strategy in addressing problems can and should take the national gene bank, whose role in the context of globalization goes beyond just storing the national gene pool of genetic resources in ex situ. This gene bank received from the Government of the special status of the National Center for Genetic Resources of Agricultural Biodiversity is responsible not only for long-term preservation of the national heritage but also for the observance of national interests in the field of monitoring the progress of genetic resources.

2.6 WILD RELATIVES OF CULTIVATED PLANTS IN TAJIKISTAN

The areas of maximum diversity, including a number of endemic forms and signs, according to Vavilov usually are the centers of formation. Vavilov showed that in the centers of origin of cultivated plants, which are found intense morphogenetic processes that manifest themselves primarily dominant traits. Recessive same symptoms, which are often associated and valuable plant quality can be stably express themselves, but going beyond these centers, where they will not be suppressed and masked signs dominant. Apart from the basic geographical patterns of the evolutionary process, Vavilov revealed a number of regularity in the distribution of individual and groups of species and varieties of plants around the globe. For example, if Europe's flora is very poor in species of woody plants, the Caucasus, the mountainous and foothills regions of Central Asia, rich in their diversity (Vavilov, 1931). The results of the geographical analysis showed that the maximum number of sections of these species are concentrated in the central and eastern parts of the range kind, namely, in the

Caucasus, in South-West Asia, and Central Asia. In the Caucasus, grows a maximum variety of subspecies, varieties, and forms described in the species *A. tauschii* and *A. cylindrica.* This is good evidence that in the South Caucasus is the center of their formation. Probably, Central Asia is a secondary center forming these species and for *A. tauschii* it captures more and northern Afghanistan.

The ancient agricultural culture of the population of Tajikistan contributed to the creation of many varieties of cultivated plants and domestic animal breeds, based on the gene pool of wild relatives, primarily native species. Currently cultivated in Tajikistan are more than 85 species and 360 varieties and hybrids of cultivated plants for different purposes. Main areas of agro-ecosystems are located below 3000 m above sea level m and under the terms of precipitation moisture is divided into two subzones: Subzone rainfed agriculture and subzone, where farming is possible with irrigation. Tajikistan now has a detailed list of species of plants, amounting to more than 9000 items of plant species. According to National Accounting, Tajikistan flora includes 9771 species of plants, 650 species are endemic, 1000 species—wild relatives of cultivated plants, 226 species listed in the Red Book of Tajikistan. National inventories of flora are a published book "Flora of Tajikistan" in 10 volumes (1957–1991), which is the species diversity of flora of Tajikistan. This catalog provided a unified description of plants of Tajikistan, with an indication of habitat types, most types of illustrations, and maps of the distribution of all taxa. Species diversity of endemic flora of Tajikistan is also described in detail in the book catalog "Flora of Tajikistan", 1957–199; and in the Red Book of Tajikistan, 1988. It established a national target for the protection and rational use of rare species. Approved Plant Protection Act (1994) published by Red Book of Tajikistan (1988). Taxonomy of the country is represented in several publications: "Flora of Tajikistan" in 10 volumes (1957–1991), "Fauna of Tajikistan" in 20 volumes (1960–1988), "the vascular plants of Tajikistan" (1999), as well as in publications on birds (Tajikistan Fauna/Birds volume 19 1, 2, 3 parts (1971, 1973, 1977), fish, insects, and mammals in paleontology. These publications describe 10, 000 species of plants and about 13,500 animal species.

Tajikistan has a very rich biodiversity. Taking into account the uniqueness of the flora and fauna of the country, a large number of rare, relict, and endemic species, a rich gene pool of wild relatives of cultivated plants and domesticated animals, as well as the traditionally high level of their application in various sectors, in situ conservation is considered by the Government of Tajikistan as the original warranty sustainable use of biodiversity. To this end, the country has established a network of specially protected natural areas, composed of 4 parks, 14 nature reserves, 1 national, and 1 historical and natural park. In 1992, a National Park had been established in the eastern part of the country to protect different types of terrain (mountainous deserts, glaciers, lakes, juniper forests, mountain steppes, and alpine meadows). In 1993 Shirkent Natural Historic Park to protect and preserve the ancient cultural landscape was created in the country, unique rocks, dinosaur footprints and juniper forests. The National Strategy for the Conservation and Sustainable Use of Biodiversity clearly defines the key policy issues in this area. The most important of these are as follows:

1. The development of national and regional networks of protected areas;
2. The development of national and regional ecological networks, which can be integrated with the Euro-Asian network;
3. Implementation of measures for the protection of rare and endemic species of flora and fauna, reprinting of the Red Book of Tajikistan;
4. Development and improvement of the existing legislative framework for the conservation of biodiversity and management of human activities;
5. The integration of the requirements of the conservation of biodiversity into sectoral policies and projects.

Tajikistan has great potential of genetic resources, including unique genetic collection, genetic collection of crops, and wild species of flora. Cereals, legumes, and industrial crops: 1 5,000 accessions, fruits, and berries: 10,000 accessions stored in a variety of scientific and educational institutions (national herbarium includes about 350,000–400,000 plant specimens of modern flora).

The main holders of the genetic collections are the following institutions: the Botanical Garden of the Academy of Sciences, the various departments of the Tajik Academy of Agriculture, which contain a number of genetic collections of trees and shrubs, shoots of fruit trees, medicinal herbs, rare and endangered species, etc. The main genetic types of agricultural floristic species, herbs, potatoes, various fruits, industrial plants collected in different scientific institutions of the Tajik Academy of Agricultural Sciences. Genetic collection of fruits and berries is about 10,000 sorts of apricot, apple, plum, almond, etc. (collector: NGO "Bogparvar" TAAS). Cereals, legumes and industrial crops, potatoes 15,000 accessions (Collector: Institute of Plant Physiology and Genetics, Academy of Sciences of the Republic of Tajikistan) and 22,000 (NGO "Ziroat" TAAS).

2.6.1 NATIONAL CENTER FOR GENETIC RESOURCES INTERNATIONAL INSTITUTE

ICARDA assisted CA region, including Tajikistan about creating centers of plant genetic resources. In 2001, exploration center of the Tajik Academy of Agricultural Sciences, the parent organization was the NGO "Ziroatkor" TAAS (now the Institute of Agriculture). Resolution of the Government of the Republic of Tajikistan December 29, 2006, No. 791 was created by the National Republican Center of Genetic Resources TAAS.

Activity: National Republican Center of genetic resources is determined by the scientific and technical selection and genetic trends, long-term plans approved by the Tajik Academy of Agricultural Sciences. At the Republican National Genetic Resources Centre have the following objectives:

1. The collection, study, and preservation of genetic resources of all types of crops and their wild relatives growing in the Republic of Tajikistan;
2. The collection, study, and preservation of unique, rare, valuable, endangered, rare accessions of cereals, legumes, fodder, oilseeds, fruits and vegetables, grapes, medicinal plants, and industrial crops;
3. Identifying the location and documentation of the local age-old accessions of fruit crops and medicinal plants;
4. The creation of "Databases" on the crops and provide the new technologies of the information system;
5. Regeneration of accessions collected in the laboratory and field conditions; definition seed germination, propagation, and preservation of seeds in the gene bank;
6. Recommendation accessions on economic grounds as a valuable source of raw material for genetic breeding purposes;
7. Collector's organization: demonstration plots of the number of unique rare, valuable, endangered, rare accessions of fruit crops, grapes, and their wild relatives.

The main activity: cereals, legumes, fodder, technical, oil, medicinal, wild relatives of cereals, and leguminous plants, horticulture, viticulture, nurseries, vegetable growing, melon, potato, and protection of plants against pests and diseases.

2.6.2 INSTITUTE OF HORTICULTURE

The history of the Institute of Horticulture begins in 1932. At this time, according to the Decision of Tajikistan Government in Khujand area in the north of the country was established grapefruit and vegetable farms. After passing a series of reorganization and redeployment in Dushanbe in 1956, he received the name of the Research Institute of Horticulture, Viticulture and subtropical crops named after I. V. Michurin. In 1978 reduced Institute Research Institute of Horticulture, Viticulture, and Horticulture. According to the Resolution of the Government of the Republic of Tajikistan on May 2, 2008, No. 215 NGO "Bogparvar" was renamed to the Institute of Horticulture of the Tajik Academy of Agricultural Sciences.

Activity: Institute of Horticulture is determined by science and technology and production lines, long-term plans approved by the Tajik Academy of Agricultural Sciences. At the Institute of Horticulture have the following objectives:

1. The development and deepening of scientific research by industry horticulture, viticulture, horticulture, and potatoes, taking into account natural and economic conditions of the Republic of zones, increasing research effectiveness;
2. The development of new high-quality and high-yielding, resistant to pests and diseases, as well as unfavorable climatic and soil conditions, varieties of fruit, grapes, vegetables, and potatoes;
3. The cultivation of planting material of fruit and grapes, high-quality seeds of vegetable crops and watermelon plantation;
4. Participation in development issues gain production, inter-farm cooperation and agro-industrial integration in horticulture, viticulture, vegetable growing, and potato;
5. Development and improvement of scientific, technical proposals and forecasts on the development of horticulture, viticulture, vegetable, and potato of the republic;

6. Performance of tasks in the production and implementation of crop production.

The main activities are: Horticulture, viticulture, nurseries, vegetable growing, melon, potato, protection of plants against pests and diseases.

2.6.3 PAMIR AGRICULTURAL EXPERIMENT STATION

Pamir Agricultural Experiment Station was founded in 2003 with the Tajik Academy of Agricultural Sciences, with the support of GBAO Hukumat and MSDSP AKF.

Activity: Pamir experimental agricultural station is determined by the scientific and technical selection and technical trends, long-term plans approved by the Tajik Academy of Agricultural Sciences.

In the Pamir experimental agricultural station has the following objectives:

1. Test and induced local varieties of crops on productivity and disease and pest resistance in the conditions of Badakhshan;
2. Development and improvement of technology of cultivation of gardens, planting and growing of seed potatoes and horticultural crops in Badakhshan;
3. Development of scientific and industrial development system yak;
4. Recommendation accessions on economic grounds as a valuable source of raw material for breeding purposes;
5. Collector's organization: the number of demonstration plots of accessions of fruit and cereals as well as potatoes.

The main activities are: breeding, cereals, legumes, fruit, nurseries, vegetable, potato, protection of plants against pests, and diseases.

2.7 BIODIVERSITY OF TAJIKISTAN

In the mountainous regions of Tajikistan contain 0.66% of the global diversity of animals and 1.8% of plants, including wild relatives of domesticated animals and cultivated plants. Tajikistan also has a rich gene pool of species that are a potential resource for the creation of highly productive and resistant cultivars of ornamental plants, medicinal, aromatic, and industrial raw materials. The richness of biodiversity manifested at the genetic, species, population, biocenosis and ecosystem levels, there are many relict and endemic species. The majority of components are vulnerable to the effects of anthropogenic factors. On the territory of modern Tajikistan are more than 9000 species of spore and flowering plants. The terrain is mostly mountainous, arable land scarce.

Some local plant species are introduced to the culture and well adapted to the conditions of their environment and make invaluable genetic resources, both national and international communities. The gene pool of cereals, legumes, oilseeds is about 3000 samples, including 510 samples of wheat, barley: 500 samples, rye: 115 samples, oats: 60 samples, chickpeas: 500 samples, lentils: 80 samples, soybean: 46 samples, peanut: 8 samples, and corn: 234 samples. Collection accessions staple cotton has more than 600 different varieties and accessions, which were tested in all the ecological zones of Tajikistan. The collection of subtropical crops are: 7 varieties of pecans, almonds: 30 varieties, 46: jujube, 43: figs and a variety of persimmon, pistachio, sea-buckthorn, hazelnut, fruit mulberry. It is breeding work with citrus: lemon, oranges, and tangerines. Gourds are represented mainly melons, watermelons, and pumpkins. In the Pamirs revealed more than 43 varieties of soft wheat, 16 of which are liguleless forms. Under the conditions of the Pamirs in nurseries and botanical collections are more than 20,000 species and varieties of plants. The collection of the Pamir Botanical Garden has more than 40 varieties of apple trees, 38 varieties of apricots, pear varieties of 15, 14 varieties of peach, 20 varieties of mulberry and different varieties of strawberry, raspberry, currant, gooseberry, and other crops. Total collection of varieties, hybrids, and various forms of cotton, cereals, legumes, oilseeds, fruits, vegetables, subtropical, citrus fruits, and berries, and other crops is more than 32,000 local and imported samples. However, in recent years due to the socio-economic difficulties poorly replenished collection materials, in poor condition are breeding station, experimental plots, nurseries, botanical gardens, and others. This increases the risk of loss of the gene pool of the national wealth of biodiversity.

Some studies in the field of agriculture were conducted in the framework of projects and grants from international financial institutions such as FAO, German Agro Action, Asian Development Bank (ADB), International Monetary Fund (IMF), the Global Environment Facility (GEF), the World Bank. International Center for Agricultural Research in the Dry Areas (ICARDA) and others. Special attention is paid to the creation of regional networks in certain cultures, in order to support scientific communication and exchange of germplasm. Extensive searches are supported by donors in Tajikistan of new nontraditional forms of cereals, fodder crops, root crops, fruits and berries for the conduct of farming practices in the country. Work is happening on the breeding program launch promising varieties of cotton, cereals, legumes, oilseeds, and maize. Experienced stations are studies on the collection of 1400 fruit trees and grapes, of which 190 varieties of apricot, 67 varieties of peach, 125 varieties of apple, 76 varieties of pear, 110 varieties of walnut, 50 varieties of grenades, 350 varieties of grapes, and others. Research works is expanding on breeding and potato seed production. Also conducted are environmental testing and more than 70 accessions of potato clones. More than 1000 tested samples of wheat, including 300 accessions of local origin are obtained. Selection and seed-growing work are done on the onions, garlic, carrots, tomatoes, cabbage, peppers, sweet and hot, cucumbers, eggplant, radishes, turnips, radish, and others.

Central Asia is one of the five most important centers of origin of cultivated plants, and rich in species and intraspecific diversity of many world-famous fruit crops. Crops such as apple, apricot, walnut, grape, peach, pomegranate, and almond ancient long grown in home gardens and on farms (in situ/on farm) in the region, while their wild relatives are still to be found in their natural habitat (in situ). The southern position of Tajikistan, its arid climate, which is similar in many respects to the Mediterranean subtropical and extremely varying with the strength of vertical zoning, determines, in one hand, a very specific ecological, and biological features of fruit plants (their love of light, drought resistance, hetero rhythmic), on the other hand, a kind of structure they form plants. As believes Vavilov (1935), here is the birthplace of many of the now cultivated fruit.

Fruit plants entered the human economy, undoubtedly, a very long time. Apparently, the emergence of the primary centers of horticultural culture associated with relict islands Tertiary forests, where one finds a more favorable climate, abundant food procured by wild plants, such as fruit. The history of many of them is of thousands of years; it is possible that fruit cultivation began earlier than many other plants, including bread. Among the literary heritage of A. Regel (Zapryagaeva, 1964) of particular interest is the characterization of wood and handicraft vegetation. Already in those early years, the researcher noted for Pamir-Alai huge variety of wild fruit and stressed their importance for the establishment of centers of origin of cultivated plants. Studies by Fedchenko, conducted in Tajikistan (1925, 1937), are also important for the knowledge of wild fruit. Shows the relationship of forests Pamir-Alai with forests of Western Tien Shan, he emphasized the great shaping of apple that is important in elucidating the origin cultivars. He noticed that the wild vine climbs in the mountains to a height where cultivars have not divorced. In future, as a result of expeditions to Darwaz Russian scientists, Baranov, Raikova, Popov in 1927–1930, have been described in this isolated outbreak of the Tajik agricultural culture walnut, apple, pear, plum, cherry plum, and many other species, have been shown their distribution border in the vertical profile (Zapryagaeva, 1964). According to Popov, human tribes living in such peculiar natural conditions were in the thick of the process of species formation, at sources of "natural gene pool." As part of the forest surrounded by fruit, man at the dawn of agricultural civilization. By selecting the best form of the wild thickets created many local varieties of pears, apples, especially pomegranate, and grapes. Enrichment assortment of fruit went on simultaneously due to the transfer of forest seedlings in the garden, wildings, and subsequent refining. Ennobled trees differ only in slightly larger fruit with juicy pulp and sladnoy. The rich nature of the Pamir-Alai Mountains: An inexhaustible source of plant biodiversity, where he meets an abundance of wild fruit, such as pistachio, walnut, almond, barberry, currants, dwarf cherry, and other species. They are an inexhaustible fund for breeding work aimed at the development of new varieties of fruit, adapted to dry and hot lowlands, anhydrous and cold highlands of Tajikistan. In addition to the direct use of wild fruit, many of them—walnut, pomegranate, and others—are medicinal plants, so there are inputs for drug prescribing can be done in a very large scale.

Maintaining the regional diversity of fruit crops and their wild relatives has global importance in the present and for the future to ensure the genetic resources of all user groups. The cheapest and promising way to increase the productivity of plants is the development and introduction of high technologies based on the use of genetic resources, adapted to local growing conditions, advances in molecular biology, and biotechnology. At the present stage of development of agricultural production as the science of selection is becoming increasingly important, sort of becomes a powerful means of increasing the productivity of fields, and a variety of assortment of crops, food security. The sources of valuable raw material for the creation of new sustainable and high-yield varieties of fruit crops and grapes is a rich diversity of wild relatives of mountain fruit forests of Central Asia, recognized by Vavilov, one of the centers of origin of cultivated plants.

The huge presence of the species in the country, high-quality and population diversity of fruit crops and grapes allows extensive research on their involvement in the selection process. Before each breeder, the task is to bring a new variety, adapted to local environmental conditions with high taste fruit and economically valuable traits like productivity, winter hardiness, drought tolerance, resistance to pests and diseases, and others. In order to fulfill such an important task required for hybridization to implement a proper selection of the starting parent forms with positive signs and adaptation properties as the phenotype and genotype. Creating a database of adaptive traits and agronomic properties of the known fruit crops and grapes, it is essential to further improve the existing assortment and increase the production of fruits in the country. With the help of directional selection, we can achieve the development of the plants, the most economically valuable biological characteristics, the creation of varieties for each soil–climatic zones with different ripening times, increased resistance to diseases and pests, and adverse climatic conditions. Ultimately, the elimination of new varieties provides a yield increase of apple, pear, apricot, and grapes, saving money and material resources. It is, therefore, important to develop a state-genetic breeding program to develop new varieties of fruit crops and grapes using local diversity of these cultures as donor's agronomic characters.

Tajik scientist Nazirov (2011) indicates the possibility of using a Sievers apple breeding due to their large polymorphism, and the use of local varieties directly in production. They studied intravariability of Sievers apple on the main economically valuable attributes and properties, such as winter hardiness, drought resistance, heat resistance, disease resistance, and yield. It has been allocated more than 120 forms and varieties of local apple Sievers, promising for breeding and recommended for orchards and plantations in the dry mountain areas of Tajikistan and other Central Asian republics. As a result of the work was compiled a schematic map of the spread of varieties and forms, taking into account climatic conditions. Collection of wild (600 samples) and local forms of apple Tajikistan (220) is in Faizabad zonal experimental station of the Institute of gardening and vegetable growing of the Tajik Academy of Agricultural Sciences.

Another example of local flora is pistachio *Pistacia vera* L., one of the main forest-forming species in the foothills of southern Tajikistan. Pistachio is the most important forest reclamation, water conservation, and soil conservation value. Exceptional drought resistance, undemanding to soil, the ability to develop a deep root system makes pistachio valuable crop for Tajikistan, with its arid climate and mountainous terrain. As a very drought-resistant breed, pistachio can grow where they cannot be grown, no other fruit trees.

Within the framework of the ISTC Project No. T-1105 "Genome analysis of cereals in Tajikistan" work of collecting, gliadins analyzing was carried out, DNA genotyping of wild species *Aegilops* L., as well as local indigenous varieties of wheat. Types of *Aegilops* L. take great attention of researchers as a promising genetic source. As shown by numerous studies (Konorev, 1979), species of the genus *Aegilops* L. directly involved in the formation of tetra- and hexaploid wheat and now many of them are used for hybridization as gene sources a number of valuable features. For example, the view is a donor *A. tauschii* drought and pathogen resistance.

The geographical location of Tajikistan, with its peculiar soil and climatic conditions and a number of extreme factors (temperature changes, high-altitude stages, drought, salinity), contributed to the formation of unique species *Aegilops* L. and soft wheat native. Depending on the habitats formed, biotypes with varying degrees of stability are found. Some species have adapted to the hot and saline habitats while others to a moderate temperature habitats. Tougher climate, acting as a selective filter, contributed to the preservation and dissemination of the fittest biotypes. We have, in order to study the biodiversity species *Aegilops* L., growing in different climatic conditions of Tajikistan, was held expeditionary collecting more than 600 samples collected in the regions of the country: Hissar, Turkestan, Zeravshan ridges, and southern Tajikistan and neighboring countries (Kyrgyzstan and Uzbekistan) (see map). Studies have shown that on the territory of Tajikistan, the most common types of four *Aegilops* L. are: *A. tauschii*, *A. triuncialis*, *A. cylindrical*, *A. crassa*. Wide range of intraspecific diversity of genetic variation suggests that allowed these types to adapt well to adverse factors such as drought, soil salinity, and have increased immunity to disease. Each type has its own high-rise limit of distribution:

1. *A. crassa*: From 400 to 500 m above sea level.

2. *A. triuncialis*: Grows in Northern and Southern Tajikistan up to 2000 m above sea level.
3. *A. cylindrica*: Grows in the middle mountains and valleys of the up to 1800 m above sea level.
4. *A. tauschii* is widespread from 360 to 2000 m above sea level.

In order to determine intraspecific diversity was held *A. tauschii* cataloging specimens, *A. triuncialis*, *A.cylindrical*, *A. crassa*, including phenotypic characterization of samples collected from various floroceno-types and electrophoretic spectra of gliadin (Nasyrova et al., 2007). The bulk of the wheat grain (80%) is gliadins, so-called gluten proteins soluble in 70% ethanol and having a sufficiently broad molecular weight range of variability. Protein markers are used to distinguish grades of wheat and biotypes, in studying polymorphism species genomic analyses. The study of polymorphism on gliadin has been extremely useful to clarify that the classification of intraspecies identification of genetically similar samples. It is known that seed storage proteins as markers of genetic systems very successfully involved in solving urgent problems of phylogeny, taxonomy, and genetics of cultivated plants. Particularly promising is the use of protein markers in the identification of genomes and identification of phylogenetic relationships of cultivated plants to wild relatives. Rapid and accurate identification of genomes are available on the electrophoretic spectra of protein markers (Konorev, 1987).

Electrophoretic analysis of gliadin showed that the type of protein composition *A. tauschii* has a number of components from 12 to 20. Intra-specific differences in the composition of proteins are mainly observed by the presence or absence of components and the degree of severity in the electrophoregrams. The findings suggest that in each zone electrophoretic spectra present stable components that determine plant belonging to a certain type, and labile, showing a variety of species *A. tauschii*. In terms of identifying species of greatest interest are β and γ electropherograms valence band in which there is intense, pronounced components 1 and γ 2 in the valence band and the valence band 1 β. By the presence of these components to form two main types of the electrophoretic spectrum can be identified. The first type is defined by the presence of component 2 in the middle region of the spectra (γ). This type is most common of the 85 samples studied, he was diagnosed in 49. The second type of electrophoretic spectrum is determined by the presence in the same zone of intensive component 1. This type, compared with the first, is less common: of the 85 tested samples revealed it only in 26 depending on habitats present one of the said components. In very rare cases, mixed type of spectrum may occur when there are 1 and 2 components, simultaneously. In 13 of the 85 samples studied, discovered a very rare, rapidly migrating component β 1 in the valence band. It should be noted that this component is present only in samples assigned to the first type of electrophoretic spectrum. The ω-zone, in contrast to γ- and β-zones present mild gliadin and differences in the samples are largely on the presence or absence of components 1 and 2. *A. triuncialis* type, the degree of extension in Tajikistan is at second place after *A. tauschii*. Studies have shown that a large range of intra-specific variation observed on phenotypic characteristics, which allows referring *A. triuncialis* adapt to adverse environmental factors and have increased immunity to disease, in particular, to yellow rust. Unlike other species of *Aegilops* L., *A. triuncialis* contains a small number of gliadin components, from 11 to 17. The nature of the protein spectra of 25 samples studied is quite diverse and very different from the kind *A. tauschii*. The boundaries of intraspecific variation are the broadest and include almost all areas of the spectrum. Only three of the samples can be identified; electrophoretic spectra contain the maximum number of gliadin components. The remaining samples are intermediate, differing in the number and intensity of the individual components. The most varied and intense γ-spectrum area where the number of components ranges from 7 to 2. As shown by the comparative analysis, the component composition of gliadin in the samples collected in the same locus is very different. As for the β- and ω-zones, intraspecific differentiation they expressed slightly. Among *A. tauschii* and *A. triuncialis* biotypes found forms that are resistant to extreme environmental factors and have increased immunity to disease. A kind of cylindrical structure found a large intraspecific polymorphism of both gliadin spectrum and by their appearance. The differences between the samples collected by external signs manifested in the color intensity of ear and stem cells. Variety spectrum gliadin *A. cylindrica* coincided with the external phenotypic diversity. This species has a characteristic different from other species of *Aegilops*, protein spectrum with a number of components from 15 to 19. The component composition of gliadin *A. cylindrica*. For this type of characteristic intraspecific differences between samples from different habitats, which is observed in all zones of electrophoretic spectra are cylindrical type characterized by a small number of stable components electrophoresis (OEP, 0.34; 0.64; 0.66; 0.68; 0.91) which can be used as species-specific protein components. According to the number of electrophoretic labile components can judge the degree of polymorphism of the species. Among the labile components are components of electric range of 0.24, 0.27, 0.30, 0.62, 0.69, 0.71, 0.74, 0.95, and 1.0 showing a variety of species, related, in all probability, to the climatic conditions of habitat. The most saturated zone β-9 includes extensive protein components. In 10 samples, present a very rare component β-7 in the valence band (OEP 0.69) and fast migrating components in this zone 1 (Environmental impact assessment, EIA 1.0). These components are completely absent in other samples. It is assumed that these components can play a significant role in the type of adapting to environmental conditions, and possibly can serve as markers of symptoms. View *A. crassa*—with a narrow distribution area—from 400 to 550 m above sea level. Sea is characterized by a small variety. Unlike other species of *Aegilops*, this kind includes α -zone electrophoretic spectrum, which is present in wheat and contains a distinct component of the CES 1.0. There are several species-specific components in each zone electrophoresis spectrum: In ω—OEP 0.21 and 0.28, and gamma—EIA zone 0.36; 0.61 and 0.64, and in β—0.90 areas EIA, regardless of which habitats are present in all the samples studied. Most are variable ω and beta range area, due to the absence of some distinct components or variability of their intensity, as well as the emergence of new components: EIA: 0.25, 0.44, 0.74, and 0.84.

Thus, we detected intraspecific variability of certain components in four species of the genus *Aegilops*, depending on growing conditions. It was found that each species has its own characteristic spectrum of a protein with a specific set of components and distribution of gliadin. Certain types of the detected number of components can be used as a protein marker in the analysis of certain characteristics. The findings create the preconditions for the use of *Aegilops* genus in the breeding as sources of genes for improving plant resistance to disease, drought and soil salinity. According to the results, it was concluded:

1. The genus *Aegilops* L., growing on the territory of Tajikistan, has a wide range of intraspecific variability in phenotypic characteristics and composition of the main component of the protein gluten-gliadin.

2. Among *A. tauschii* and *A. triuncialis* biotypes isolated form, resistant to extreme environmental factors and has increased immunity to disease.

It is known that the cluster analysis as pattern recognition method is a highly informative tool for a wide range of biochemical studies, beginning with applications in the classification of genotypes and ending with the recognition of the fundamental problems of biopolymers, as well as establishing the factors regulating gene expression (Savić, 1991). Dendrological electrophoretic analysis of the components of gliadin proteins of the distribution zones in the four studied species *Aegilops* allowed dividing them into groups and subgroups, depending on their genetic similarity to the different climatic conditions in the places of sampling. As the data matrix of genetic distances (schemes 1–3), detectable levels of intra- and interspecies genomic polymorphism on gliadin proteins vary considerably. In this case, the maximum diversity within species was observed for *A. taucshii*, and the smallest in *A. crassa*, *A. triuncialis,* and *A. cylindrica* on the same value occupied an intermediate position. Thus, the analysis of electrophoretic spectra of gliadin (alcohol soluble) protein 4 species of the genus *Aegilops* L. showed that each of the studied species has its own characteristic spectrum (i.e., their genotype passport) with a number of components and their relative mobility. At the same time, it revealed the intraspecific and interspecific variability of certain components of protein, depending on growing conditions. For individual types it found a number of components that can be used as markers of the protein responsible for certain symptoms. A dendrological analysis of these proteins contributes to a deeper level and a comparative study of intraspecific differences and the degree of similarity, as well as gives an opportunity to capture the genetic differences between different species, ecotypes, and even individual instances of the same species in different ecological zones. In addition, the data creates the prerequisites for successful use of *Aegilops* species in breeding as a gene source with increased resistance to various diseases and adverse environmental factors.

Modern methods of genetics, based on the use of molecular markers allow the study of the organization and variability of plant genome at the level of DNA polymorphism. In this work we studied genotyped species *Aegilops* L., growing in Tajikistan, using different markers (RAPD, SNP, and SSR). The results showed that using RAPD-primers between species and within species, there is a clear polymorphism. All RAPD-primers were informative and, therefore, they can be used to study inter-polymorphism and establishing phylogenetic relationships between species and genera of cereals.

Currently, little studied comparative characteristic levels of intraspecific polymorphism and analysis of phylogenetic relationships of all kinds of *Aegilops* L. with the D-genome. For molecular analysis 15 samples were selected from each species *Aegilops* L., growing on the territory of Tajikistan. The sampling was guided by their geographical confinement so as to cover the maximum range of the species and, thus, try to more fully characterize the genetic variability of each of the analyzed species. The analysis included polyploid species *Aegilops* L., having as part of its genome gene alloploidy D: *A. crassa* (gene HcrDcr1 and HcrDcr1Dcr2), *A. cylindrica* (CcDc) and diploid form of *A. tauschii* (D), which is the D-genome donor of data types. Therefore, the aim of this study is RAPD-analysis of the variability of the diploid species *Aegilops* and alloploidy the genome D. In addition, the analysis was taken by representatives of the other diploid species *A. triuncialis* genome (C and U).

For the RAPD-analysis were taken 17 primers of 10–11 nucleotides in length, selected from the collection of RAPD-primers based on the identification of a larger number of PCR fragments when used to amplify genomic DNA *Aegilops* L. and wheat. To construct a dendrogram for cluster analysis (UPGMA) using a computer program NTSYS-pc, 2.0 used built binary matrix. To assess the level of intraspecific differences for each of the analyzed species *Aegilops* by RAPD-marking 21 selected primers were used, that showed highly informative in analyzing the genomes of cereal. Fifty-three species of four species *Aegilops*, which are found in all 1417 RAPD-fragments, were analyzed. By results of the analysis, it was determined the genetic distance and the method of hierarchical cluster analysis dendrogram constructed. The maximum intraspecific similarity was noted for *A. culindrica* (genetic similarity coefficient, 0.92). In drawing dendrogram *A. triuncialis* discovered that formed two major clusters: The first cluster brings together mainly the samples grown in the northern regions of Tajikistan, and the second cluster—in the south. The dendrogram shows that this species has the highest level of genetic diversity among all studied species *Aegilops*. Within the species most distinguished representatives of the genetic diversity of samples collected from Faizabad district (town Bodom). Between samples of Rudaki and

Varzob (Lohur and Hodzhaobigarm) significant differences were found. In drawing dendrogram *A. cylindrica*, it turned out that this species has a high level of intraspecific variation. Types of *A. crassa* and A. *tauschii* studied by primers showed approximately 50% polymorphism. The dendrogram reveals not a very high level of genetic diversity of these species. View *A. crassa* forms two main clusters. In the first cluster patterns, growing in Rudaki district, and the second cluster is divided into two subclusters, where samples of salt source Hissar grouped. Species *A. tauschii* has a high level of genetic similarity and forms two clusters, the first cluster consists of eight samples were hardly observed having genetic diversity, whereas in the second cluster had a lot of genetic diversity. The task of the research also included the analysis of phylogenetic relationships in the group of species of *Aegilops*, with D-genome, as well as the relationships of species that do not have D-genome. Scheme 4 shows a dendrogram of various types of *Aegilops*. It is well known that specie of diploid *A. tauschii* is D-gene, and species *A. crassa* and *A. cylindrica* are polyploid species but also have the D-genome. Therefore, molecular analysis to establish the phylogenetic relationships of species *Aegilops* with the D-genome is relevant. According to the literature, the diploid form of *A. tauschii* using the RAPD-analysis is the most polymorphic. As for polyploid species *A. crassa* and *A. cylindrica* with the D-genome, they have lower levels of intra-species diversity. Our data show that the samples of the form *A. crassa* have a closer level of genetic similarity with *A. tauschii*, than *A. cylindrica* that, apparently due to the fact that the species *A. cylindrica*, except for D-genome has also a C-genome. Sometimes samples of *A. cylindrica* may form one cluster with other species; in particular, our data show that samples *A. cylindricaform* with samples of the type *A. triuncialis* of Hodzhaobigarma and Lohura general subclusters. Polyploid form of *A. triuncialis* has no D-genome, but has a C-genome and U a genius by which this species forms a single cluster with *A. cylindrica*. Unlike species of D-gene, a species with a U-genome (*A. triuncialis*) are characterized by a significant level of intraspecific diversity. Type *A. triuncialis* among the species studied by us is one of the most common and has a high level of genetic diversity. Type *A. triuncialis* highlighted several examples of cluster growing in the different regions of Tajikistan. Based on our results, we can assume multiple independent origins of polyploid *Aegilops* species from different diploid species. Our findings on the basis of molecular genotyping on 21 RAPD-primer showed varying degrees of intra- and interspecific polymorphism in different species *Aegilops*.

The aim of our work was to compare the genotypes of age-old local and selected varieties of bread wheat (*Triticum aestivum* genome AAVVDD—hexaploid) using molecular DNA markers. It is shown that molecular genetic analysis makes it possible to identify the specific wheat genome RARD markers that can be used for varietal identification of genotypes and establishing phylogenetic relationships between different varieties of wheat. The study is important for local varieties and gene geographic research as it allows not only to get an idea about the main characteristics of the Aboriginal material of a species but also to restore its phylogenetic relationships. These forms are of paramount importance for breeding. It is believed that the age-old varieties and local forms, as a result of a long process of natural and artificial selection are better than others are adapted to local growing conditions and they are not unique to the optimum length of the growing season for the area, but also a set of other economically important traits. During the RAPD-PCR 17 decameric effective arbitrary primers were used, which were evaluated using intravarietal and intraspecific genetic diversity of *Triticum aestivum*. As such phenotypic characteristics as the color of the grain, the color of the ear, and awns, ear, and stalk breakage observed diversity within species. The main zone of separation of fragments is in the range 2000–200 bp. In general, amplified fragments accounted 273, 12–23 (average 16) on one RAPD primer. Of the 273 PCR fragments of 54% were polymorphic between the genotypes studied, 46% had the same length for all samples studied. All primers used in various proportions established the presence of both monomorphic and polymorphic fragments in test samples. Monomorphic fragments can be considered RARD markers to members of the species. The greater the genetic distance between the studied species, the less they have common amplification products. Detectable electrophoresis monomorphic band in different varieties suggest the common structural and functional organization of genomes. Each of the varieties has a certain range of RAPD amplified products, which differs from other quantity fragments, their size, and severity. Some primers revealed unique to one particular class amplicons and are therefore species specific. Significant differences in the number of fragments RAPD between studied varieties were noted. The samples also varied according to the number of unique, characteristic only for one type of amplicons.

Dendrogram of genetic similarity between the studied samples of wheat has been built on the basis of cluster analysis. Dendrogram revealed that the samples are characterized by a relatively high level of polymorphism intravarietal. Thus, the molecular genetic analysis makes it possible to identify specific genomic markers that can be used for varietal identification of genotypes. It is shown that the method of molecular marking of the genome based on RAPD-PCR allows the identification of members of the species *Triticum aestivum* and establishes the phylogenetic relationships between different varieties. The purposeful use of species-specific primers will allow researchers to reduce labor costs and resources required to analyze the collection of samples.

2.8 CONCLUSIONS

The genus *Aegilops* L., growing on the territory of Tajikistan, has a wide range of intraspecific variability in phenotypic characteristics and composition of the main component of the protein gluten-gliadin. Among *A. tauschii* and *A. triuncialis* biotypes isolated form, resistant to extreme environmental factors and has increased immunity to disease. Undoubtedly, the widespread use of DNA markers to assess the polymorphism of the genomes of different species of the genus Triticum will make it possible to build their genetic classification, which will reveal the history of the origin, distribution, and adaptation of wheat. Therefore, this line of research should be further developed to assess the genetic diversity of different types of this culture. In general, the assessment of genetic diversity and breeding wheat and its wild relatives with the genealogical analysis can be useful to clarify the strategies and breeding programs, selection of priority directions in the study of the collection, as well as sampling the formation of a variety of samples for analysis.

KEYWORDS

- **domestication**
- **food security**
- **plant biodiversity**
- **genotyping**
- **environmental factors**

REFERENCES

Alexanyan, S. *Cooperation Strategy of Peace Genebanks in Conditions of Globalization.* Bulletin of Applied Botany, Genetics and Breeding, 2007, 11–33.

Bezrukov, V. F. Dominant Monogenic Markers of Quantitative Traits. *Genetics* **1995,** *31* (10), 1438–1444.

Fourth National Report of the Republic of Tajikistan on Biodiversity. With the Financial Support (GEF), the Global Environment Facility, prepared as part of their obligations under the CBD (Article 26) and on the basis of the COP decision 8/14. N.

Fowler, C. Global Efforts to Save Crop Diversity. SES Edition; 30-31 State of World Report on PGRFA FAO Rome 1998, 2005, p 510.

Graham, J. S. D.; Morton, R. K.; Raison, J. K. Isolation and Characterization of Protein Bodies from Developing Wheat Endosperm. *Austral. J. Biol. Sci.* **1963,** *16*, 375–383.

Karimov, H. H. *The State, Problems and Prospects of Development of Biological Sciences in Tajikistan;* Donish Publisher: Dushanbe, 2008.

Khavkin, E. E. Molecular Markers in Plant Breeding. *Agri. Biol.* **1997,** *5,* 3–21.

Konarev, V. G. Protein Markers in Varietal Identification and Registration of Plant Genetic Resources. Collection of Scientific Works of Applied Botany, Genetics and Breeding, Leningrad, 1987, *114*, 3–11.

Konarev, V. G. Proteins as Genetic Markers in the Study of the Nature and Origin of the Genomes of Crop Plants. Works of Applied Botany, Genetics, and Breeding 1979, *63* (3), 3–8.

Koval, S. F. What Is a Model of Sorts? Monograph Omsk; 1999, 200.

Kumak, V. A. Physiological Substantiation of Models of Wheat M, Agropromizdat, 1985, 268.

Merezhko, A. F. Donor Problem in Plant Breeding. 1994, 126.

Mitrofanova, O. P. Wheat Genetic Resources in Russia: State and Study predselektsionnoe. Vavilov *J. Gen. Breed.* **2012,** *16* (1): 10–20.

Nasyrova, F. U.; Naimov, S.; Nigmonov, M. Catalogue of Species and Genus Aegilops Specimens Collected in Different Ecological Geographic Areas of Tajikistan. Dushanbe, 2007, 56.

Nazirov, H. N. Selection and Hozyaystvnnaya Value of Local Varieties and forms Sievers Apple *Malus sieversii* (Ledeb.) M Roem in Tajikistan. Abstract of dissertation for the degree of Doctor of Agricultural Sciences: Dushanbe, 2011.

Proceedings of Applied Botany, Genetics and Breeding. Including C-164, Petersburg, 2007.

Safarov. Dushanbe, The National Center for Biodiversity and Biosafety, 2009, 117.

Savchenko, A. D.; Imamkulova, Z. A., Ahmad, H. M. Horticultural Crops Pistachios in Tajikistan, Dushanbe, 2009, 19.

Savich, I. M. Izoperoksidazy and Polypeptides Storage Proteins as Biochemical Markers of Maize, Rice, and Sorghum. Abstract of Dissertation for the Degree of doc Biol Science Moscow, 1991, 45.

Second National Report on biodiversity conservation in Tajikistan Safarov. Dushanbe: The National Center for Biodiversity and Biosafety, 2005, 116.

Sozinov, A. A. *Polymorphism of Proteins and its Importance in Genetics and Breeding;* Nauka: Moscow, 1985, 272.

Strategic Framework for the Implementation of a European Genebank Integrated System (2006); (Discussion Paper). IPGRI Italy, pp 1–0.
The Convention on Biological Diversity. Text and Applications Geneva, Switzerland: The CBD Secretariat 2002, 34.
The Genetic Resources of Cultivated Plants in the XXI Century. Vavilov II International Conference Status, Problems and Prospects. Abstracts, St Petersburg, 2007.
Vavilov, N. I. Breeding as a Science. Theoretical Basis of Plant Breeding M L, 1935, 17–74.
Vavilov, N. I. Global Centers of High-quality Resources (genes) of Cultivated Plants. In Report at the General Meeting of the V International Congress of Genetics, 1987, pp 146, 1927 Sep; Berlin. (In the book "The Origin and Geography of Cultivated Plants").
Virchow, D. Efficient Conservation of Crop Genetic Diversity. Theoretical Approaches and Empirical Studies//Springer, 2003, 247.
World Resource Institute. UNEP//Conservation International, 2006, 10.
Zapryagaeva, V. I. *Wild Fruit of Tajikistan;* Science Publishing House: Moscow, 1964.

CHAPTER 3

Phytodiversity and Conservation Status of the Nara Desert, Pakistan

R. QURESHI

Department of Botany, Pir Mehr Ali Shah Arid Agriculture University, Rawalpindi, Pakistan

Corresponding author. E-mail: rahmatullahq@uaar.edu.pk; rahmatullahq@yahoo.com

ABSTRACT

Nara Desert is an arid tract located in Sindh Province, Pakistan between latitudes 26°–28° north and longitude 68°–70° east. Phytogeographically, most of the area falls under the Saharo-Sindian Region, with a small patch of western portion represented by the Indian elements. The area typically possesses xerophytic vegetation. This is a hot sandy desert with the highest temperature with very low and diurnal fluctuation in precipitation. The area is presenting a unique landscape with 267 species including 3 ferns, 1 gymnosperm, and 64 monocots. The present work is thus a contribution of more than 10 years exploration of this xeromorphic landscape. This fragile ecosystem supports human and livestock population by providing fuel, food, fodder, sheltering, ethnomedicine, etc., as natural resources. Furthermore, the increasing incidence of lingering drought, desertification, deforestation, and soil erosion posed serious threats to this area. These factors have exerted an enormous burden on the ecology of the region. Floristic biodiversity, human livelihood, threats, and conservation issues and their some possible measures are discussed in this chapter.

3.1 INTRODUCTION

The Nara Desert, Pakistan is located at 26°–28° N and 68°–70° E (Fig. 3.1), with 50- and 115-m elevations. This desert is bordered by eastern side by Rajasthan, India, that is, Jodhpur and Jaisalmer. Topographically, the area possessed three physical features sandy hills, steep slopes, and vast low-lying flat areas. The accretion of sand in the gigantic mass outlines a hill which is known as sand dune/ridges. The dune has three permanent features crest, the topmost portion; Swale (flank)/sloes, the middle portion, and foot; the base of the dune. The area flanked by two succeeding dunes is known as an interdunal valley (Qureshi and Bhatti, 2005a). Most of the area (approximately 80%) is gentle undulating sandy plain with frequent sand dunes, however, the number and size of which increases from West to East.

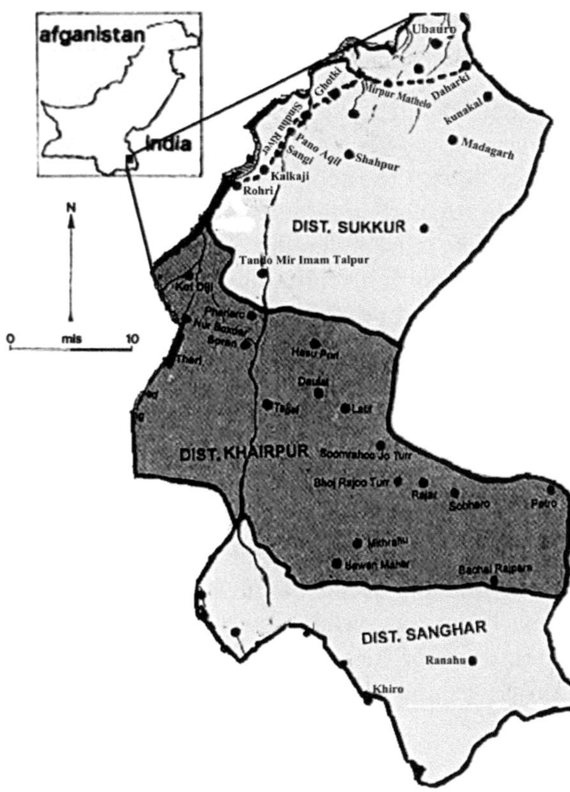

FIGURE 3.1 Location map of the Nara desert, Sindh, Pakistan.

This is a scorching sandy desert with the temperature fluctuations of 20–45°C. Majority of the area falls in hyperarid to arid region, typified by tremendous temperature (a long dry summer), extreme drought adjunct by high wind velocity, low humidity, high evapotranspiration, and too sparse rainfall. The southern part (Khipro, Sanghar) is relatively semiarid in nature that supports somewhat dense vegetation. The annual rainfall varies from 88 to 135 mm, typically received throughout the monsoon. The water is scarce in the Nara desert and the groundwater resources are lying at the depth of 50–300 feet from the surface (Qureshi and Bhatti, 2005b).

The Nara Desert is blessed with a range of flora. Keeping in view, a field survey was conducted to document the flora along with the identification of problems/threats to this unique landscape and their possible suggestion for the rehabilitation of the degraded lands. Previously, the author contributed some papers on the flora and vegetation of this unique landscape (Qureshi, 2008a, 2008b; Qureshi and Bhatti, 2008a, 2008b, Qureshi, 2011, Qureshi and Bhatti, 2014). The present communication is updated information on the flora and their status that can be utilized by the scientific community for the planning and management and their sustainable utilization.

3.2 MATERIALS AND METHODS

The whole study area was extensively surveyed to record the flora from September 1998 to December 2008. Based on physiognomic features, six distinct habitat types were delineated linking with the typical flora. During the survey plant specimens along with their local names and ecological features were also recorded. The voucher specimens were prepared using the standard method and identified with the help of various floristic materials (Cooke, 1903–1906; Jafri, 1966; Nasir and Ali, 1970–1989; Ali and Nasir, 1989–1993; Ali and Qaiser, 1993–1995, 2000–2004; Batanouny, 1981; Matthew, 1981–1983; Shetty and Singh, 1987, 1991; Bhandari, 1987; Boulos, 1991). The determined specimens were placed in the Shah Abdul Latif University Botanical Garden and Herbarium, Khairpur. The ecological features, land/vegetation degradation was observed personally as well as discussed with the local inhabitants and recorded in the field notebook. Based on observations, some remedial measures are proposed.

3.3 RESULT AND DISCUSSION

3.3.1 PHYTOGEOGRAPHY

Phytogeographically, most of the study area comes under the Saharo-Sindian region, whereas its small patch of the western portion is also represented by the species from the Indian Region (Ali and Qaiser, 1986). All of the areas possess typically the xerophytic vegetation. The hot dry deserts of Cholistan, Tharparkar, and Thal represent similar vegetation and are, therefore, included in the Saharo-Sindian region.

3.3.1.1 FLORA AND VEGETATION

Most of the floristic elements are of Saharo-Sindian origin. However, a small patch of the western portion may represent Indian elements (Ali and Qaiser, 1986). The whole area represents typically the xerophytic vegetation. Floristically, the area possesses 267 species distributed across 167 genera and 59 families (Table 3.1). The floristic composition revealed that the flora includes 3 ferns, 1 gymnosperm, and 64 monocots (Table 3.2). There were four monocotyledon families and amongst them, the Poaceae family contributed the largest number of taxa (28 genera, 47 spp.), followed by Cyperaceae (4 genera and 14 spp.). Out of 51 dicotyledonous families, Fabaceae contributed highest taxa (12 genera and 26 spp.), followed by Asteraceae (10 genera, 13 spp.), Boraginaceae (4 genera, 9 spp.), Capparidaceae (6 genera, 9 spp.), Euphorbiaceae (4 genera, 8 spp.), and Amaranthaceae, Chenopodiaceae, Solanaceae (5 genera, 7 spp. each) and Convolvulaceae (4 genera and 5 spp.).

There were 11 life forms/habits of the flora recorded from the Nara Desert. Table 3.3 depicted the life form of the existing vegetation that reveals that most of the area was dominated by herbs (114 spp., 42.70%), followed by grasses (47 spp., 17.60%), shrubs (35 spp., 13.11%), subshrubs (29 spp., 10.86%), sedges (14 spp., 5.24%), and trees (13 spp., 4.87%). The desert generally takes a green tinge after the rains in the early spring. The ephemerals cannot sustain generally the burning heat and the long-lasting drought and escape soon, after producing seed for the next favorable season. The vegetation composition comprised on common trees and shrubs, such as *Acacia nilotica, Prosopis cineraria, Tamrarix aphylla, Salvadora oleoides, Zizyphus nummularia, Calligonum polygonoides, Aerva javanica, Crotalaria burhia, Dipterygium glaucum*, and *Leptadenia pyrotechnica.*

TABLE 3.1 Inventory of the Flora Along Their Families, Local Names, Life Form/Habit, and Their Conservation Status Recorded from the Nara Desert, Pakistan.

S. No.	Family	Plant species	Local name	Life form/habit	Conservation status
1	Acanthaceae	*Barleria acanthoides* Vahl	–	Subshrub	Rare
2	Acanthaceae	*Blepharis sindica* Stocks ex. Anders	Utangan	Herb	Rare
3	Adiantaceae	*Adiantum capillus-veneris* L.	Pursha	Fern	Rare
4	Aizoaceae	*Gisekia pharancoides* L.	Buto	Herb	Rare
5	Aizoaceae	*Limeum indicum* Stocks ex T. Anderson.	Dhoor Chhapri	Herb	Very common
6	Aizoaceae	*Sesuvium sesuvioides* (Fenzl) Verdc.	Lonak	Herb	Common
7	Aizoaceae	*Trianthema portulacastrum*L.	Waho	Herb	Very common
8	Aizoaceae	*T. triquettra* Rottl. and Willd.	Wasanh/Waho	Herb	Rare
9	Aizoaceae	*Zelya petandra* (L.) Jeffrey	Wasanh/Waho	Herb	Very common
10	Amaranthaceae	*Achyranthes aspera* L.	Ubat Kandri	Herb	Rare
11	Amaranthaceae	*Aerva javanica* var. *bovei* Webb.	Booh	Subshrub	Rare
12	Amaranthaceae	*A. javanica* var. *javanica* (Burm.f.) Juss. ex. Journ. A. Schultes.	Booh	Subshrub	Very common
13	Amaranthaceae	*Amaranthus graecizans* subsp. *thllungianus* (Nevski) Gusev.	Mariro	Herb	Common
14	Amaranthaceae	*A. virdis* L.	Mariro	Herb	Common
15	Amaranthaceae	*Celosia argentea* L.	Boobak	Herb	Common
16	Amaranthaceae	*Digera muricata* (L.) Mart.	Lulur	Herb	Common
17	Arecaceae	*Phoenix sylvestris* (L.) Roxb.	Jhangli khaji	Tree	Rare
18	Aristolochiaceae	*Aristolochia bracteolata* Lamk.	Zeerawund	Herb	Vulnerable
19	Asclepiadaceae	*Calotropis procera* (Willd.) R. Br.	Ak	Shrub	Common
20	Asclepiadaceae	*Leptadenia pyrotechnica* (Forssk.) Decne.	Khip	Shrub	Common
21	Asclepiadaceae	*Oxystelma esculentum* (L.f.) R. Br.	Phuli	Liana	Rare

TABLE 3.1 *(Continued)*

S. No.	Family	Plant species	Local name	Life form/habit	Conservation status
22	Asclepiadaceae	*Pentatropis nivalis* (J.F.Gmel.) Field & J. R. I. Wood	Wan Wehri	Climber	Rare
23	Asteraceae	*Amberboa ramosa* (Roxb.) Jafri.	–	Herb	Vulnerable
24	Asteraceae	*Carthamus oxycantha* Bieb.	Pohli	Bushy subshrub	Common
25	Asteraceae	*Cirsium arvense* (L.) Scop.	Leh	Herb	Very common
26	Asteraceae	*Echinops echinatus* DC.	Kanderi Bhattar	Bushy subshrub	Rare
27	Asteraceae	*Eclipta prostrate* (L.) L.	Bhangrio	Herb	Rare
28	Asteraceae	*Iphiona grantioides* (Boiss.) Anderb.	Gandraf	Herb or subshrub	Common
29	Asteraceae	*Launaea procumbens* (Roxb) Ramayya & Rajagopal.	Bhattar	Herb	Rare
30	Asteraceae	*L. resedifolia*(L.) Kuntze	Bhattar	Herb	Vulnerable
31	Asteraceae	*Pluchea lanceolata* (DC.) Oliv. & Hiern.	Phaar Buti	Shrub	Common
32	Asteraceae	*P. wallichiana* DC.	–	Herb	Vulnerable
33	Asteraceae	*Sonchus asper* (L.) Hill.	Bhattar	Herb	Common
34	Asteraceae	*S. oleraceus*L.	Bhattar	Herb	Common
35	Asteraceae	*Xanthium strumarium* L.	Bhurt	Shrub	Very Common
36	Bignoniaceae	*Tecomella undulata* (Sm.) Seem.	Lohiro	Shrub/Small tree	Threatened
37	Boraginaceae	*Arnebia hispidissima* (Sieber) DC.	Harnub	Herb	Threatened
38	Boraginaceae	*Cordia gharaf* (Forssk.) Ehren. ex Asch.	Lesuro	Small tree	Vulnerable
39	Boraginaceae	*C. myxa* L.	Lesuro	Small tree	Vulnerable
40	Boraginaceae	*Heliotropium crispum* Desf.	Kharsan	Herb	Very Common
41	Boraginaceae	*H. curassavicum* L.	Kharsan	Herb	Rare
42	Boraginaceae	*H. europeum* L.	Uth Charo	Herb	Common
43	Boraginaceae	*H. strigosum* Willd.	Kharsan	Herb	Rare
44	Boraginaceae	*H. subulatum* (Hochst. ex DC.) Vatke.	Kharsan	Herb	Rare

TABLE 3.1 *(Continued)*

S. No.	Family	Plant species	Local name	Life form/habit	Conservation status
45	Boraginaceae	*Trichodesma indicum* (L.) R.Br.	Gaaozaban	Herb	Rare
46	Brassicaceae	*Farsetia hamiltonii* Royale.	Lathio	Herb	Common
47	Brassicaceae	*F. jacquemontii* Hook. f.& Thoms.	Lathio	Herb	Common
48	Brassicaceae	*Coronopus didymus*(L.) Smith	Jhangli halon	Herb	Very Common
49	Burseraceae	*Commiphora stocksiana* (Engl.) Engl.	Gugur	Shrub	Threatened
50	Burseraceae	*C. wightii* (Arn.) Bhandari	Gugur	Shrub	Threatened
51	Caesalpiniaceae	*Caesalpinia bonduc* (L.) Roxb.	Pahar Wal	Liana	Rare
52	Caesalpiniaceae	*Senna holosericea*(Fresen.) Greuter	Sindhi senna	Herb	Common
53	Caesalpiniaceae	*S. italica* Mill.	Ghora Wal	Herb	Very Common
54	Capparidaceae	*Cadaba farinosa* Forssk.	Khabri Wal	Shrub	Rare
55	Capparidaceae	*Capparis decidua* (Forssk.) Edgew.	Kirar	Shrub	Rare
56	Capparidaceae	*C. spinosa* L.	Golaro	Sub-shrub	Rare
57	Capparidaceae	*Cleome brachycarpa* Vahl.	Dhanaar Khathoori	Herb	Rare
58	Capparidaceae	*C. scaposa* D.C.	Khathoori	Herb	Very common
59	Capparidaceae	*C. viscosa* L.	Kinni Buti	Herb	Common
60	Capparidaceae	*Dipterygium glaucum* Decne.	Phair	Subshrub	Common
61	Capparidaceae	*Gynandropsis gynandra* (L.) Briq.	Kinro	Herb	Rare
62	Capparidaceae	*Maerua crassifolia* Forssk.	Khabar	Shrub	Threatened
63	Caryophyllaceae	*Vaccaria hispanica* (Miller) Rauschert	Taklo	Herb	Common
64	Caryophyllaceae	*Spergula arvensis*L.	Lani	Herb	Common
65	Chenopodiaceae	*Arthrocnemum indicum* (Willd.) Moq.	Lano	Shrub	Common
66	Chenopodiaceae	*Chenopodium album* L.	Chil	Herb	Very common
67	Chenopodiaceae	*C. murale* L.	Chil	Herb	Common

TABLE 3.1 *(Continued)*

S. No.	Family	Plant species	Local name	Life form/habit	Conservation status
68	Chenopodiaceae	*Haloxylon salicornicum* (Moq.) Bunge ex Boiss.	Lano	Shrub	Common
69	Chenopodiaceae	*H. stocksii* (Boiss.) Benth. & Hook.	Sacho Lano	Shrub	Rare
70	Chenopodiaceae	*Salsola imbricata* Forssk.	Lano	Subshrub	Very common
71	Chenopodiaceae	*Saueda fruticosa* (L.) Forssk.	Lani	Subshrub	Very common
72	Convolvulaceae	*Cressa cretica* L.	Oin	Herb	Rare
73	Convolvulaceae	*Convolvulus arvensis* L.	Naro	Climber	Common
74	Convolvulaceae	*C. glomeratus* Choisy.	Kirhanj	Prostrate herb	Rare
75	Convolvulaceae	*C. prostratus* Forssk.	Kirhanj	Herb	Vulnerable
76	Convolvulaceae	*Ipomoea aquatica* Forssk.	Naro	Herb	Rare
77	Convolvulaceae	*I. carnea* (Martius ex Choisy) D. F. Austin	Akri	Shrub	Common
78	Convolvulaceae	*Seddera latifolia*Hochst. et Steud.	–	Shrub	Threatened
79	Cucurbitaceae	*Citrullus colocynthis* (L.) Schrad.	Trooh	Prostrate subshrub	Very common
80	Cucurbitaceae	*Ctenolapis cerasiformis* (Stocks.) Hooker.f.	–	Prostrate herb	Rare
81	Cucurbitaceae	*Cucumis melo* var. *agrestis* Naudin.	Mitero	Prostrate herb	Common
82	Cucurbitaceae	*C. prophaterum* L.	Kando Mitero	Prostrate herb	Rare
83	Cucurbitaceae	*Luffa echinata* Roxb.	Jhangli toori	Climber	Threatened
84	Cucurbitaceae	*Mukia maderaspatana* (L.) M. J. Roem.	Chibhri Wal	Prostrate herb	Common
85	Cuscutaceae	*Cuscuta chinensis* Lam.	Bepari	Stem parasite	Common
86	Cyperaceae	*Bolboschoenus affinis* (Roth.) Drobov	Kaluro	Sedge	Common
87	Cyperaceae	*B. glaucus* (Lam.) S. G. Smith	Kaluro	Sedge	Common
88	Cyperaceae	*Cyperus arenarius* Retz.	Moniah Gaah	Sedge	Vulnerable
89	Cyperaceae	*C. articulatus* L.	Kal	Sedge	Common
90	Cyperaceae	*C. bulbosus* Vahl	–	Sedge	Rare

TABLE 3.1 *(Continued)*

S. No.	Family	Plant species	Local name	Life form/habit	Conservation status
91	Cyperaceae	*C. conglomeratus* Rottb.	Moniah Gaah	Sedge	Vulnerable
92	Cyperaceae	*C. difformis* L.	–	Sedge	Common
93	Cyperaceae	*C. iria*L.	–	Sedge	Common
94	Cyperaceae	*C. laevigatus* L.	–	Sedge	Very common
95	Cyperaceae	*C. longus* L.	–	Sedge	Very common
96	Cyperaceae	*C. rotundus* L.	Kabah	Sedge	Very common
97	Cyperaceae	*Fimbristylis bisumbellata* (Forssk.) Bubani	Kal	Sedge	Common
98	Cyperaceae	*F. cymosa* R. Br.	Kal	Sedge	Common
99	Cyperaceae	*Schoenoplectus litoralis* (Schrad.) Palla	Kaluro	Sedge	Common
100	Elatinaceae	*Bergia aestivosa* Wight & Arn.	–	Sub-shrub	Rare
101	Ephederaceae	*Ephedra ciliata* Fisch. & Mey. ex C. A. Meyer.	Santh	Liana	Vulnerable
102	Equisetatceae	*Equisetum arvense*L.	–	Fern	Rare
103	Euphorbiaceae	*Chrozophora plicata* (Vahl) A. Juss. ex Spreng.	Nil Kanthi	Subshrub	Rare
104	Euphorbiaceae	*Euphorbia caducifolia* Haines	Thor	Shrub	Rare
105	Euphorbiaceae	*E. clarkeana* Hook. f.	Kher Wal	Herb	Very common
106	Euphorbiaceae	*E. hirta* L.	Kheer Wal	Herb	Common
107	Euphorbiaceae	*E. thymifolia* L.	Kheera Wal	Herb	Common
108	Euphorbiaceae	*Phyllanthus maderaspatensis* L.	Hazar dani	Herb	Rare
109	Euphorbiaceae	*P. reticulatus* Pair.	Kamho	Shrub	Rare
110	Euphorbiaceae	*Ricinus communis* L.	Heran buti	Shrub	Common
111	Fabaceae	*Alhagi maurorum* Medic.	Kandero	Subshrub	Very common
112	Fabaceae	*Alysicarpus monilifer* (L.) DC.	–	Herb	Rare
113	Fabaceae	*A. scariosus* Grah. ex Thwaites	–	Herb	Rare

TABLE 3.1 *(Continued)*

S. No.	Family	Plant species	Local name	Life form/habit	Conservation status
114	Fabaceae	*Argyrolobium roseum*ssp. *Ornithopodioides* (Jaub. & Spach) Jafri & Ali	–	Herb	Rare
115	Fabaceae	*Crotolaria burhia* Ham. ex Benth.	Chagg	Subshrub	Common
116	Fabaceae	*C. medicaginea* var. *medicaginea*Lamk.		Herb	Rare
117	Fabaceae	*Indigofera argentea* Burm.f.	Lathio	Herb	Rare
118	Fabaceae	*I. cordifolia* Heyne ex Roth	Lathio	Herb	Common
119	Fabaceae	*I. hochstetteri* Baker	Bheekar	Herb	Rare
120	Fabaceae	*I. linifolia* (L.) Retz.	Vekar	Herb	Rare
121	Fabaceae	*I. oblongifolia* Forssk.	Jhil	Sub-shrub	Common
122	Fabaceae	*I. sessiliflora* DC.	Vekar	Herb	Rare
123	Fabaceae	*Lathyrus aphaca* L.	Jhangli matri	Herb	Common
124	Fabaceae	*Medicago polymorpha* L.	Mena	Herb	Very common
125	Fabaceae	*Melilotus alba* Desr.	Sinjhi	Herb	Common
126	Fabaceae	*M. indica*(L.) All	Sinjhi	Herb	Common
127	Fabaceae	*Rhynchosia minima* (L.) DC.	Wan Wehri	Climber	Common
128	Fabaceae	*R. pulverulenta*Stocks	Wan vehri	Climber	Rare
129	Fabaceae	*R. schimperii* Hochst. ex Boiss.	–	Subshrub	Vulnerable
130	Fabaceae	*Sesbania bispinosa* (Jacq.) W. F. Wight	Jantur	Subshrub	Common
131	Fabaceae	*Tephrosia falciformis* Romaswami.	Drebarr	Herb	Vulnerable
132	Fabaceae	*T. purpurea*(L.) Pers.	Andhri/Siringh	Subshrub	Rare
133	Fabaceae	*T. strigosa* (Dalz.) Sant. & Mahesh.	–	Herb	Rare
134	Fabaceae	*T. uniflora* var. *petrosa* (Blatter and Hallberg) Gillett and Ali	Andhri/Siringh	Subshrub	Very common
135	Fabaceae	*T. villosa* (L.) Pers.	Andhri/Siringh	Subshrub	Very common

TABLE 3.1 *(Continued)*

S. No.	Family	Plant species	Local name	Life form/habit	Conservation status
136	Fabaceae	*Vicia sativa*L.	Matri	Herb	Common
137	Frankeniaceae	*Frankenia pulveulenta*L.	–	Herb	Rare
138	Fumariaceae	*Fumaria indica* (Hausskn.) H. N. Pugsley	Doonhi Buti/ Shahtaro	Herb	Very common
139	Geraniaceae	*Monsonia heliotropoides* (Cav.) Boiss.	–	Herb	Rare
140	Hydrocharitaceae	*Hydrilla verticillata* (L. f.) Royale	Machhi Gaah	Herb	Rare
141	Lamiaceae	*Leucaus aspera* (Willd.) Link.	Goomi Buti	Herb	Rare
142	Lamiaceae	*Salvia santolinifolia*Boiss.	–	Bushy	Threatened
143	Liliaceae	*Asphodelus tenuifolius* Cavan	Basri	Herb	Very common
144	Malvaceae	*Abutilon fruticosum* Guill. & Perr.	Pat Teer	Subshrub	Rare
145	Malvaceae	*A. indicum* (L.) Sweet	Pat Teer	Shrub	Common
146	Malvaceae	*A. muticum* (Del.ex DC.) Sweet	Pat Teer	Shrub	Common
147	Malvaceae	*Hibiscus lobatus* (J. A. Murray) O. Ketze.	–	Herb	Rare
148	Malvaceae	*Malvastrum coromendelianum* (L.) Garcke	–	Subshrub	Very common
149	Malvaceae	*Pavonia glechomaefolia* S. Abedin	–	Herb	Rare
150	Marsileaceae	*Marsilea minuta*L.	Lusani	Fern	Rare
151	Menispermaceae	*Cocculus hirsutus* (L.) Diels	Fareed Buti	Climber	Rare
152	Menispermaceae	*C. pendulus* (J. R. and Forst) Diels	–	Liana	Rare
153	Mimosaceae	*Acacia jacquemontii* Benth.	Banwar	Shrub	Common
154	Mimosaceae	*A. nilotica* ssp. *cupressiformis* (J. L. Stewart) Ali and Faruqi	Sindhi Babur	Tree	Rare
155	Mimosaceae	*A. nilotica* subsp. *indica* (Benth.) Brenan	Sindhi Babur	Tree	Very common
156	Mimosaceae	*A. senegal* (L.) Willd.	Khaunr	Tree	Rare
157	Mimosaceae	*Prosopis cineraria* (L.) Druce	Kandi	Tree	Common

TABLE 3.1 *(Continued)*

S. No.	Family	Plant species	Local name	Life form/habit	Conservation status
158	Mimosaceae	*P. glandulosa* Torr.	Kando	Shrub	Rare
159	Mimosaceae	*P. juliflora* (Swartz) DC.	Devi	Shrub	Very common
160	Molluginaceae	*Glinus lotoides* L.	Kotak	Herb	Rare
161	Molluginaceae	*Mollugo cerviana* (L.) Ser.	Hazar Dani	Herb	Common
162	Molluginaceae	*M. nudicaulis* Lamk.	Hazar Dani	Herb	Rare
163	Molluginaceae	*M. pentaphylla* L.	Hazar Dani	Herb	Rare
164	Moraceae	*Ficus religiosa* L.	Pipul	Tree	Rare
165	Moraceae	*F. bengalensis* L.	Barr	Tree	Common
166	Moringaceae	*Moringa oleifera* Lamk.	Suhandaro/ Suhanjro	Tree	Rare
167	Neuradaceae	*Neurada procumbens* L.	Chhipri	Herb	Vulnerable
168	Nyctaginaceae	*Boerhavia diffusa* L.	Dakhri	Herb	Rare
169	Nyctaginaceae	*B. procumbens* Banks ex Rxb.	Dakhri	Herb	Common
170	Nyctaginaceae	*B. rubicunda* Steud.	–	Herb	Threatened
171	Nyctaginaceae	*Commicarpus boisseiri* (Heimer) Cufod.	Dakhri	Herb	Common
172	Orobanchaceae	*Cistanche tubulosa* (Schrenk) Hook. f.	Bhunjro	Root parasite	Common
173	Oxalidaceae	*Oxalis corniculata* L.	Khatti buti	Herb	Very common
174	Poaceae	*Aeluropus lagopoides* (L.) Trin. ex. Thw	Kalar Gaah	Grass	Common
175	Poaceae	*Aristida adscensionis* L.	Lumb Gaah	Grass	Very common
176	Poaceae	*A. hystricula*Edgew.	Lumb Gaah	Grass	Common
177	Poaceae	*A. funiculata* Trin. and Pupr.	Lumb Gaah	Grass	Common
178	Poaceae	*A. mutabilis* Trin. and Rupr.	Lumb Gaah	Grass	Common
179	Poaceae	*Avena fatua* L.	Jangli Jai	Grass	Common

TABLE 3.1 *(Continued)*

S. No.	Family	Plant species	Local name	Life form/habit	Conservation status
180	Poaceae	*Brachiaria eruciformis* (J. E. Smith) Griseb.	Sawri	Grass	Common
181	Poaceae	*B. ramosa* (L.) Stapf	Sawri	Grass	Common
182	Poaceae	*B. ovalis* (L.) Stapf	Sawri	Grass	Common
183	Poaceae	*Cenchrus biflorus* Roxb.	Mohabbat buti	Grass	Very common
184	Poaceae	*C. ciliaris* L.	Bhurt	Grass	Very common
185	Poaceae	*C. pennisetiformis* Hochst. and Steud. ex Steud.	Bhurt	Grass	Common
186	Poaceae	*C. prieurii*(Kunth) Maire	Bhurt	Grass	Rare
187	Poaceae	*C. setigerus* Vahl.	Bhurt	Grass	Rare
188	Poaceae	*Chloris barbata* Sw.	Gaah	Grass	Common
189	Poaceae	*Cymbopogon jwarancusa* (Jones) Schult.	Kattan	Grass	Very common
190	Poaceae	*Cynodon dactylon* (L.) Pers.	Chhabar	Grass	Very common
191	Poaceae	*Dactyloctenium aegyptium* (L.) P. Beauv.	Gandheer Gaah	Grass	Very common
192	Poaceae	*D. aristatum* Link	Gandheer Gaah	Grass	Common
193	Poaceae	*D. scindicum* Boiss.	Gandheer Gaah	Grass	Common
194	Poaceae	*Desmostachya bippinata* (L.) Stapf.	Drabh	Grass	Very common
195	Poaceae	*Dichanthium annulatum* (Forsskal) Stapf.	Dinohi	Grass	Common
196	Poaceae	*Digitaria sanguinalis* (L.) scop.	Karash	Grass	Common
197	Poaceae	*Dinebra retroflexa* (Vahl.) Panzer.	Gaah	Grass	Rare
198	Poaceae	*Echinochloa colona* (L.) Link	Sanwak	Grass	Common
199	Poaceae	*Eragrostis barrelieri* Dav.	Gaah	Grass	Common
200	Poaceae	*E. cilianensis* (All.) Lut. ex F.T. Hubbard	Gaah	Grass	Common
201	Poaceae	*E. ciliaris*(L.) R.Br.	Gaah	Grass	Common
202	Poaceae	*E. minor* Host.	Makhni Gaah	Grass	Common

TABLE 3.1 *(Continued)*

S. No.	Family	Plant species	Local name	Life form/habit	Conservation status
203	Poaceae	*E. tenella*(L.) P. Beauv. ex Roem. & Schult.	Gaah	Grass	Common
204	Poaceae	*Imperata cylindrica* (Linn) Raeuschel.	Drabh	Grass	Rare
205	Poaceae	*Lasiurus sindicus* Henr.	Booro	Grass	Vulnerable
206	Poaceae	*Leptothrium senegalense* (Kunth) W.D. Clayton	Gaah	Grass	Rare
207	Poaceae	*Ochthochloa compressa* (Forssk.) Hilu.	Mandhano	Grass	Very common
208	Poaceae	*Panicum turgidum* Forsk.	Sewan	Grass	Common
209	Poaceae	*Paspalidium geminatum* (Forssk.) Stapf in Prain	Gaah	Grass	Common
210	Poaceae	*Paspalum paspalodes* (Michx.) Scribner	Gaah	Grass	Common
211	Poaceae	*P. scrobiculatum*L.	Gaah	Grass	Common
212	Poaceae	*Phalaris minor* Retz.	Banak	Grass	Very common
213	Poaceae	*Phragmites australis* (Cav.) Trin. ex Steud.	Nar/Naro	Grass	Rare
214	Poaceae	*P. karka* (Retz.) Trin. ex Steud.	Nar/Naro	Grass	Very common
215	Poaceae	*Polypogon fugax* Nees ex Steud.	Nor puchh	Grass	Common
216	Poaceae	*Saccharum bengalense* Retz.	Booro	Grass	Common
217	Poaceae	*S. griffithii* Munro ex Boiss..	Booro	Grass	Rare
218	Poaceae	*S. spontaneum* L.	Kanh	Grass	Common
219	Poaceae	*Setaria pumila* (Pouir.) Roem. and Schult.	Sawri	Grass	Very common
220	Poaceae	*Stipagrostis plumosa* (L.) Munro ex T. Anders.	Lumb Gaah	Grass	Very common
221	Polgonaceae	*Calligonum polygonoides* L.	Phog	Shrub	Very common
222	Polgonaceae	*Persicaria barbata*(L.) Hara	–	Herb	Rare
223	Polgonaceae	*P. glabra* (Willd.) M. Gómes	–	Herb	Rare
224	Polgonaceae	*Polygonum plebejum* R.Br.	Kheer Wal	Herb	Very common
225	Polgonaceae	*Rumex dentatus* subsp. *klotzschianus* (Meisn.) Rech. f.	Jhangli palak	Herb	Very common

TABLE 3.1 *(Continued)*

S. No.	Family	Plant species	Local name	Life form/habit	Conservation status
226	Polygalaceae	*Polygala erioptera* DC.	Hiran Buti	Herb	Common
227	Polygalaceae	*P. irregularis* Boiss.	Hiran Buti	Herb	Rare
228	Portulacaceae	*Portulaca oleracea* L.	Lonak	Herb	Common
229	Potamogetonaceae	*Potamogeton perfoliatus* L.	Machhi Gaah	Herb	Common
230	Primulaceae	*Anagallis arvensis* var. *coerulea* (L.) Gouan	Bili buti	Herb	Common
231	Rhamnaceae	*Zizyphus nummularia* (Burm. F.) Wt. &Arn.	Jhangoori Ber	Shrub	Very common
232	Salicaceae	*Populus euphratica* Olivier.	Bahan	Tree	Rare
233	Salvadoraceae	*Salvadora oleoides* Decne.	Jar/Peroon	Shrub	Common
234	Salvadoraceae	*S. persica* L.	Khabbar	Shrub	Rare
235	Scrophulariaceae	*Anticharis glandulosa* Aschers.	Gaamesh	Herb	Rare
236	Scrophulariaceae	*A. linearis*(Benth.) Hochst. ex Aschers.	–	Herb	Threatened
237	Scrophulariaceae	*Bacopa monnieri*(L.) Wettestein in Engl. & Prantl.	Kuari lonak	Herb	Rare
238	Scrophulariaceae	*Kickxia ramosissima* (Wall.) Janchen	–	Herb	Threatened
239	Scrophulariaceae	*Schweinfurthia papilionacea* (Burm. f.) Boiss.	Akri/Paneer Wal	Subshrub	Threatened
240	Solanaceae	*Lycium edgeworthii* Dunal.	–	Subshrub	Threatened
241	Solanaceae	*Datura fastuosa* L.	Charyo Dhatooro	Herb	Common
242	Solanaceae	*Physalis minima*L.	Khat mithro	Herb	Rare
243	Solanaceae	*Solanum surattense* Burm.f.	Kanderi Wal	Subshrub	Common
244	Solanaceae	*S. nigrum* var.*nigrum* L.	Kanwal Buti	Herb	Common
245	Solanaceae	*Withania coagulans* (Stocks) Dunal	Paneer	Shrub	Threatened
246	Solanaceae	*W. somnifera* (L.) Dunal	Akri	Shrub	Threatened
247	Sterculiaceae	*Melhania denhamii* R. Br.	–	Herb	Threatened
248	Tamaricaceae	*Tamarix aphylla* (L.) Karst.	Lawo	Tree	Very common

TABLE 3.1 *(Continued)*

S. No.	Family	Plant species	Local name	Life form/habit	Conservation status
249	Tamaricaceae	*T. dioica* Roxb. ex Roch	Lai	Shrub	Common
250	Tamaricaceae	*T. indica* willd.	Lai	Shrub	Common
251	Tamaricaceae	*T. pakistanica* Qaiser	Lai	Shrub	Rare
252	Tamaricaceae	*T. passernioides* Del ex. Desv var. *macrocarpa* Ehrenb	Lai	Shrub	Common
253	Tiliaceae	*Corchorus aestuans*L.	Datehri	Herb	Common
254	Tiliaceae	*C. depressus* (L.) Stocks.	Mundheri	Herb	Rare
255	Tiliaceae	*C. tridens* L.	Datehri	Herb	Common
256	Tiliaceae	*C. trilocularis* L.	Datehri	Herb	Common
257	Tiliaceae	*Grewia tenax* (Forssk.) Aschers and Schweinf.	–	Shrub	Rare
258	Typhaceae	*Typha domingensis* Pers.	Pan	Shrub	Very common
259	Typhaceae	*T. elephantina* Roxb.	Deer/Pan	Shrub	Very common
260	Verbenaceae	*Phyla nodiflora* (L.) Greene.	Bukkan	Herb	Common
261	Zygophyllaceae	*Fagonia bruguieri* DC. var. *rechingeri* Had.	Dramaaho	Bushy subshrub	Common
262	Zygophyllaceae	*F. indica* var. *schweinfurthii* Hadidi	Dramaaho	Bushy subshrub	Common
263	Zygophyllaceae	*Seetzania lanata* (Willd.) Bullock.	–	Herb	Threatened
264	Zygophyllaceae	*Tribulus longipetalus* Viv.	Bakhro/Bhurt	Herb	Very common
265	Zygophyllaceae	*T. ochroleucus* (Maire) Ozenda and Quezel	Bakhro/Bhurt	Herb	Rare
266	Zygophyllaceae	*T. terrestris* L.	Bakhro/Bhurt	Herb	Very common
267	Zygophyllaceae	*Zygophyllum simplex* L.	Jand Laani	Herb	Common

TABLE 3.2 Contribution of Groups/Families in Terms of Genera and Species of the Flora of Nara Desert.

S. No.	Group/plant family	No. of genera	No. of species
	PTERIDOPHYTES		
1	Adiantaceae	1	1
2	Equisetatceae	1	1
3	Marsileaceae	1	1
	GYMNOSPERM		
1	Ephederaceae	1	1
	MONOCOTYLEDONS		
1	Poaceae	28	47
2	Cyperaceae	4	14
3	Typhaceae	1	2
4	Liliaceae	1	1
	DICOTYLEDONS		
1	Fabaceae	12	26
2	Asteraceae	10	13
3	Boraginaceae	4	9
4	Capparidaceae	6	9
5	Euphorbiaceae	4	8
6	Amaranthaceae	5	7
7	Chenopodiaceae	5	7
8	Convolvulaceae	4	7
9	Solanaceae	5	7
10	Zygophyllaceae	4	7
11	Aizoaceae	5	6
12	Cucurbitaceae	5	6
13	Malvaceae	4	6
14	Mimosaceae	2	7
15	Polygonaceae	4	5
16	Scrophulariaceae	4	5
17	Tamaricaceae	1	5
18	Tiliaceae	2	5
19	Asclepiadaceae	4	4
20	Molluginaceae	2	4

TABLE 3.2 *(Continued)*

S. No.	Group/plant family	No. of genera	No. of species
21	Nyctaginaceae	2	4
22	Brassicaceae	2	3
23	Caesalpiniaceae	2	3
24	Acanthaceae	2	2
25	Burseraceae	1	2
26	Caryophyllaceae	2	2
27	Lamiaceae	2	2
28	Menispermaceae	1	2
29	Moraceae	1	2
30	Polygalaceae	1	2
31	Salvadoraceae	1	2
32	Arecaceae	1	1
33	Aristolochiaceae	1	1
34	Bignoniaceae	1	1
35	Cuscutaceae	1	1
36	Elatinaceae	1	1
37	Frankeniaceae	1	1
38	Fumariaceae	1	1
39	Geraniaceae	1	1
40	Hydrocharitaceae	1	1
41	Moringaceae	1	1
42	Neuradaceae	1	1
43	Orobanchaceae	1	1
44	Oxalidaceae	1	1
45	Potamogetonaceae	1	1
46	Primulaceae	1	1
47	Portulacaceae	1	1
48	Rhamnaceae	1	1
49	Salicaceae	1	1
50	Sterculiaceae	1	1
51	Verbenaceae	1	1
		167	267

TABLE 3.3 Life Form of the Flora of Nara Desert, Pakistan.

S. No.	Life form	No. of spp.	Percentage
1	Climber	6	2.25
2	Fern	3	1.12
3	Grass	47	17.60
4	Herb	114	42.70
5	Liana	4	1.50
6	Root parasite	1	0.37
7	Sedge	14	5.24
8	Shrub	35	13.11
9	Stem parasite	1	0.37
10	Subshrub	29	10.86
11	Tree	13	4.87

3.3.2 PROBLEMS AND THREATS

3.3.2.1 CONSERVATION STATUS OF ETHNOBOTANICAL TAXA

The conservation status of plant species was recorded/observed from the study area based on extensive field exclusion and presented in Table 3.1. Overall, 102 species (38.20%) were commonly recorded from the project area, followed rare (85 spp., 31.84%), very common (49 spp., 18.35%). It is very important to note that 17 species (6.37%) were threatened and 14 species were vulnerable (Fig. 3.2). The threatened species include *Anticharis linearis, Arnebia hispidissima, Boerhaviarubicunda, Commiphora stocksiana, C. wightii, Kickxia ramosissima, Luffa echinata, Lycium edgeworthii, Maerua crassifolia, Melhania denhamii, Tecomella undulate, Salvia santolinifolia, Schweinfurthia papilionacea, Seddera latifolia, Seetzania lanata, Withania coagulans,* and *W. somnifera.* In case of vulnerable species, *Aristolochia bracteolata, Amberboa ramosa, Cordia gharaf, C. myxa, Convolvulusprostratus, Cyperus arenarius, C. conglomeratus, Ephedra ciliata, Launaea resedifolia, Pluchea wallichiana,* and *Rhynchosia schimperii* were recorded.

It is worthwhile to note that annual species, like *Mollugo nudicaulis, Tephrosia falciformis, Neurada procumbense,* and *Schweinfurthia papilionacea,* once present, could not be find in the following field trips. The reason could be the dry spell. Furthermore, chopping/lopping of tree, like *Prosopis cineraria,* has been found at large scale. This tree is being used for coal making. As a result, its population is decreasing at an alarming scale.

Excessive use of some of the species resulted in constant threat and their population is gradually declining. Among them important plant species which are extensively used in various purposes are *Capparis decidua, Senna italica, Citrullus colocynthis, Corchorus depressus, Fumaria indica, Leptadenia pyrotechnica, Neurada procumbens, Prosopis cineraria, Trichodesma indicum,* etc.

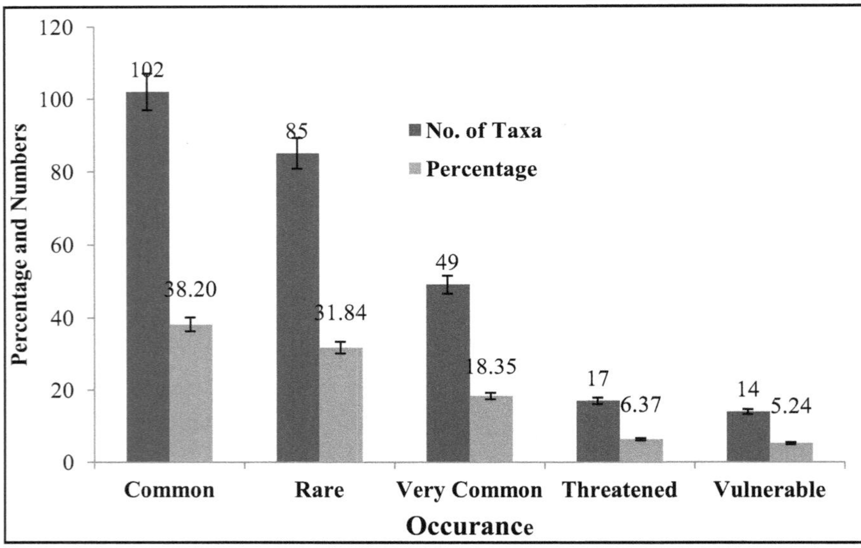

FIGURE 3.2 (See color insert.) Summary of conservation status of the taxa.

3.3.2.2 THREATS

Many factors are involved in exerting huge pressure on the vegetation that results in infinite desertification. These may include incompatible and inconsistent rainfalls, low humidity, extreme temperatures coupled with anthropogenic activities.

3.3.2.3 HUMAN NEEDS VERSUS THE FLORA

Local inhabitants use native flora to meet their daily life requirements such as traditional medicine, fuel wood, forage, food, vegetables/

wild fruits, roof thatching, agricultural implements, timber/furniture, matrices making, ropes, basketry, sitting chair, broom making, tooth sticks, clothes washing, leather tanning, detergent, fencing/hedge, shade, ornamental/recreational, soil binders, windbreaker, and poisonous (Qureshi, 2004).

In the Hilly tract, most of the pastures are grazed by transient cattle. These cattle are brought to pasture as soon as the sufficient grasses and other annual plants begin their spring growth. These pastures growing in interdunal valleys of the desert and in the depressions or in the foothills are smaller and are often more severely overgrazed than those of large plants, shrubs, and trees because of easy approach. The most desirable of the forage species are grazed first and, as a result, area depleted almost to the point of complete elimination in some pastures. In the period of scarcity less palatable plants took place and it is observed that *Tephrosia* spp. are grazed when they completely dried and there is no other green plant easily available in the area. The southwest area is observed almost complete elimination by *Cyperus arenarius* and *C. conglomeratus* since these plants are a high level of palatability and more desirable than other plants. During study period no rainfall observed and likewise, highly palatable sedges could not emerge. Severe climatic conditions during the last five years caused severe depletion of the grasses and sedge population is gradually decreasing at an alarming rate. The population of grasses and high palatable annual plants is declining due to excessive utilization of forbs at a very early stage before the seed formation. This practice will lead to removing the herbage cover from the soil surface.

Dun lands and slopes are dominated by the tussock grasses *Panicum turgidum* and *Lasiurus sindicus* were grazed by flocks of goat, cows, and camels at medium and high rates of utilization in either the early wet, late wet or dry seasons. Effects of grazing on species composition were greatest in the early wet season.

3.3.2.4 ECONOMIC USE

The native flora is rich in fodder/forage source for their livestock, so people use them for enhancing livestock rearing which is the main source of their livelihood. The other use categories include fuelwood and construction wood for roofing and small implements, wood, and branches for livestock fencing, and medicinal plants for people and livestock. The use of natural medicines for livestock and people is also very common. Some plant materials used for special purposes, such as latex "*Gugul*" obtained from *Commiphora wightii,* which is sold in regional markets. This gummy material is used in the herbal pharmaceutical industry, as well as fumigants in temples.

3.3.2.5 MEDICINAL PLANT DIVERSITY

Medicinal plants are of special interest to the inhabitants of the desert due little or sometimes no access to the modern healthcare facilities. The availability of effective herbal drug resources at the doorstep associated with the generation blocked traditional knowledge retain the traditional communities to depend for more than 95% of their primary healthcare medicine from phytomedicine. Qureshi and Bhatti (2008) reported 51 plant species distributed across 43 genera and 28 families that are being used by the inhabitants for treating various diseases.

3.3.2.6 GAS EXPLORATORY ACTIVITIES

Three gas companies/fields namely, Sawan, Miano (OMV Pakistan), and Qadanwari (LASMO Oil Pakistan) are operating in the study area for consistent searching fossil fuels since the decade. Various activities, such as the construction of roads, well drilling, and camping result in clearing centuries-old climax vegetation. Such activities have exerted immense pressure on the natural habitat as well on vegetation.

3.3.2.7 HOUBARA BUSTARD (TALOOR) HUNTING

The area is also fragmented by the human activities for developing access road in-game purpose (Houbara bustard or *Taloor* hunting). For this purpose, the network of the access road is developed in the Nara Desert to assist hunting, ecotourism as well as area development. Such activities destroyed hundreds of plant species. All the discussed human activities coupled with natural calamities have displaced countless perennial and centuries-old germplasm and increased the desertification in the Nara desert. These factors resulted in a gradual decrease in the numbers of rare and endemic species (Qureshi, 2011).

3.3.3 CONSERVATION/RESTORATIVE MEASURES

In order to mitigate the severity of the landscape, the following are some of the suggested measures for the betterment of the desert ecosystem:

3.3.3.1 REFORESTATION OF NATIVE TREE SPECIES

Nine native trees are recorded from the Nara desert (Qureshi and Bhatti, 2005a), although their distribution is very sparse. Therefore, there is a need to introduce these species from the depleted areas. *Prosopis cineraria* can be regarded as the lifeline of the inhabitants since it is profoundly exploited to meet their various needs. Efforts are required by the planners/managers to plant this species. Another important species is *Acacia senegal* that is comparatively rapid growing and also used in various needs.

3.3.3.2 SAND DUNE STABILIZATION

The native trees are capable of sand dune stabilization. Trees such as *Acacia senegal*, *Albizzia lebbeck*, *Cordia rothii*, *Dalbergia sissoo*, *Prosopis cineraria*, *Salvadora oleoides*, *Zizyphus jujuba* and *Z. mauritiana*are naturally adapted to the existing climatic conditions and are extremely drought tolerant. In addition, these species are highly palatable and used as a fuelwood by the local communities. Some of the shrubs, such as *Calligonum polygonoides* and *Zizyphus nummularia* and grasses for instance *Lasiurus sindicus* and *Panicum turgidum*, are also valuable in checking soil erosion.

3.3.3.3 GERMPLASM CONSERVATION

Since the people of the study area are solely dependent on the livestock keeping, there is a trend to increase their herds. A large number of grasses, herbs, and shrubs supply a rich source of feed for their livestock. Keeping in view, there is a need to protect some interdunal valleys for conserving germplasm/seed banks.

3.4 CONCLUSIONS

Plants enrich rangelands in shaping the pasture lands. In addition to feeding the livestock, local inhabitants utilize native plants for meeting their basic requirements, like fuel, human diet, and medicine (Bhatti et al., 2001; Qureshi, 2004). The whole complex ecosystem under environmental stress exerted by grazers and browsers, increasing land for agriculture, developmental activities by various discussed agencies, fuel and timber, and range resource collectors for medicinal and residential uses. The process of deterioration is gaining momentum with more and more thorough anthropogenic activities by various exploratory agencies as well as increasing human and livestock pressure. Though this unique landscape, possessed sufficient plant diversity, however, the population of valuable species is rapidly on decline stage. Therefore, this is the need of hour to launch urgent rehabilitation measures for controlling excessive grazing, deforestation, and habitat destruction. The degraded areas may be rehabilitated through planting fodder tree species and reseeding of palatable grasses. Furthermore, there is a need to collect and preserve the genetic diversity of range and forage grasses, legumes, and other forbs.

ACKNOWLEDGMENTS

I thankfully acknowledge Pakistan Science Foundation, Islamabad for providing funds under the research project (45) for the baseline study. Thanks go to Indus For All Programme (IFAP), a subsidiary organization of World Wide Fund for Nature (WWF), Pakistan for providing an opportunity to work on the vegetation of Chotiari wetlands complex located in the southern extreme of Nara desert called *Achhro Thar* (white desert). While conducting my field expedition in the northern part of the desert, I would like to express my thanks to Sardar Mohammad Bux Khan Maher, who provided base camp facilities at Dilkusha Bungalow at Sardar Garh (Ghotki) and to LASMO oil Pakistan for providing accommodation in their Campsite during plant expedition.

KEYWORDS

- **Nara Desert**
- **natural resources**
- **sustainable utilization**
- **Saharo-Sindian**
- **xerophytic vegetation**
- **fragile ecosystem**

REFERENCES

Ali, S. I., Qaiser, M., Eds.; *Flora of Pakistan;* (Fascicle series), 1993–1995; 2000–2004.

Ali, S. I., Nasir, Y. J., Eds.; *Flora of Pakistan* (Fascicle series), 1990–1991.

Batanouny, K. H. Ecology and Flora of Qatar. Centre for Scientific and Applied Research, University of Qatar, 1981.

Bhandari, M. M. *Flora of Indian Desert;* Scientific Publishers, 1978.

Bhatti, G. R.; Qureshi, R.; Shah, M. Ethnobotany of Qadan Wari of Nara Desert. *Pak. J. Bot.* **2001,** *33* 801–812.

Boulos L. *Flora of Egypt;* Al Hadara Publishing Cairo, 1, 1991.

Jafri, S. M. H. *The Flora of Karachi;* The Book Corporation: Pakistan, 1966.

Matthew, K. M. Flora of Tamil Nadu Carnatic. The Rapinat Herbarium. St. Joseph's College, Tiruchirapalli 620002, India, 1981–1983; pp 1–3.

Nasir, E.; Ali, S. I.; Eds.; *Flora of Pakistan;* (Fascicle series), Department of Botany, University of Karachi: Pakistan, 1970–1989.

Qureshi, R. Floristic and Ethnobotanical Study of Desert Nara Region, Sindh. Ph.D. Thesis, Department of Botany, Shah Abdul Latif University, Khairpur, Sindh, Pakistan, 2004, pp 1–454.

Qureshi, R. Preliminary Floristic List of Chotiari Wetland Complex, Nawab Shah, Sindh, Pakistan. *Pak. J. Bot.* **2008,** *40* (6), 2281–2288.

Qureshi, R. Vegetation Assessment of Sawan Wari of Nara desert, Pakistan. *Pak. J. Bot.* **2008,** *40* (5), 1885–1896.

Qureshi, R. Floral Diversity, Threats and Conservation Measures: A case study from Nara Desert, Pakistan. In *Deserts: Fauna, Flora and Environment;* Guevara. J., Grünwaldt, E., Eds.; Nova Science Publishers: Inc Hauppauge, New York, USA, 2011; pp 53–77.

Qureshi, R.; Bhatti, G. R. Nara Desert, Pakistan: Part I: Soils, Climate and Vegetation. *Rangelands* **2005a,** *27* (5), 27–31.

Qureshi, R.; Bhatti, G. R. Nara Desert, Pakistan: Part II: Human Life. *Rangelands* **2005b,** *27* (5), 32–35.

Qureshi, R.; Bhatti, G. R. Diversity of Micro-habitat and its Plant Resources in Nara Desert, Pakistan. *Pak. J. Bot.* **2008a,** *40* (3), 979–992.

Qureshi, R.; Bhatti, G. R. Ethnobotany of Plants Used by the Thari people of Nara Desert, Pakistan. *Fitoterapia* **2008b,** *79*, 468–473.
Qureshi, R.; Bhatti, G. R. Floral Diversity of Nara Desert, Pakistan. In *Biodiversity of Pakistan;* Rafique, M., Ibupoto, K. A., Hussain, S. A., Mahmood, K, Eds; Pakistan Museum of Natural History: Islamabad, Pakistan, 2014; pp 413–431.
Shetty, B. V.; Singh, V. Flora of Rajasthan, Botanical Survey of India, Old Connaught Place, Dehra Dun1991, pp 1–2.
Singh, Pandey. Economic and Medicinal Plants of Indian Desert. In *Desert Resources and Technology;* Singh A., Ed.; Scientific Publishers: India, 1983; pp 307–366.

CHAPTER 4

Acute and Biological Effects of Some Synthetic Organic Compounds on Spiny Bollworm

R. M. FIKRY[1*], S. A. RASLAN[2], N. A. ISMAIL[1], H. S. EL-TAHAWE[2], and H. NANGYAL[3]

[1]*Department of Chemistry, Faculty of Science, Zagazig University, Zagazig, Egypt*

[2]*Plant Protection Research Institute, Dokki, (Sharkia Branch), Agricultural Research Center, Giza, Egypt*

[3]*Department of Botany, Hazara University, Mansehra, Khyber Pakhtunkhwa, Pakistan*

[*]*Corresponding author. E-mail: hasnain308@gmail.com*

ABSTRACT

Three synthetic compounds namely: A = 3-amino-5-chloro-4- phenyl azo-1H-pyrazole, B = 3-amino-5-mercapto-4- phenyl azo-1H-pyrazole, C = 3-amino-5- hydrazino-4-phenylazo-1H-pyrazole were tested on neonate larvae of Spiny bollworm, *Earias insulana* (Boisd.) at constant conditions of 27±2°C and 65±4% R.H. Acute toxicity as well as their latent effects on the different stages of the survived larvae were studied.

Corresponding to the acute toxicity, at LC_{50} level the toxicity effect of the four tested chemicals were in descending order as follows: A, C, and B compounds. During larval stage, phenyl pyrazoles caused insignificant prolongation in duration of the larval stage of spiny bollworm. All tested chemicals showed highly significant decreased in the larval weight of spiny bollworm. All compounds caused insignificant increases on larval mortality percentage of spiny bollworm compared with control.

All tested chemicals resulted in pupation percentages less than control. Pupal weights were not affected significantly with the tested chemicals. All compounds caused insignificant decreases on pupal durations of spiny bollworm compared with control. Significant differences were found between the tested compounds and untreated check on pupal mortality.

Percent of adult's emergence was significantly affected by the tested compounds. The sex ratio of emerged adult's spiny bollworm was insignificantly affected by the three tested compounds compared with untreated ones. A, B, and C compounds shortened the male longevity compared with the untreated. A and B compounds shortened the female longevity compared with untreated. But C compound elongated the female longevity compared with the untreated check.

All treatments decreased number of deposited eggs, which proved significant with A, B, and C chemicals. Incubation periods did not affect. Significantly B, A, and C compounds caused reduction in the hatchability of deposited eggs by spiny bollworm female survived from treated newly hatched larvae.

4.1 INTRODUCTION

Cotton considers an important crop all over the world hence it is strategic crop by sharing in Egyptian national income. Cotton plants like other field crops are liable to attack by several species of insect pests during its growing season. In Egypt, bollworms were considered the most destructive pests infesting cotton plants causing severe damage resulting in high loss in both quality and quantity of cotton (Khidr et al., 1996).

The pesticides were used to control the spiny bollworm. The level of resistance to some insecticides has increased and new group of chemicals was needed to manage resistant populations in cotton. Insecticidal pyrazoles was a new family of insecticides that act on the gamma–amino butyric acid (GABA) receptors of insects by blocking the passage of chloride ions, inducing disruption of the central nervous system (Cole et al., 1993). The fipronil is the first compound of phenyl pyrazole insecticide class to be registered for commercial use which is toxic to piercing–sucking, and chewing insects (Eric et al., 2001). Pyrazoles has shown excellent activity against *Spodoptera litura*, tarnished plant bug and European corn borer larvae (Scott and Sondgrass, 2000).

The first objective of the present study is to evaluate the toxicity of some synthetic pyrazoles against the newly hatched larvae of the spiny bollworm *Earias insulana* (Boisd.), under the constant conditions of 26±1°C and 80±5% R H. The second objective of this work is to study the latent effects of these compounds on some biological aspects of spiny bollworm such as larval mortality, larval durations, weights of full-grown larvae and pupae resulted from treated newly hatched larvae, percent of pupation, pupal period, moth's emergence percentage, and sex ratio. The longevity of female and male were recorded. The numbers of deposited eggs and hatchability percentage were also considered.

4.2 MATERIALS AND METHODS

4.2.1 *ARTIFICIAL DIET*

The artificial diet for maintaining a mass culture of spiny bollworm larvae was followed according to the method described by Rashad et al. (1993). Diet medium consists of 860 g dried kidney beans (boiled in water), 128 g dried active yeast, 10 g ascorbic acid, 5 g methyl-P-hydroxy benzoate, 5 g sorbic acid, and 48 g agar to which 600 mL distilled water was added.

4.2.2 *REARING TECHNIQUE OF PINK BOLLWORMS*

Active and healthy full-grown larvae of spiny bollworm *Earias insulana* (Boisd.) were collected from infested cotton bolls at the end of cotton growing season. These larvae were reared in electrical incubation at constant conditions of 27±2°C and 65±5% R.H, on artificial diet previously described in bollworms laboratory, Plant Protection Research Institute (Sharkia branch) at Zagazig.

The emerged adults were continually reared for about 30 successive generations under the previously mentioned conditions to obtain individuals of a susceptible strain, which their larvae were used for the present toxicological studies.

4.2.3 CHEMICAL USED

Three synthetic chemical compounds namely:

A = 3-amino-5-chloro-4- phenyl azo-1H-pyrazole
B = 3-amino-5-mercapto-4- phenyl azo-1H-pyrazole
C = 3-amino-5- hydrazino-4-phenylazo-1H-pyrazolewere synthesized at the laboratory of faculty of Science, Zagazig University.

4.2.4 ACUTE EFFECT AGAINST SPINY BOLLWORM NEONATE LARVAE

The surfaces of petri dishes containing layer of artificial diet (5 g) were treated with different concentrations of the three tested chemicals by adding 3 mL of each concentration over the artificial diet and shake the petri dish to distribute the chemical solution over all the surface of the artificial diet, and immediately the excess of chemical solution was removed from the dish, and the petri dish was left on air to dryness.

Using a camel hair brush, 20 newly hatched larvae (0–6 h old) were transferred over treated diet, and the Petri dishes were covered with fine and soft toilet paper below the glass cover to prevent larvae from escape, and that represent one replicate. Three replicates for every tested chemical as well as control (treated with water only) were incubated under control conditions.

The dead larvae were counted after 48 h and the percent of mortality of larvae was estimated. The LC_{50}, LC_{90}, and slope values of each tested compound were calculated according to the method described by Finney (1971).

Toxicity index (TI) and relative toxicity were determined by using Sun's equation (1950) as follows:

$$\text{Toxicity index (TI)} = \frac{LC_{50} \text{ or } LC_{90} \text{ of the chemical (A)}}{LC_{50} \text{ or } LC_{90} \text{ of the chemical (B)}} \times 100$$

where

A = the most effective chemical and
B = the other tested chemical.

Relative potency (RP) values were measured according to the method described by Zidan and Abdel-Megeed (1988).

$$\text{Relative potency (RP)} = \frac{LC_{50} \text{ or } LC_{90} \text{ of the lowest toxic insecticide}}{LC_{50} \text{ or } LC_{90} \text{ of the tested insecticide}}$$

4.2.5 CHRONIC EFFECT AGAINST PINK BOLLWORM NEONATE LARVAE

To study the latent effects of tested chemicals on certain biological aspects of spiny bollworm *Earias insulana* (Boisd.), the alive larvae after 48 h of one concentration which caused less than LC_{20} for each tested chemical and the control treatment were transferred individually into glass tube (2×7.5 cm), containing 2 g of untreated diet. Afterward, glass tubes were covered with a piece of absorbent cotton and held under the same conditions as mentioned before, the weight of the 4th instar larvae, time of entering the last instar in the prepupal period and date of adult emergence were estimated. From these records, some biological aspects such as durations of larvae, pre-pupae and pupae as well as weight of both 4th instar larvae and pupae, and percent of pupation were considered. The insects were sexed in the fourth larval stage. After the insect pupation, pupae were transferred individually to clean vials covered with cotton stopper and incubated till moth emergence. The emergence percentage and sex ratio were calculated.

The emerged moths were caged in one pair (male and female) in glass jar (7.00 cm in diameter of lower 11.5 cm in height) under the previously mentioned rearing conditions. For each concentration, the newly emerged moths of each treatment and control were placed in Jar for mating. A piece of cotton wool, previously soaked in 10% sugar solution was hung inside the jar near its upper opening for moth feeding and changed by new one every 48 h. Upper opening of jar was covered by paper for egg-laying response of females, thin covered with muslin cloth and tightly secured with rubber bands. Each jar was examined daily to count the number of eggs deposited per female and date of adult death. Hatchability of all deposited eggs per female was also determined.

At the beginning of the female oviposition period, 100 eggs in each replicate were taken and incubated under the previously mentioned optimum conditions and the date of hatching was recorded. Several biological aspects such as oviposition periods, female fecundity of eggs, longevity of adult males and females, hatchability percentage and incubation period of deposited eggs were also estimated.

4.3 STATISTICAL ANALYSIS

The obtained results of mortality each and biological measurements were subjected to analysis of variance to clear the significantly of the toxic and latent effects of the different tested materials against the tested insect. The analysis of variance was computed using Costat computer program 3.03 copyright 1986 Cohort Software. P. O. Box 1149, Berkeley CA 9471 (Costat program methods).

4.4 RESULTS AND DISCUSSION

4.4.1 ACUTE EFFECT AGAINST SPINY BOLLWORM NEONATE LARVAE

The following three synthesis chemical compounds were evaluated against the spiny bollworm neonate larvae:

A = 3-amino-5-chloro-4- phenyl azo-1H-pyrazole;
B = 3-amino-5-mercapto-4- phenyl azo-1H-pyrazole; and
C = 3-amino-5- hydrazino-4-phenylazo-1H-pyrazole.

4.4.2 LC_{50} AND LC_{90} VALUES

Data in Table 4.1, show that at LC_{50} level the toxicity effect of the three tested chemicals can be arranged in descending order as follows: A, C, and B, where their values were, 5891.216, 20,086.71, and 24,055.03 ppm, respectively.

Corresponding the LC_{90} values data in Table 4.1 show that chemical A was the most toxicant one, followed by B and C chemicals, which recorded LC_{90} of 17,897.83, 210,663.1, and 226,502.9 ppm, respectively.

Generally, from LC_{50} and LC_{90} values, it could be notice that chemical A was the most toxicant one, and chemical B and C were in par to each other in their toxicant effect and strongly less in their toxicant action than chemical A against the spiny bollworm neonate larvae; and it could be concluded that chemical A is considered one of the most promising compound for controlling *E. insulana*.

4.4.3 TOXICITY LINES, SLOPE, AND LC_{50}/LC_{90} RATIO

According to the results obtained in Table 4.1, it could be proved that chemical A has the steepest toxicity line, where it has the highest slope value and lowest LC_{90}/LC_{50} ratio, while chemical B has the flattest one, where it has the lowest slope value and highest LC_{90}/LC_{50} ratio and the compound C was in intermediate between A and B chemicals.

The slope values in descending order were 1.2556, 1.3159, and 2.6556, while LC_{90}/LC_{50} ratio was 3.038, 9.416, and 10.487 for B, A, and C, respectively.

4.4.4 TOXICITY INDEX

According to Sun's equation (1950) the relative toxicity of the four tested chemicals against the spiny bollworm neonate were evaluated.

At the LC_{50} and LC_{90} levels, chemical A taken as the standard (which resulted in the least LC_{50} of all tested chemicals) and given the arbitrary value of 100 units, the toxicity indices of the other chemicals calculated by the ratio of (LC50 or LC90 of the standard chemical/LC_{50} or LC_{90} of the standard chemical × 100), which at LC_{50} were 29.328 and 24.490 for B and C, respectively and at LC_{90} level, toxicity indices were 8.4959 and 7.9018 for B and C, respectively.

4.4.5 RELATIVE POTENCY

The potency levels of the tested chemicals are expressed as the number of folds at the required toxicity level, compared with the least effective chemical in the four chemicals tested against the spiny bollworm neonate larvae. Hence the number of folds representing the relative potency level at Table 4.1 was obtained by dividing the LC_{50} and LC_{90} of chemical C, which was considered the standard chemical at the LC_{50} and LC_{90} level by the corresponding figures of each tested chemicals. At the LC_{50} level, the relative potency levels expressed as number of folds indicate that A and B were 4.0832 and 1.1975 times as toxic as C. At the LC_{90} level, the relative potency levels for A and B were 12.655 and 1.07521 times as compared with C compound against the spiny bollworm neonate larvae.

TABLE 4.1 Toxicity of Different Tested Chemicals Against the 1st Instarneonate Larvae of Laboratory Strain of the Spiny Bollworm, *Earias insulana* (Boisd.).

Chemicals	LC_{50} (ppm)	LC_{90} (ppm)	Relative potency (folds) at		Toxicity index at		Slope	LC_{90}/ LC_{50}
			LC_{90}	LC_{50}	LC_{90}	LC_{50}		
A	5891.216	17,897.83	12.655	4.0832	100.00	100.00	2.6556	3.038
B	20,086.71	210,663.1	1.07521	1.1975	8.4959	29.328	1.2556	10.487
C	24,055.03	226,502.9	1.00	1.00	7.9018	24.490	1.3159	9.416

A = 3-amino-5-chloro-4- phenyl azo-1H-pyrazole; B = 3-amino-5-mercapto-4- phenyl azo-1H-pyrazole; and C = 3-amino-5- hydrazino-4-phenylazo-1H-pyrazole.

4.4.6 LATENT EFFECT

The main objective of these experiments is to study the latent effect of phenyl pyrazoles on some biological aspects of the Survived larvae and subsequent all development stages of spiny bollworm. The results could be discussed as the follows:

4.4.7 LARVAL DURATION

Data in Table 4.2 indicate that it can be concluded that the phenyl pyrazoles caused prolongation in duration of the larval stage of spiny bollworm, which was significant only with compound C as compared with untreated larvae. The mean larval durations were 15.26, 15.67, and 16.10 days for A, B, and C, respectively, compared with 15.00 days for control.

4.4.8 LARVAL WEIGHT

Statistical analysis of data presented in Table 4.2 shows that all tested chemicals highly significant decreased larval weight of spiny bollworm. The mean larval weight was 0.0675, 0.0623, and 0.0640 g for A, B, and C, respectively, while control revealed 0.0834 g.

4.4.9 LARVAL MORTALITY PERCENTAGE

Data presented in Table 4.2 indicate that all compounds caused insignificant increase on larval mortality percentage of spiny bollworm compared with control. The highest average percentage of larval mortality (28.86%) was obtained with compound B while the lowest percentage (13.33%) was recorded for C chemical as compared with control which was 10.00% larval mortality.

4.4.10 PUPATION PERCENTAGE

The effect of the four tested compounds on the pupation percentage of spiny bollworm was shown in Table 4.2. Statistical analysis of data indicated that the effect of tested compounds on pupation percentage was insignificant. The highest average of pupation percentage was 86.67% for C Compound. While the lowest one was 71.14% fcr B compound compared with 91.00% for untreated one. Generally, all tested chemicals resulted in pupation percentages less than control.

4.4.11 PUPAL WEIGHT

The result concerning the effect of the tested compounds on the pupal weight indicated insignificant effect as shown in Table 4.2. The mean pupal weights were 0.0548, 0.0523, 0.0524, and 0.0205 g for A, B, and C, respectively, while it was 0.0630 g for control. The highest average pupal weight of spiny bollworm was 0.0548 g for A compound. While the lowest one was 0.0523 g for B compound.

4.4.12 PUPAL DURATION

Data presented in Table 4.2 indicate that all compounds caused insignificant decreases on pupal durations of spiny bollworm compared with control. The longest pupal period was 10.38 days for B compound, while the lowest one was 8.73 days for C compound compared with 10.00 days for untreated.

4.4.13 PUPAL MORTALITY PERCENTAGE

Data in Table 4.2 indicate that it can be concluded that the phenyl pyrazoles caused increased in pupal mortality of spiny bollworm, which was significant only with compound C as compared with untreated pupal. Mortality: the highest average percentage of pupal mortality (22.23%) was recorded for C compound while the lowest one (5.16%) was recorded for B compound compared with untreated check.

4.4.14 ADULT EMERGENCE

Statistical analysis of the data in Table 4.2 indicates that percentage of adult's emergence in case of compound C was significantly affected by the tested compounds. The average percentages of adult's emergence were 90.27%, 94.84%, and 77.78% for A, B, and C compounds, respectively compared with untreated (96.00%). The highest adults' emergence percentage was 94.84% for B compound while the lowest one was 77.78% for C compound.

Generally, it was noticed that phenyl pyrazole compounds caused lower adult emergence than control.

4.4.15 SEX RATIO

Date in Table 4.2 show the sex ratio calculated as percent of females from the total number of emerged adult. The sex ratio of emerged adult's spiny bollworm was insignificantly affected by the three tested compounds compared with untreated ones. The percent of females differed slightly from compound to another. The percentages of females were 39.07%, 41.67%, 47.03%, and 45.33 % for A, B, C compounds and untreated check, respectively.

4.4.16 ADULT LONGEVITY

4.4.16.1 MALE LONGEVITY

Data in Table 4.3 indicate that the three compounds had insignificant effect on the spiny bollworm male longevity. Treated neonate larvae with A, B, and C compounds shortened the male longevity compared with the

TABLE 4.2 Effect of Tested Compounds on Biological Aspects of Larval and Pupal Stages of Spiny Bollworm Treated as Newly Neonate Larvae.

Chemicals	Larval duration (days)	Larval weight (grams)	Larval mortality %	% Pupation	Pupal weight (grams)	Pupal duration (days)	Pupal mortality %	% Adult emergence
A	15.26b	0.0675b	26.67	73.34	0.0548	9.74	9.73ab	90.27ab
B	15.67ab	0.0623b	28.86	71.14	0.0523	10.38	5.16b	94.84a
C	16.10a	0.0640b	13.33	86.67	0.0524	8.73	22.23a	77.78b
Control	15.00b	0.0834a	10.00	91.00	0.0630	10.00	4.00b	96.00a
F-test	*	***	N.S.	N.S.	N.S.	N.S	*	*
$LSD_{0.05}$	0.76	0.01	–	–	–	–	13.27	13.27

*Means followed by the same letter vertically did not differ significantly.
***Significance level at 0.001.

A = 3-amino-5-chloro-4- phenyl azo-1H-pyrazole; B = 3-amino-5-mercapto-4-phenyl azo-1H-pyrazole; and C = 3-amino-5- hydrazino-4-phenylazo-1H-pyrazole.

untreated. The longest male longevity was 11.53 days recorded with B compound while the shortest one was 8.94 days for A compound compared with the control of 15.40 days.

4.4.16.2 FEMALE LONGEVITY

The obtained data in Table 4.3 revealed that the female longevity resulted from compound A was significantly shorter compared with untreated one. The mean female longevities were 14.50, 16.33, and 18.07 days for A, B, and C compounds, while the untreated was 17.77 days. The A and B compound shortened the female longevity (14.50, 16.33 days) compared with untreated (17.77 days). But C compound elongated the female longevity compared with the untreated check.

4.4.17 PRE-OVI POSITION PERIOD

Data presented in Table 4.3 show that the means of time required for maturation of the spiny bollworm female ovaries and starting the eggs deposition were 5.40, 6.60, and 5.63 days for A, B, and C compounds. While the untreated one was 5.20 days. The compounds B was elongated the pre-oviposition periods, but A and C compounds were in par with control. No significant differences were found between the tested compounds and untreated check.

4.4.18 OVIPOSITION PERIOD

The egg-laying period of the emerged female from treated newly hatched larvae were insignificant shortened compared with the untreated Table 4.3. The mean oviposition periods were 5.27, 4.60, and 5.78 days for A, B, and C compound while the untreated control was 7.00 days.

4.4.19 POSTOVIPOSITION PERIOD

Date in Table 4.3 indicate that the postoviposition period of the spiny bollworm female developed from newly hatched larvae treated with the three

TABLE 4.3 Effect of Tested Compounds on Biological Aspects of Adults of Spiny Bollworm Treated as Newly Neonate Larvae.

Chemicals	Sex ratio % females/ total adults	Preoviposition period (days)	Oviposition period (days)	Postoviposition period (days)	Male longevity (days)	Female longevity (days)
A	39.07	5.40	5.27	3.87	11.53	14.50c
B	41.67	6.60	4.60	5.13	13.33	16.33b
C	47.03	5.63	5.78	6.67	14.53	18.07a
Control	45.33	5.20	7.00	5.57	15.40	17.77ab
$LSD_{0.05}$	–	–	–	–	–	1.57
F-test	N.S.	N.S.	N.S.	N.S.	N.S.	*

*Means followed by the same letter vertically did not different significantly.

A = 3-amino-5-chloro-4- phenyl azo-1H-pyrazole; B = 3-amino-5-mercapto-4-phenyl azo-1H-pyrazole; and C = 3-amino-5- hydrazino-4-phenylazo-1H-pyrazole.

compounds was insignificant compared with control. It was clear that these periods were 3.87, 5.13, and 6.67 days for A, B, and C compounds while the untreated was 5.57 days.

4.4.20 FEMALE FECUNDITY

The number of eggs deposited by female was insignificantly affected by treated of C compound compared with untreated but it was significant in case of A and B compounds compared with untreated Table 4.4. The mean number of deposited eggs by one female was reduced by treating the three compounds compared with untreated. The mean number of eggs was 41.70, 37.77, and 68.93 eggs/female for A, B, and C compounds, respectively, compared with 107.87 eggs/female. Generally, the fecundity of female produced from newly hatched larvae treated with phenyl pyrazole compounds was greatly reduced than that obtained from control.

4.4.21 INCUBATION PERIOD

Data in Table 4.4 show insignificant effects between tested compounds and untreated.

The mean incubation period of spiny bollworm eggs lasted 4.13, 4.09, and 4.08 days for A, B, and C compounds, respectively, while, the untreated check revealed 4.16 days.

4.4.22 HATCHABILITY PERCENTAGE

Data presented in Table 4.4 show that B, C, and D compounds caused significant reduction in the viability of eggs deposited by spiny bollworm female survived from treated newly hatched larvae. The highest hatchability percentage was 85.17% for B compound, while, the lowest one was 68.67% and 77.67% recorded for A and C compounds, compared with 95.67% for control. Generally, the phenyl pyrazole compounds caused higher reduction than that of control on the hatchability rate.

The present results are in accordance with those obtained by Arthur (2002) working on stored grain insects; adult red flour beetles and maize weevils treated with 10 ppm Ethiprole (pyrazole compound) alone or in combination treatments. The mortalities of maize weevils ranged from 77.9% to 100%, and mortalities of red flour beetles were from 46.2% to 94.2%. Eric et al. (2001) found that phenylpyrazole insecticide fipronil was very toxic to neonate European corn borer larvae in feeding bio assays (LC_{50} = 3.34 ng a.i./cm^2 of treated diet) and to fifth instars in topical bio assays (LD_{50} = 18.78 ng/insect). Ester et al. (1997) mentioned that seeds of winter leeks (*Allium porrum*) coatings with fipronil and imidacloprid gave a good control against *Thrips tabaci*. Krushelnycky and Reimer (1998) showed that phenyl pyrazoles were active against many species of ants. Recently, Chandler et al. (2004) found that phenyl pyrazoles have been highly toxic to Esturaine copepod (*Amphias custenuiremis*) with a 96-h acute LC_{50} of 6.8 mg/L.

TABLE 4.4 Effect of Tested Compounds on Fertility of Spiny Bollworm Treated as Newly Neonate Larvae.

Chemicals	No. of eggs/Female	Incubation period (days)	% Hatchability
A	41.70b	4.13	68.67d
B	37.77b	4.09	85.17b
C	68.93ab	4.08	77.67c
Control	107.87a	4.16	95.67a
$LSD_{0.05}$	41.13	–	6 391
F-test	*	N.S.	***

*Means followed by the same letter vertically did not differ significantly.
***Significance level at 0.001.

A = 3-amino-5-chloro-4- phenyl azo-1H-pyrazole; B = 3-amino-5-mercapto-4-phenyl azo-1H-pyrazole; and C = 3-amino-5- hydrazino-4-phenylazo-1H-pyrazole.

KEYWORDS

- acute
- biological
- effects
- synthetic
- compounds
- bollworm
- toxicity

REFERENCES

Arthur, F. H. Efficacy of Ethiprole Applied Alone and in Combination With Conventional Insecticides for Protection of Stored Wheat and Stored Corn. *Protect. Sci.* **2002,** *11* (1), 134–148.

Chandler, G. T.; Cary, T. L.; Volz, D. C. Fipronil Effects on Estuarine Copepod (*Amphiascus tenuiremis*) Development, Fertility, and Reproduction; a Rapid Life–cycle Assay in 96-well Mircoplate Format. *Environ. Toxical. Chem.* **2004,** *23* (1), 117–124.

Cole, L. M.; Nicholson, R. A.; Casida, J. E. Action of Phenyl Pyrazole Insecticides at the GABA Gated Chloride Channel. *Pest. Biochem. Physiol.* **1993,** *46*, 47–54.

Eric, W. D.; Scharf, M. E.; Siegfried, B. D. Toxicity and Neurophysiological Effects of Fipronil and its Oxidative Sulfone Metabolite on European Corn Borer Larvae (*Lepidoptera Crambidae*). *Pest. Biochem. Physiol.* **2001,** *71*, 97–106.

Ester, A.; Devogel, R.; Bouma, E. Controlling *Thrips tabaci* (Lind.) in Leek by Film–coating Seeds with Insecticides. *Crop Prot.* **1997,** *16*, 673–677.

Finney, D. J. Probit Analysis, 3rd ed.; Cambridge University Press: London, UK, 1971.

Giles, D. P.; Copping, L. G.; Willis, R. J. Benzophenone Hydrazones, A Group of Novel Insecticides. *Crop Prot. Coun.* **1984,** *2*, 405–412.

Holan, G.; Smith, D. R. J. A New Selective Insecticidal Uncoupler of Oxidative Phosphorylation. *CSIRO Div. Appl. Org. Chem.* **1986,** *42* (5), 558–560.

Khidr, A. A.; Desuky, W. M. H.; El-Sheakh, A. A.; Raslan, S. A. A. Sequential Use of Some Insecticides Against Cotton Bollworms in Control Trials. *Egypt. J. Agric. Res.* **1996,** *74* (2), 321–332.

Krushelnycky, P. D.; Reimer, N. J. Bait Preference by the Argentine Ant (*Hymenoptera : formicidae*) in Haleakala National park, Hawaii in Haleakala National Park, Maui, *Hawaii Environ. Entomol.* **1998,** *27* (6), 1473–1481.

Rashad, A. M.; Abd EL-Salam, N. M. IGRs as Bioactive Agents Against the Pink and Spiny Bollworms. *Egypt. J. Appl. Sci.* **1993,** *8* (1), 303–315.

Scott, W. P.; Sondgrass, G. L. A Review of Chemical Control of the Tarnished Plant Bug in Cotton. *South Western Entomol.* **2000,** *23*, 67–81.

Sun, Y. P. Toxicity Index, An Improved Method of Comparing the Relative Toxicty of Insecticides. *J. Econ. Entomol.* **1950,** *43*, 45–53.

Zidan, Z. H.; Abdel-Megeed, M. I. *New Approaches in Pesticides and Insect Control*; Arabic Publishing House and Delivery: Cairo, 1988; Vol. 605.

CHAPTER 5

The Simplest Way to Determine the Most Appropriate Diversity Indices in a Study Area

J. PARCHIZADEH*

Third Floor, Number 24, Zartoshtian Alley, Hafez Street, Tehran City, Tehran Province, Iran

Corresponding author. E-mail: Jamshid.Parchizadeh@gmail.com

ABSTRACT

Determining the most appropriate diversity indices in a study area is one of the necessary things to preserve animal species. Therefore, the simplest way to do it seems to be quite important in ecological communities. Here, with the help of a case study in Iran, I explained the simplest and easiest way to determine the most sensitive biological indices in a study area. My survey showed that Menhinic's species richness, Shannon–Weaver diversity, and Simpson's measure of evenness were the most appropriate indices to calculate the richness, the diversity, and the evenness in the study area.

5.1 INTRODUCTION

Lack of information regarding the ecology of animal species is the most significant cause of their extinction. The goal of wildlife conservation strategy is to employ expert management techniques for the ongoing preservation of species (Sutter et al., 2005). Assessment of ecosystems seems to be one of the most important concepts in ecology (Mirzaie et al., 2013). Thus, biological indices are used to facilitate the interpretation of ecological data. Determining the most appropriate diversity indices in a study area is necessary to preserve animal species. However, the elusive nature of large and medium-sized mammals make measurements difficult. There are so many biological indices that the calculation of all of them is both time-consuming and difficult (Shirood Mirzaie et al., 2013). In general, diversity is defined as the number of different items and their relative abundance. For biodiversity, these items are organized from complete ecosystems to the chemical structures (Aslam, 2009). Species richness, which is the simplest way to describe community and regional diversity, is defined as the number of species in a given area. However, diversity takes into account how individuals are distributed among those species. Evenness is also defined as the measure of equality of abundances in a community (Alatalo, 1981; Aslam, 2009; Gotelli and Colwell, 2001). Richness, diversity and the relative abundance of target species (how relative abundance is distributed among species) is important concepts in ecological communities (Wilsey and Stirling, 2007).

Therefore, with the help of a case study in Iran, we will discuss in brief the simplest way to determine the most appropriate biological indices in ecological communities in this chapter.

5.2 MATERIALS AND METHODS

Camera traps are useful tools to provide valuable information about both individuals and populations. The information collected using camera traps can influence the management of conservation and direct the managers toward making more appropriate decisions regarding conservation of different species (Stein et al., 2008). Therefore, from June 2013 to August 2014 in a survey that took place in Lar Protected Area in northeastern Iran measuring 735.42 km^2 in size, I gathered information regarding large and medium-sized mammals from 278.2 km^2 that included mountains, canyons, and plains to measure species richness, diversity, and evenness. Based on both published official documents of the Iran Department of the Environment (DoE) and the valuable experience of wildlife rangers, key spots were identified and camera traps (n = 10) (Bushnell, Overland Park, Kansas City, MO, USA) were installed. The sites of all camera traps were located by GPS Garmin etrex 10 and plotted on the map in ArcGIS 10.3 for measurements. A total of 47 camera-trap stations were chosen. I programed the cameras to operate round-the-clock 24 h/day to take only photographs instead of motion pictures in order to save battery usage.

All cameras were labeled by numbers and positioned 40–60 cm from the ground (Rios-Uzeda et al., 2007). The cameras were camouflaged so as to prevent them from being stolen. I did not use lure (bait) or any other attractants in order to obtain pure information of the populations. I checked the cameras monthly and sometimes twice a month to change batteries and download taken photographs. In order to avoid repetition in the counting process, I based identification on natural markings and environmentally induced characteristics, such as color, size, presence of scars, and having cubs or not (Ball, 1975; Foster and Harmsen, 2011). Individual leopards were recognized via their unique spot and rosette patterns on their flanks and limbs (Ghoddousi et al., 2008), but the individuals of other six target species were identified via color, size, scars, and height of horns. If any individual from one species resembled another individual of the same species in two different photographs from the same station within a period of 1 h, that individual was counted only once assuming that the individual had been counted earlier.

Using a variety of indexes to compare species richness, diversity, and evenness in three different habitat types in the study area the following were chosen:

1. $D_{Mg} = (S - 1)/\ln N$
2. $D_{Mn} = S/\surd N$
3. $H' = -\Sigma^{s}_{i=1} P_i(\ln P_i)$
4. $1 - D = 1 - \Sigma\, n_i(n_i - 1)/N(N - 1)$
5. $J' = H'/H'_{max} = H'/\ln S$
6. $V' = (1/D)/D_{max} = (1/D)/S$

(1) Margalef's species richness index, (2) Menhinic's species richness index, (3) Shannon–Weaver diversity index, (4) Simpson's diversity index, (5) Pielou's index of evenness, and (6) Simpson's measure of evenness. Where D_{Mg} represents Margalef's species richness index, S equals the total number of species, and N is the total number of population in the sample. D_{Mn} is Manhinic's species richness index. H' represents Shannon–Weaver diversity index, and P_i is the proportion of population in species i to total sample. D is the Simpson's diversity index, and n_i represents the total number of individuals of each species. J' represents Pielou's species evenness index, and H'_{max} is the theoretical maximum value for H' if all species in the sample were equally abundant. V' is Simpson's Measure

of Evenness, and D_{max} represents the maximum amount of the species diversity index (Sharifi et al., 2013; Shirood Mirzaie et al., 2013).

5.3 RESULTS AND DISCUSSION

I sampled 2783 camera-trap nights. Total number of photographs was 1080. Out of 1080 photographs, 433 (40%) were of birds, wildlife rangers, poachers, and other non-target species that were excluded from the analysis, and 647 (60%) were of target species. I identified seven large and medium-sized terrestrial mammal species from two orders (Cetartiodactyla and Carnivora) and five families (Bovidae, Ursidae, Felidae, Canidae, and Suidae) in the study area during my survey.

The mountain area contained seven species, followed by the canyon, and the plain areas with six and three species, respectively (Table 5.1). All six indices in all three graphs showed similar patterns with different gradients (Fig. 5.1). The canyon area had the highest Menhinic's species richness index with 1.55, followed by the plain, and the mountain areas, with 0.65, and 0.47, respectively. Higher value of Shannon–Weaver diversity index means heterogeneity in the study area, whereas the lesser value indicates the homogeneity (Altaf et al., 2014). Hence, the canyon area showed more heterogeneity with 1.64, while the plain area was more homogeneous with 0.51. In case of Simpson's measure of evenness, the canyon, the plain, and the mountain areas showed amounts of 0.98, 0.46, and 0.38, respectively. This meant that the species were distributed more evenly in the canyon and less evenly in the mountain (Table 5.2).

Shirood Mirzaie et al. (2013) stated that biological indices were used to assess species interaction, fundamental community ecology, environmental stress, and biological factors, and the index with the most gradient was the best and the most sensitive one for that concept of ecological estimations. According to them, the aim of comparing different indices was to find out the most appropriate index for a study area. In my study, Menhinic's species richness index showed more gradient than Margalef's species richness index, and this index seemed more sensitive. As a result, Menhinic's index was considered a more suitable measurement to use in calculating the richness of the study area. Also, the more gradient of the Shannon–Weaver diversity index compared to Simpson's diversity index made it a better estimator to calculate the diversity of the study area. The more gradient Simpson's measure of evenness compared to the Pielou's index of evenness in the graph indicated that Simpson was more precise and sensitive for evenness investigations in the study area (Fig. 5.1).

TABLE 5.1 Abundance of the Large and Medium-sized Mammals in Three Different Habitat Types of the Study Area, Iran.

Common names	Scientific names	Mountain	Canyon	Plain	Total
Wild goat	*Capra aegagrus*	97	5	–	102
Alborz red sheep	*Ovis orientalis*	96	2	–	98
Brown bear	*Ursus arctos*	3	4	–	7
Persian leopard	*Panthera pardus*	5	1	1	7
Red fox	*Vulpes*	15	2	2	19
Golden jackal	*Canis aureus*	1	1	–	2
Wild boar	*Sus scrofa*	4	–	18	22
Total	–	221	15	21	257

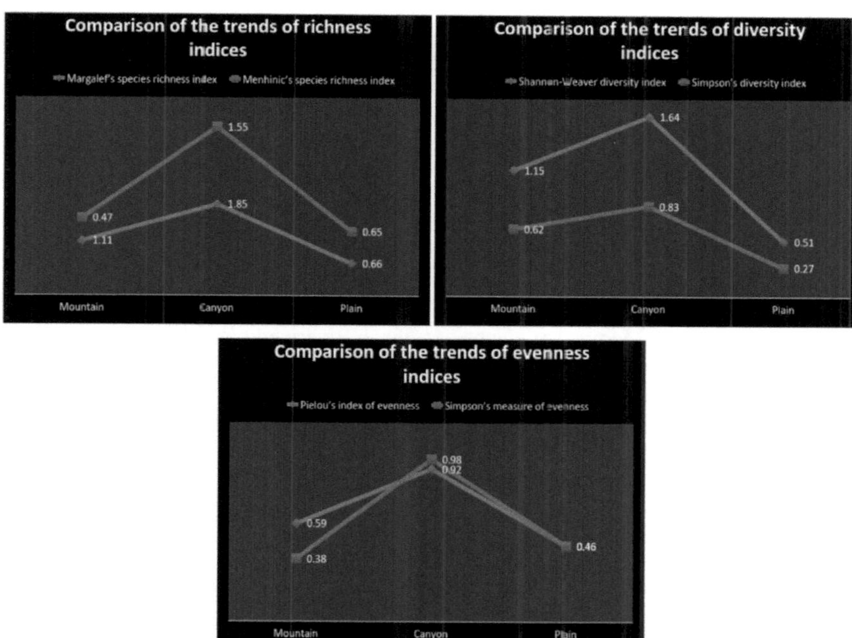

FIGURE 5.1 (See color insert.) Comparison of the trends of richness, diversity, and evenness indices among three different habitat types of the study area in Iran.

TABLE 5.2 Comparison of Margalef's Species Richness (D_{Mg}), Menhinic's Species Richness (D_{Mn}), Shannon–Weaver diversity (H'), Simpson's Diversity ($1 - D$), Pielou's Index of Evenness (J'), and Simpson's Measure of Evenness (V') of Three Different Habitat Types in the Study Area, Iran.

Habitat type	D_{Mg}	D_{Mn}	H'	$1 - D$	J'	V'
Mountain	1.11	0.47	1.15	0.62	0.59	0.38
Canyon	1.85	1.55	1.64	0.83	0.92	0.98
Plain	0.66	0.65	0.51	0.27	0.46	0.46

KEYWORDS

- **species richness**
- **diversity**
- **evenness**
- **Iran**
- **indices**

REFERENCES

Alatalo, R. V. Problems in the Measurement of Evenness in Ecology. *Oikos* **1981,** *37*, 199–204.

Altaf, M.; Irfan, A. J.; Munir, M. A.; Ashraf, S.; Umair, M.; Iqbal, K. J.; Khan, A. M.; Ali, Z. Diversity of Wild Mammalian Fauna of Chenab Riverine Forest, Punjab, Pakistan. *J. Animal Plant Sci.* **2014,** *24*,1342–1347.

Aslam, M. Diversity, Species Richness and Evenness of Moth Fauna of Peshawar. *Pak. Entomol.* **2009,** *31*, 99–102.

Ball, R. E. Time-lapse Cameras as an Aid in Studying Grizzly Bears in Northwest Wyoming. *Int. Conf. Bear Res. Manag.* **1975,** *4*, 331–335.

Foster, R. J.; Harmsen, B. J. A Critique of Density Estimation From Camera-trap Data. *J. Wildlife Manag.* **2011,** *9999*, 1–13.

Ghoddousi, A.; Hamidi, A. H. K.; Ghadirian1, T.; Ashayeri, D.; Moshiri, H.; Khorozyan, I. The Status of the Persian Leopard in Bamu National Park, Iran. CAT *News* **2008,** *49*,10–13.

Gotelli, N. J.; Colwell, R. K. Quantifying Biodiversity: Procedures and Pitfalls in the Measurement and Comparison of Species Richness. *Ecol. Lett.* **2001,** *4*, 379–391.

Rıos-Uzeda, B.; Gomez, H.; Wallace, R. B. A Preliminary Density Estimate for Andean Bear Using Camera-trapping Methods. *Ursus* **2007,** *18*, 124–128.

Sharifi, A.; Sepehry, A.; Barani, H. Investigation Biodiversity of Vegetation Species in Roadside (Case Study: Incheh-broun (Golestan Province). *Int. J. Agron. Plant Prod.* **2013,** *4*, 856–862.

Shirood Mirzaie, F.; GhorbaniR.; Montajami, S. A Comparative Study of Different Biological Indices Sensitivity: A Case Study of Macroinvertebrates of Gomishan Wetland, Iran. *World J. Fish Marine Sci.* **2013,** *5*, 611–615.

Stein, A. B.; Fuller, T. K.; Marker, L. L. Opportunistic Use of Camera traps to assess Habitat-specific Mammal and Bird Diversity in Northcentral Namibia. *Biodiv. Conserv.* **2008,** *17*, 3579–3587.

Sutter, J. V.; Andersen, M. E.; Bunnell, K. D.; Canning, M. F. Clark, A. G.; Dolsen, D. E.; Howe, F. P. *Purpose of the Comprehensive Wildlife Conservation Strategy*. Utah Comprehensive Wildlife Conservation Strategy (CWCS), 2005; pp 1–1.

Wilsey, B.; Stirling, G. Species Richness and Evenness Respond in a Different Manner to Propagule Density in Developing Prairie Microcosm Communities. *Plant Ecol.* **2007,** *190*, 259–273.

CHAPTER 6

Diversity of Pollen Grain Fertility and Crossing of Potato in Tajikistan

K. PARTOEV[1*], K. MELIKOV[2], A. NAIMOV[1], H. NANGYAL[2], and F.A. KHAN[3]

[1]*Institute of Botany, Plant Physiology and Genetics, Academy of Sciences, Republic of Tajikistan, Tajikistan*

[2]*Department of Botany, Hazara University, Mansehra, Khyber Pakhtunkhwa, Pakistan*

[3]*Dr. A. Q. Khan Institute of Biotechnology and Genetic Engineering (KIBGE), University of Karachi, Karachi, Pakistan*

[*]*Corresponding author. E-mail: pkurbonali@yahoo.com*

ABSTRACT

Potato tubers as a valuable food product for most people in the world are considered "second bread." Agricultural environment of mountainous areas of Tajikistan situated at an altitude of more than 1800 m above sea level makes it possible to grow good and quality harvest of potato. Under these cool mountainous conditions, potato plants are of little affection, have intensive bloom, and form a lot of berries and botanic seeds.

Thus, on basis of above researchers work analysis and our experience on potato crossbreeding obtained in India on 2009, for the first time in the highland of Tajikistan we have begun to research the rate of pollen grain fertility and to cross different grades and hybrids of potato and the results of this researchers brining in this article.

Among cultivated potato varieties Cardinal grade has the least percent of fertile pollen grains—26.5% and the new perspective Dusti grade has the greatest percent of those—95.2%. As a result of potato varieties and hybrids crossings, 6 kg of hybrid berries has been obtained which seed will be sown and studied in the future.

Our research works in Tajikistan direct to selected new clones of potato tolerant to high temperature and drought weather in the future for support of rural farmers in global climate change situation.

6.1 INTRODUCTION

During the researches made in different soil and climatic conditions a number of scientists (Perlova, 1958; Balashev, 1968; Lebedeva, 1970; Bukasov and Kameraz, 1972; Yashina et al., 1983; Mumindjanov, 2003; Shpaar, 2004; Kiru, 2007; Partoev et al., 2007, 2008; Carli et al., 2008) have ascertained the significance of breeding pair selection and locality of crossbreeding so as to get good results of potato hybridization.

According to the following breeders: Mendiburu and Peloquin (1976), Frankel and Galun (1977), Kaushik et al. (1997), Gopal (1994), Kumar and Gopal (2003), Pandey et al. (2003), Gupta et al. (2004), Luthra et al. (2006) from India, success of selective and genetic work aimed at nurture of new perspective grades depends to a large extent on pollen grain fertility and vitality when conducting different crossing between grades and species of potato.

6.2 METHODOLOGY AND SUBJECTS

To study the methods and ways of identification of pollen grain fertility and conducting potato crossbreeding, during the first two weeks of July 2009, with the assistance of the International Potato Center (CIP) we succeeded in visiting the Central Potato Research Institute (CPRI) in Shimla, India (K. Partoev). During this visit, such experienced Indian selectionists like Dr. Pandey S. Kumar, Jay Gopal, Vinod Kumar, Niradj Sharma, Lutra Kumar, Shambu Kumar, S. Chakrabari, and others accessibly trained us in methods of estimating pollen grain vitality and conducting potato cross. The above methods have been used in our research work in Tajikistan.

To identify vitality of potato pollen grains the following types of work should be prepared and carried out:

1. To make acetocarmine solution, dropping bottle, forceps, and needle ready for potato pollen release.
2. To remove flower pistil using forceps.
3. To pour out the pollen on the slide in the acetocarmine drop by shaking the stamen filament of the flower.
4. To cover the acetocarmine and pollen mixture with a glass sheet.
5. The ready preparation is to be examined through the microscope in 3–5 min.
6. The quantity of acetocarmine colored pollen grains (both fertile—vital pollen grains and yellow pollen grains—sterile or unviable grains) should be counted in three or five microscopic fields.
7. The counted quantity of colored and uncolored pollen grains is to be recorded into the logbook.

In some cases, when shaking the stamen filament pollen grains are not poured out. It is observed when the blooms or anthers are not mature enough or can be also connected with genetic character of potato grades and species. In that case, one or two anthers are taken and by means of forceps and needle, pollen grains are released and mixed with an acetocarmine drop on the slide, following which anthers' remains are removed and the preparation is getting ready to be examined on the glass sheet cover.

Pollen vitality was analyzed in three terms: on July 20 (the beginning of blooming period), on August 1 (mass blooming period) and on August 10 (the end of blooming period) in 2009–2013. In each term, the pollen of five plants of each sample was taken, 15 plants in total. Fertile (vital) and sterile (unviable) pollen grains were counted in three microscope fields with increasing of 8×7 on the microscopes MBC-9 and MBC-10 and with increasing of 20×7 on the microscope MBU-4A.

As a material for research 62 clones and varieties of potato that were kindly given by the International Center of Potato Investigation (CIP and CGIAR) in the network of plant gene pool investigation in the republics of Central Asia and the Caucasus were served. Also, the collection of potato samples obtained by the Institute of Plant Physiology and Phytogenetics of the Academy of Sciences of the Republic of Tajikistan and Social Organization "Tukhmiparvar" from VIR in 90 years of the last century. Sample planting was made on April 2009–2013 at an altitude of 2700 m above sea level on the field station of Social Organization "Tukhmiparvar" situated in Jirgital region, Sari-Kendja village, Muksu jamoat, Republic of Tajikistan.

Here, we also carried out straight and backcrossing of different potato grades and species for obtaining first-generation hybrids. For conducting potato grades crossbreeding after 3 p.m. emasculation of maternal plant flowers (removal of flower stamen filament with forceps) was made. After 4 p.m. we plucked paternal plant flowers, removed their pistils and kept them in the room temperature environment from 4 p.m. to 9 a.m. of the next day in Petri dish.

In the morning, inside the room, we shook stamen filaments with anthers using forceps and needle and picked the pollen into the glass Petri dishes of small size. The picked pollen was used for pollination of emasculated flowers by touching the pistil head with the picked pollen of paternal flowers. After pollination, we attached a paper label indicating hybrid combination (maternal form × paternal form) and the date of crossbreeding written in pencil to the stem. The number of pollinated flowers was recorded into the potato crossbreeding logbook in compliance with the dates of pollination.

6.3 RESULTS AND DISCUSSIONS

Investigations concerning identification of potato pollen grain fertility have been conducted in Jirgital region at an altitude of 2700 m above sea level where an experimental plot of the Social Organization "Tukhmiparvar" and the Institute of Botany, Plant Physiology, and Genetics of the Academy of Sciences of the Republic of Tajikistan is situated. In the course of the investigations, we managed to ascertain that fertile (vital) potato pollen grains placed in the acetocarmine drop are colored and turned red. As to sterile pollen grains, they are not colored and will have yellow coloring. Their number can be easily counted through the microscope (Fig. 6.1).

Thus, by means of the above methodology for the first time in our republic, we have estimated pollen grain fertility of potato grades and species. According to our researches, pollen grain fertility of potato clones and grades in highland environment of our republic is genotypically specific. Most of the examined clones and grades have more than 80–97% of fertile (vital) pollen grains. At the same time, it is necessary to note that pollen grain fertility of some potato grades makes only 5–10% (Table 6.1).

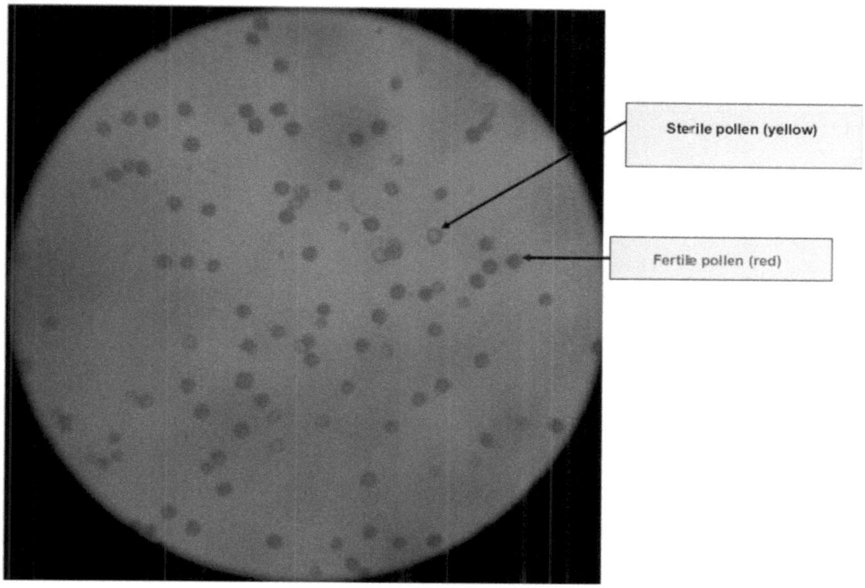

FIGURE 6.1 Sterile (yellow) and fertile (red) potato pollen grains through the microscope.

TABLE 6.1 Rate of Fertile Pollen Grains of Potato Grades in Highland Environment. (Jirgital Region at an Altitude of 2700 m above sea level).

S.No.	Varieties of potato	Analysis implementation dates (2019–2013)			Mean
		July 20	August 1	August 10	
1	Cardinal	28.0	24.4	27.1	26.5
2	Zarina	79.2	83.0	84.0	82.1
3	Dusti	95.0	96.6	94.0	95.2
4	Faizabad	85.9	86.0	84.1	85.3
5	Zhukovski	85.7	93.3	90.3	89.8
6	Condor	83.0	86.3	85.0	84.8
7	Picasso	86.4	85.2	84.1	85.2
8	Hybrid 22	84.5	87.5	90.4	87.5
9	Hybrid 23	64.7	74.9	98.2	79.3
10	Clone 27/5	64.7	63.5	63.6	63.9
11	Clone 30/9	76.7	77.0	78.9	77.5
12	Clone 36/6	66.7	68.3	70.3	68.4
13	Clone 37/2	84.1	86.2	93.8	88.0

TABLE 6.1 *(Continued)*

S.No.	Varieties of potato	Analysis implementation dates (2019–2013)			Mean
		July 20	August 1	August 10	
14	Clone 40/1	79.3	81.0	87.3	82.5
15	Clone 47/4	93.4	91.5	85.9	90.3
16	Clone 47/8	83.3	85.1	89.8	86.1
17	Clone 47/11	91.7	90.2	89.7	90.5
18	Clone 50/7	85.9	85.5	86.7	86.0
19	Clone 50/9	87.4	87.6	84.1	86.4
20	Clone 52/6	75.8	73.6	74.6	74.7
21	Clone 1	94.4	88.4	97.5	93.4
22	Clone 2	5.6	7.3	6.7	6.5
23	Clone 3	5.3	6.0	8.0	6.4
24	Clone 4	62.0	63.0	60.5	61.8
25	Clone 7	5.7	5.2	6.3	5.7
26	Clone 8	95.0	97.9	98.0	97.0
27	Clone 11	11.5	12.0	11.3	11.6
28	Clone 13	95.7	97.5	90.0	94.4
29	Clone 14	10.9	11.6	8.0	10.2
30	Clone 15	5.6	7.2	7.3	6.7
31	Clone 18	81.0	80.5	75.7	79.1
32	Clone 21	87.2	94.7	82.7	88.2
33	Clone 22	85.8	84.9	91.5	87.4
34	Clone 24	47.5	38.8	42.6	43.0
35	Clone 25	94.5	98.0	90.9	94.5
36	Clone 26	75.4	73.1	75.1	74.5
37	Clone 27	90.0	88.6	92.1	90.2
38	Clone 40	32.0	35.9	33.8	33.9
39	Clone 42	73.0	70.8	72.9	72.2
40	Clone 43	83.5	84.8	75.6	81.3
41	Clone 45	96.8	88.9	86.2	90.6
42	Clone 48	5.7	4.8	5.6	5.4
43	Clone 50	9.2	10.0	9.0	9.4
44	Clone 53	95.7	97.1	84.2	92.3
45	Clone 54	89.3	87.8	94.0	90.4

TABLE 6.1 *(Continued)*

S.No.	Varieties of potato	Analysis implementation dates (2019–2013)			Mean
		July 20	August 1	August 10	
46	Clone 55	90.7	89.9	90.5	90.4
47	Clone 56	95.2	97.5	96.8	96.5
48	Clone 58	10.6	9.1	9.3	9.7
49	Clone 59	88.4	86.0	93.0	89.1
50	Clone 60	87.6	89.4	88.1	88.4
51	Clone 63	88.8	82.8	92.1	87.9
52	Clone 64	5.0	6.6	7.0	6.2
53	Clone 65	5.2	4.7	5.4	5.1
54	Clone 66	96.6	95.0	94.0	95.2
55	Clone 67	6.0	7.1	7.1	6.7
56	Clone 68	23.6	25.0	25.0	24.5
57	Clone 71	23.9	21.0	19.8	21.6
58	Clone 73	78.0	76.0	76.5	76.8
59	Clone 75	95.4	94.9	91.9	94.1
60	Clone 76	97.9	85.6	81.1	88.2
61	Clone 79	95.3	89.0	85.4	89.9
62	Clone 80	13.0	11.8	13.0	12.6
X		64.9	64.7	64.9	64.8
S		34.3	34.2	34.3	34.2
S_x		4.3	4.2	4.2	4.2
V (%)		52.9	52.8	52.6	52.5
CD ($P \leq 0.05$)		8.6	8.4	8.6	8.5

Among potato varieties cultivated in the mountainous area Cardinal grade had the least per cent of fertile pollen grains—26.5%. The new perspective Dusti grade had the greatest per cent of those—95.2%. This grade was raised by the scientists of the Institute of Plant Physiology and Phytogenetics of the Academy of Sciences of the Republic of Tajikistan, the Social Organization "Tukhmiparvar" and the International Potato Center (CIP). This year the grade has been given to the Tajik State Commission on quality testing and grade protection.

It should be noted that some clones sampled individually from hybrid swarm F_1 and propagated from one plant by culture method (in-vitro plants) had stamen filaments of different shapes. For example, clones 43, 48, and 54 had different morphological attributes of stamen filaments (Fig. 6.2). Clone 43 had plants with two types of stamen filament shape. One part of the plant had a regular stamen filament; the other part of the plant had a modified stamen filament when stamen filaments are not pressed to the pistil stem as it is with regular plants. They are detached from it and look friable.

Clone #43

Clone # 48

Clone # 54

FIGURE 6.2 Different shapes of flowers, stamen filaments, and pistils of potato clones.

Thus, according to the stamen filament shape among these new potato clones, we can observe definite variation. It should be noted that the rest morphological signs (color of flowers and plants, leaf shape, evolution phenological stage and tuber color of the examined clones) remained stable. In addition, the researches indicated that the plants with modified shape of the stamen filament have weak yellow color of the stamen filament and low fertility of pollen grains compared to regular plants (Table 6.2).

TABLE 6.2 Pollen Grain Fertility Subject to the Shape of the Flower Stamen Filament, %.

S. No.	Varieties of potato	Analysis implementation dates (2019–2013)			Mean
		July 20	August 1	August 10	
1	Clone 43 (plants with regular stamen filament)	83.5	84.8	75.6	81.3
2	Clone 43 (plants with modified shape of stamen filament)	45.7	50.1	47.5	47.8
3	Clone 48 (plants with modified shape of stamen filament)	5.7	4.8	5.6	5.4

As you can see from above data (Table 6.2), plants with modified shape of stamen filament have considerably less fertile pollen grains compared to those with regular stamen. The sign of cytoplasmic male sterility is greatly detected with clone 48 where pollen grain fertility was only 5.4%. Along with it, we have ascertained that although many potato clones have regular stamen filament shape and bright yellow coloring of the stamen filament and anther their index of fertility is very low. They are clones 2, 3, 7, 15, 50, 58, 64, 65, and 67 that have only 5.1–9.7% of fertile pollen grains. These clones can serve a good baseline for potato breeding through straight crossing (without anther removal) and obtaining botanic seeds without flower emasculation. Thus, specific potato clones with male sterility sign have been defined.

As a result of potato hybridization carried out in Jirgital region from July 19 to August 11 (1650 crossings) by early September we have got regular berries F_1 (Fig. 6.3), which seeds will be studied in the future (Table 6.3).

TABLE 6.3 The Number of Pollinated Flowers and Set Berries Obtained Through Potato Hybridization.

S. No.	Maternal form	Paternal form	Number of pollinated flowers	Number of gathered berries	Rate of berry formation
1	Clone 75	Kufri Sadabakhar*	18	0	0
2	Clone 76	Kufri Jirgari*	48	18	37.5
3	Clone 66	Kufri Jiota*	30	10	33.3
4	Khahadin	Kufri Arunda*	38	3	7.9
5	Clone 75	Kufri Anchan*	15	0	0
6	Clone 75	Kufri Suraya*	34	6	17.6
7	Clone 59	GTH/107	26	0	0
8	Clone 66	Kufri Suraya*	7	2	28.6
9	Clone 76	Mix of pollen	29	5	17.2
10	Clone 68	Mix of pollen	14	0	0
11	Clone 63	Mix of pollen	12	3	13.6
12	Clone 75	Clone 76	25	6	24.0
13	Clone 76	Clone 75	22	8	36.4
14	Clone 79	Clone 66	7	0	0
15	Zhukovski ranniy	Dusti	72	0	0
16	Dusti	Zhukovski raniy	15	0	0
17	Cardinal	Dusti	70	43	61.4
18	Dusti	Cardinal	20	0	0
19	Dusti	Condor	191	83	43.5
20	Zarina	Dusti	40	0	0
21	Dusti	Picasso	44	17	38.6
22	Hybrid 23	Dustï	23	0	0
23	Clone 40/1	Dusti	22	1	4.5
24	Clone 40/1	Cardinal	17	0	0
25	Hybrid 22	Dusti	11	0	0
26	Hybrid 23	Picasso	53	0	0
27	Faizabad	Picasso	127	41	32.3
28	Cardinal	Condor	125	88	70.4
29	Cardinal	Picasso	105	84	80.0
30	Clone 48	Dusti	65	2	3.1
31	Clone 40/1	Dusti	22	16	72.7

TABLE 6.3 *(Continued)*

S. No.	Maternal form	Paternal form	Number of pollinated flowers	Number of gathered berries	Rate of berry formation
32	Cardinal	Dusti	50	0	0
33	Hybrid 23	Dusti	85	0	0
34	Clone 40/1	Dusti	30	6	20.0
35	Zarina	Hybrid 23	25	2	8.0
36	Dusti	Condor	85	53	63.4
37	Clone 48	Condor	28	14	50.0
	Total		1650	471	28.5

*The pollen of potato for crossing was brining from Central Potato Research Institute (CPRI) in Shimla, India in 2009 (K. Partoev).

FIGURE 6.3 Potato hybrid berries, obtained through Dusti × Picasso grades crossing.

6.4 CONCLUSIONS

For the first time in the mountainous area of Tajikistan study of pollen grain fertility and hybridization of potato varieties (*Solanum tuberosum L.*) have been carried out. The degree of pollen grain fertility and great variability of this genetic sign among 62 potato samples (5–97%) has been identified. It has been stated that some potato clones have great variability according to flower and anther coloring and flower stamen filament shape. Ten potato clones with male sterility sign and less quantity of fertile pollen grains (5.1–9.7%) have been picked out. Such clones can be a good material for selection and genetic work aimed at raising of new potato hybrids and grades. Among cultivated potato varieties Cardinal grade has the least percent of fertile pollen grains—26.5% and the new perspective Dusti grade has the greatest percent of those—95.2%. As a result of potato varieties and hybrids crossings, 6 kg of hybrid berries have been obtained which seed will be sown and studied in the future.

Our research works in Tajikistan direct to selected new clones of potato tolerant to high temperature and drought weather in the future for support of rural farmers in global climate change situation.

KEYWORDS

- **selection**
- **pollen vitality**
- **hybridization**
- **clones**
- **hybrids**

REFERENCES

Balashev, N. N. *Growing of Potato and Vegetables*; M Kolos, 1968; p 366.

Bukasov, S. M.; Kameraz, A. Y. *Selection and Seed Farming of Potato*; Leningrad Kolos, 1972; p 359.

Carli, C.; Khalikov, D.; Yuldashev, F. In *Recent Advances in Potato Research and Development in Central Asia*. Abstracts Global Potato Conference, Delhi, 2008; pp 31–32.

Frankel, R.; Galun, E. *Pollination Mechanisms, Reproduction and Plant Breeding*; Springer Verlag: Berlin-Heidelberg- New York, 1977; pp 35–78.

Gopal, J. Flowering Behavior, Male Sterility and Berry Setting in Tetraploid *Solanum tuberosum* Germplasm. *Euphytica* **1994,** *72*, 133–142.

Gupta, V. K.; Thakur, K. C.; Shantanu, K.; Pandey, S. K.; Uma, Sah. *True Potato Seed – An Alternative Technology for Potato Production in North-eastern Hill Region*; CPRI: Shimla, 2004; pp 1–21.

Kaushik, S. K.; Birhman, R. K.; Singh, B. P.; Gopal, J. In *Combining Ability and Heterosis for Field Resistance to Late Blight in Potato.* National Symposium on Molecular Approaches in Plant Disease Management, November 14–15; CPRI: Shimla, 1996.

Kiru, S. D. Outcomes and Prospects of the World Potato Collection Research. Genetic Resources of Cultivated Plants in XXI Century. The II Vavilov International Conference Reports Theses, St. Petersburg, 2007; pp 474–476.

Kumar, R.; Gopal, J. Combining Ability of Andigena Accessions for Yield Components and Tuber Dry Matter in Third Clonal Generation. *J. Ind. Potato Assoc.* **2003,** *30*, 3–4.

Lebedeva, N. V. To the Question of Mountainous Potato Seed Farming in the Northern Tajikistan. *Newslett. Sci. Tech. Inf.* **1970,** *8*, 47–50.

Luthra, S. K.; Pandey, S. K.; Singh, B. P. *Potato Breeding in India*; Central Potato Research Institute, 2006; pp 3–71.

Mendiburu, A. O.; Peloquin, S. J. Sexual Polyploidization and Deploidization. Some Terminology and Definitions. *Ther. Appl. Genet.* **1976,** *48*, 137–143.

Mumindjanov, K. A. *Physiological and Biotechnological Approach to Selection and Seed Farming of Potato*: Dushanbe, 2003; p 126.

Pandey, S. K.; Gupta, P. K. Genetic Divergence and Combining Ability Studies on True Potato Seeds (TPS) in Potato *(Solanum tuberosum L.). J. Ind. Potato Assoc.* **1997,** *24*, 1–16.

Pandey, S. K.; Singh, S. V.; Chakrabarti, S. K.; Manivel, P. *New Potato Hybrids*; Central Potato Research Institute: Shimla, 2005; pp 3–44.

Partoev, K.; Karimov, B.; Sulangov, M.; Melikov, K. Study of Potato Varieties in Different Ecological Conditions of Tajikistan. Genetic Resources of Cultivated Plants in XXI Century. The II Vavilov International Conference Reports Theses, St. Petersburg, 2007; pp 329–331.

Partoev, K.; Naimov, S.; Melikov, K.; Jumakhmadov, A.; Abdurakhimov, S. *The Hybridization of Potato (Solanum tuberosum L.) in Tajikistan*: Dushanbe, 2010; p 38.

Partoev, K.; Sulangov, M.; Melikov, K. In *Potato Research and Development in Tajikistan.* Abstracts Global Potato Conference, Delhi, 2008.

Perlova, R. L. *Potato Species Habit in Different Regions of the USSR*; M Publising House of the Academy of Sciences: USSR, 1958; p 238.

Shpaar, D. *Potato. Torzhok OOO Variant*, 2004; p 461.

Yashina, I. M.; Sklyarova, I. P.; Kiryukhin, V. P. Recommendations on Development of Potato Grade Model for Non-Black Soil Area of the Country and Physiological and Biochemical Methods of Selection Material Assessment in Practical Selection; M. Kolos, 1983; pp 48–54.

CHAPTER 7

Impact of Arsenic on Structural and Functional Composition of Dominant Bacterial Populations Associated with Various Natural Ecosystems

SOMA GHOSH[1*], MEENAKSHI MUKHERJEE[1], and PINAKI SAR[2]

[1]*School of Environmental Studies, Jadavpur University, Kolkata, West Bengal, India*

[2]*Department of Biotechnology, Indian Institute of Technology Kharagpur, Kharagpur, West Bengal, India*

**Corresponding author. E-mail: ghoshsoma1987@gmail.com*

ABSTRACT

Arsenic (As), a carcinogenic element is abundant in earth's crust, is increasingly contaminating different ecosystems due to various natural and anthropogenic reasons. The presence of this toxic element has been known to influence the bacterial community composition of natural ecosystems. This study, therefore, reviews the major bacterial populations found to be associated with As contaminated aquifers, agricultural fields, mines and mine tailings, and marine environments. Bacterial genera belonging to *Alphaproteobacteria, Betaproteobacteria, Gammaproteobacteria, Deltaproteobacteria, Epsilonproteobacteria, Bacteroidetes, Firmicutes, Nitrospira, Aquificeae, Acidobacterium, Verrucomicrobium, Ignavibacterium,* etc., were found to be associated with different As-impacted ecosystems globally, among which *Proteobacteria* were found to be the most abundant. Metabolic capabilities of these dominant bacterial genera with respect to various electron-donor oxidation and electron-acceptor reduction are reviewed. An unweighted paired group mean analysis suggests that bacterial genera of the different phylogenetic lineage may act in consortia. Besides, As (V) reduction is found to be better associated with potential chemoautolithotrophs, nitrate reduction, and ammonia oxidation. This review shall provide essential information regarding As contamination of various environments and the usual consortia associated with that particular ecosystem. Therefore, this study would act as a useful resource for designing of strategies pertaining to as bioremediation of impacted sites without disturbing the natural bacterial flora.

7.1 INTRODUCTION

Arsenic (As), a teratogenic and carcinogenic toxin, ranking 1st on the Priority List of Hazardous Substances from the US Agency for Toxic Substances and Disease Registry (https://www.atsdr.cdc.gov/spl/index.html), is abundantly present on earth's crust. Elevated concentrations of As in groundwater, air, water, and soil due to various natural and anthropogenic reasons, finally enters the human body through direct ingestion, inhalation, or through the food chain. Apart from carcinogenic effects, diabetes, obesity, cardiovascular diseases, allergies, inflammatory bowel disease, and others are commonly associated with the As exposed populations. Arsenic can be easily solubilized in groundwaters and remain unpredictable as most of the As compounds are colorless, odorless, and tasteless. Dissolution of As compounds in water generally depends on the nature of sorbents available, pH, redox conditions (Eh), and microbial activity. Both inorganic and organic species of As exist in one of the four oxidation states, that is, −3, 0, +3, and +5. Elemental As hardly exists in nature and As in −3 oxidation states exist in highly reducing conditions. Bacterial populations associated with the As contaminated environment are the major players in oxidation, reduction, and methylation of As.

Various studies on bacterial diversity predominating As-impacted environments have been carried out and their predictive role have been established. For example, As-contaminated paddy fields have been found to be dominated by dissimilatory Fe reducing bacteria (FeRBs). These have been known to reduce Fe(III) utilizing the soil organic carbon during their anaerobic respiration causing the dissolution of Fe(III) and simultaneous solubilization of As. Role of sulfate reducing bacteria (SRBs) in such fields have also been indicated (Liesack et al., 2000; Wind et al., 1999) resulting in incorporation of As(III) in Fe(II)-sulfide minerals (Bostick and Fendorf, 2003), potentially reducing As bioavailability to the plant being cultivated. Role of As metabolizing bacterial diversity has been found to be predominant in As contaminated aquifers (Oremland and Stolz, 2005; Ghosh and Sar, 2013, Paul et al., 2015). Arsenic oxidizers oxidize As (III) to As(V) in the aqueous phase whereas As reducers may reduce As(V) either dissimilatorily, under anaerobic condition, or may simply undergo a cytosolic reduction of As(V) under aerobic or anaerobic condition. Depending upon the bacterial metabolic interactions with As, that is, whether the organisms use As in their energy metabolisms or only transforming it for detoxification, and whether the organisms are heterotrophic or lithotrophic, As transforming micro-organisms could be categorized into five groups: (1) dissimilatory As(V)-reducing prokaryotes (DARPs), (2) As-resistant micro-organisms (ARMs), (3) chemoautotrophic As(III) oxidizers (CAOs), (4) heterotrophic As(III) oxidizers (HAOs), and (5) phototrophic As(III) oxidizers (PAOs) (Sar et al., 2016). As(III) oxidation in both CAOs and HAOs is found to be catalyzed by a membrane-bound respiratory As(III) oxidase enzyme encoded by *aio* gene. Cytosolic As(V) is reduced and effluxed by the *ars*operon, where *arsC* encoded protein reduces As(V) to As(III) and As(III) is extruded out of the cell via *arsB*or *acr3* gene product in ARMs. As(V) is also reduced anaerobically via *arr*operon where As(V) is being used as a terminal electron acceptor for dissimilatory energy acquisition by DARPs (Fig. 7.1) (Mukhopadhyay et al., 2002; Silver and Phung, 2005; Cavalca et al., 2013). On the other hand, phylogenetically diverse microbes have been reported to harbor abundant and diverse As methylating genes (*ars*M) in the rice rhizosphere soil and roots. These bacteria can methylate As(III) or As(V) to their organic counterparts. Methylated As(III) compounds are the most toxic compounds and those of As(V) are the least toxic depending on their abilities to degrade DNA (Dopp et al., 2004). The DNA damage potential of As compounds decreases in the order dimethylarsinic acid (DMA)(III) > monomethylarsonic acid (MMA) (III) > [As(III), As(V)] > MMA(V) > DMA(V) > trimethylarsine (TMA)(V) (Wang and Mulligan, 2006). Another study on As resistance potential of bacterial communities in soils and sediments of gold tailings showed As(V) and As(III) metabolism to be coupled with denitrification (Guan et al., 2017).

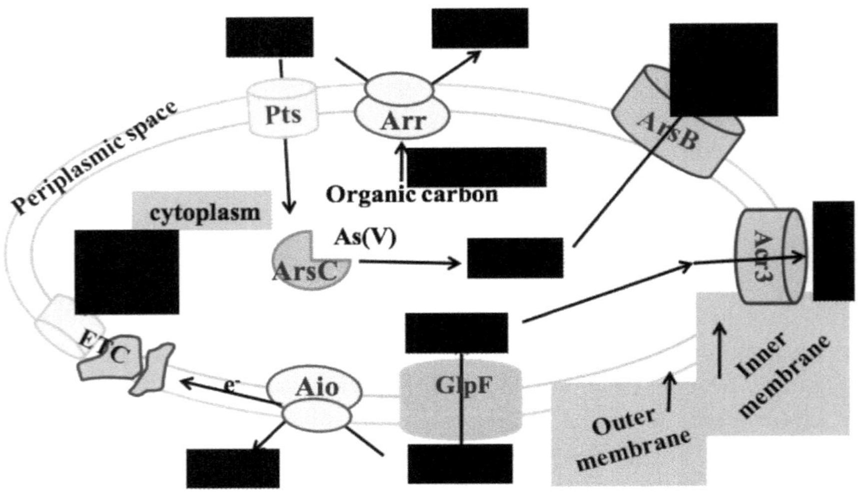

FIGURE 7.1 Schematic diagram depicting mechanisms involved in bacterial interaction with inorganic species of As and functional genes responsible for their biotransformation.

Numerous such studies have been performed in various As contaminated environments. However, there is a need to understand the basic role of the predominant microbial communities and interrelationship of various biogeochemical cycles undergoing in such environments. This review provides a clear understanding of the predominant bacterial groups in As contaminated sites, their roles, and expected bioremediation potential of these bacterial communities.

7.2 ARSENIC CONTAMINATED SITES AROUND THE WORLD

Arsenic contamination in the groundwater aquifers beyond the consumable limit is a growing global concern. Along with anthropogenic sources of contamination, natural geological settings coupled with the biogeochemical condition of the aquifers provide various sources of As release into the groundwater. The environment of the aquifers affected by As contamination globally can be grouped as either oxidizing or reducing. Groundwater aquifers tainting various degrees of As concentrations have been reported in parts of North America as well as South America, Australia, few countries of the Middle East and a major part of South and South East Asia. A detailed global study on distribution of As in different ecosystems have been tabulated in Table 7.1. Aquifers of reducing nature of India, Bangladesh, Nepal, China, Taiwan, Vietnam, and Cambodia are the worst affected countries within South East Asia with geogenic As calamity. Arsenic in these regions are mainly derived from the erosions of the Himalayas carried by the river channels and concomitant reduction to the more mobile form in the aquifers which are generally reducing in nature. Aquifers in these regions namely Bengal Delta Plain, Hetao Plain, Red River Plain, etc., are heavily contaminated with As. Microbial populations dealing with the toxic metalloid As and geochemical conditions of the aquifers are majorly responsible for As release and contamination of aquifers. Sediments of volcanic origin or in mines are generally of oxidizing nature found in UK, USA, Ghana, Chile, Argentina, Mexico, and Canada, etc. (Nordstrom, 2002). Arsenic contamination in these regions occurs in response to different geochemical conditions, including the release of As to groundwater through the reaction of iron oxide with either natural or anthropogenic (i.e., petroleum products) organic carbon. Iron oxide also can release As to alkaline groundwater in some felsic volcanic rocks and alkaline aquifers. Sulfide minerals act as both source and sink for As. Extremely high concentrations of As (10g/L) in ground- and surface water, particularly in the west have also been found to be associated with geothermal water and high evaporation rates (Welch et al., 2000). Release of As in aquifers of these areas are again the consequences of the major bacterial populations inhabiting these oxidized environments. Evidently, the role of bacterial populations in reducing and oxidizing environments in the release of As are different depending on other physical and environmental factors influencing As speciation.

7.3 EFFECT OF PHYSICAL AND ENVIRONMENTAL FACTORS ON RELEASE OF ARSENIC SPECIES

7.3.1 REDOX PROPERTIES

Redox condition of the environment is the most important factor in regulating microbial and geochemical behavior toward As speciation and availability. The respiratory oxidoreductase enzymes found in microbes are bidirectional and act depending on the prevailing redox potential. Therefore, it has been to be one of the triggers of the environment for As speciation.

TABLE 7.1 Worldwide Distribution of As (ppb) in Various Ecosystems.

Ecosystem	Range of As contamination	Location	References
Agricultural field			
Paddy field soil	46	Bangladesh	Meharg et al., 2003
Soil and agricultural plants	6,100–16,700	Nepal	Dahal et al., 2008
Irrigational water, agricultural land, soil, and plants	41240	West Bengal	Roychowdhury et al., 2005
Agricultural soil	35,000–29,3000	Japan	Arao et al., 2010
Food chain	10,000–12,600	West Bengal	Bhattacharya et al., 2010
Agricultural soil	500	Bangladesh	Saha et al., 2007
Aquifer			
Groundwater	50–3200	Bangladesh	Alam et al., 2002
Groundwater	400–600	South and southeast Asia	Brammer and Ravenscroft, 2009
Groundwater	50–1400	California, United States	Welch et al., 2000
Groundwater	139–430	Vietnam	Berg et al., 2001
Soil and groundwater	738–2180	Canada	Wang and Mulligan, 2006
Soil and sediment	52–1045	Argentina	Mukherjee et al., 2006
Soil and water	40–860	Chile	Rivara et al., 1997
Well water	68	China	Sun et al., 2004
Groundwater	100–200	Assam, India	Ghosh and Sar, 2013
Mining and acid mine drainage			
Tailing water of abandoned Uranium mines	–	Germany	Mkandwire and Duden, 2005
Draining water of sulfur mines	1,00,000–3,50,000	Carnoules, France	Duran et al., 2012

TABLE 7.1 *(Continued)*

Ecosystem	Range of As contamination	Location	References
Tailing water of Iron mines	40-50	France	Bruneel et al., 2005
Streams in catchment of coal mines	8500	South Island, New Zealand	Lear et al., 2009
Tailing water of mines	850,000	Iron mountain	Nordstrom and Alpers, 1999
Tailing water of Tungsten–Tin mines	–	Spain	Valverde et al., 2011
Tailing water of Copper smelter	–	Montana	Macur et al., 2001
Tailing water of mines	34,000	USA	Plumlee et al., 1999
Tailing water of mines	400,000	Ural mountain	Gelova, 1977
Tailing impoundment	300–1,00,000	Ontario, Canada	McCreadie et al., 2000
Mining tailing water	50–360	British Columbia	Azcue et al., 1994
Surface water			
River flood-plains	100–122	Germany	Overesch et al., 2007
River water	400–21,800	Northerm Chile	Sancha, 1999
River water	100–114	Argentina	Lerda and Prosperi, 1996
River water	300–370	Madison and Missouri, USA	Nimick at al., 1998
River water	500–583	Ron Phibun, Thailand	Williams et al., 1996
River water	7900	Ashanti, Ghana	Smedley et al.,1996
Lakes	900–1000	Western USA	Benson and Spencer, 1983
Lakes	70–530	Canada	Bright et al., 1996
Lakes	35–100	Ontario, Canada	Azcue and Nriagu, 1995

7.3.2 ROLE OF SORPTION

Adsorption reactions, precipitation, and coprecipitation reactions with oxides- hydroxides severely limit As mobilization at near neutral pH (Edmunds et al., 2015). Both As(V) and As(III) sorb strongly onto amorphous iron oxides, goethite, lepidocrocite, and hematite, however, the behavior of As is dependent on its oxidation state and the mineralogy of the iron oxides. Competitive binding takes place between As and other sorbates such as phosphate, silicate, and bicarbonate. Since the phosphate concentrations in groundwater exceed those of As, it acts as a potent at competitor of As(V) for sorption sites on iron oxide minerals. Slow diagenesis of hydrous ferric oxide transforms them into stable iron oxides with low specific sorbing sites resulting in increased As mobilization (Dixit and Hering, 2003).

7.3.3 ROLE OF OTHER ANIONS

Various studies have reported the effect of sulfate, carbonate, bicarbonate, borate, chromate, molybdate, and nitrate on As release. Kirk et al. (2004) has reported that sulfate reduction available in the sediments by bacterial population limits As mobilization. Another study suggested that the release of As was strongly related to the bicarbonate concentration in the leaching solution. Carbonation of As-sulfide minerals, including orpiment (As_2S_3) and realgar (As_2S_2), has been found to be an important process in leaching As into groundwater under anaerobic conditions (Kim et al., 2000). Chromate, molybdate, and nitrate enhanced reduction reactions of oxyanions of As in presence of zerovalent iron (Su and Puls, 2001).

7.3.4 ROLE OF ORGANIC MATTER

Natural organic matter influences the solubility and mobility of As mainly through competitive adsorption, complexation, and redox reactions (Wang and Mulligan, 2006). Redox conditions and biogeochemical processes related to As enrichment in the aquifers largely depend on the interactions between Fe minerals and dissolved organic matter. The labile and bioavailable DOM can serve as an energy source for the indigenous microbes thereby, mediating reductive dissolution of Fe(III) oxides/hydroxides and reduction of SO_4^{2-}. Moreover, the quinone-like humic substances may even act as an electron shuttle between the labile DOM and Fe(III), SO_4^{2-} and As(V)for their reduction (Islam et al., 2004; Tufano and Fendorf, 2008). Therefore, it was argued that mobilization of As occurs by slow reduction of iron oxyhydroxides or sorbed As(V) by detrital organic carbon (Nickson et al., 1998).

7.4 BIOGEOCHEMICAL CYCLE OF ARSENIC

Arsenic, the historic infamous poison, is ubiquitous on earth's crust and has become the most notorious natural contaminant. The biological effects of the element As is diverse; As(V), a chemical analog of PO_4^{3-}, can uncouple mitochondrial oxidative phosphorylation. On the other hand, As(III), possessing high affinity for thiol groups, affects respiration by binding to the vicinal thiols in dehydrogenases. It also affects the function of glucocorticoid receptors. Methylated forms of As and organoarsenicals are thought to be much less toxic than the inorganic forms. Thereby, indigenous environmental microbes, coupled with environmental factors, have evolved with various mechanisms to transform As from one form to the other causing biogeochemical cycling of the element As. Figure 7.2 attempts to provide an overview of natural and anthropogenic contributors of As and its biogeochemical cycling. Arsenic present in earth's crust is always complexed with some ores as sulfides or oxides or hydroxides. Geological/natural activities, such as volcanic eruption, forest fires, erosion, and surface runoff, are considered to be the major contributors of As to the atmosphere. Anthropogenic activities such as smelting, industrial wastes incineration, extensive use of fertilizers, and pesticides, mining activities also contribute majorly in the addition and biogeochemical cycling of As in the environment. Wet and dry deposition of the accumulated As in the atmosphere is then precipitated to the soil and water bodies where microbial activities transform these into sorbable and mobilizable forms, depending on geochemical and microbial factors.

Microbial oxidation–reduction activities transform As(III)to As(V) and vice versa leading to their mobilization in the aquifers and thereby drinking water contamination. Inorganic As(V) entering the cells through phosphate transporters is eventually reduced to As(III)by their cytosolic detoxifying reductases or respiratory reductases in DARPs, which is then extruded out to the environment causing As(III)dissolution (Rosen., 2002;

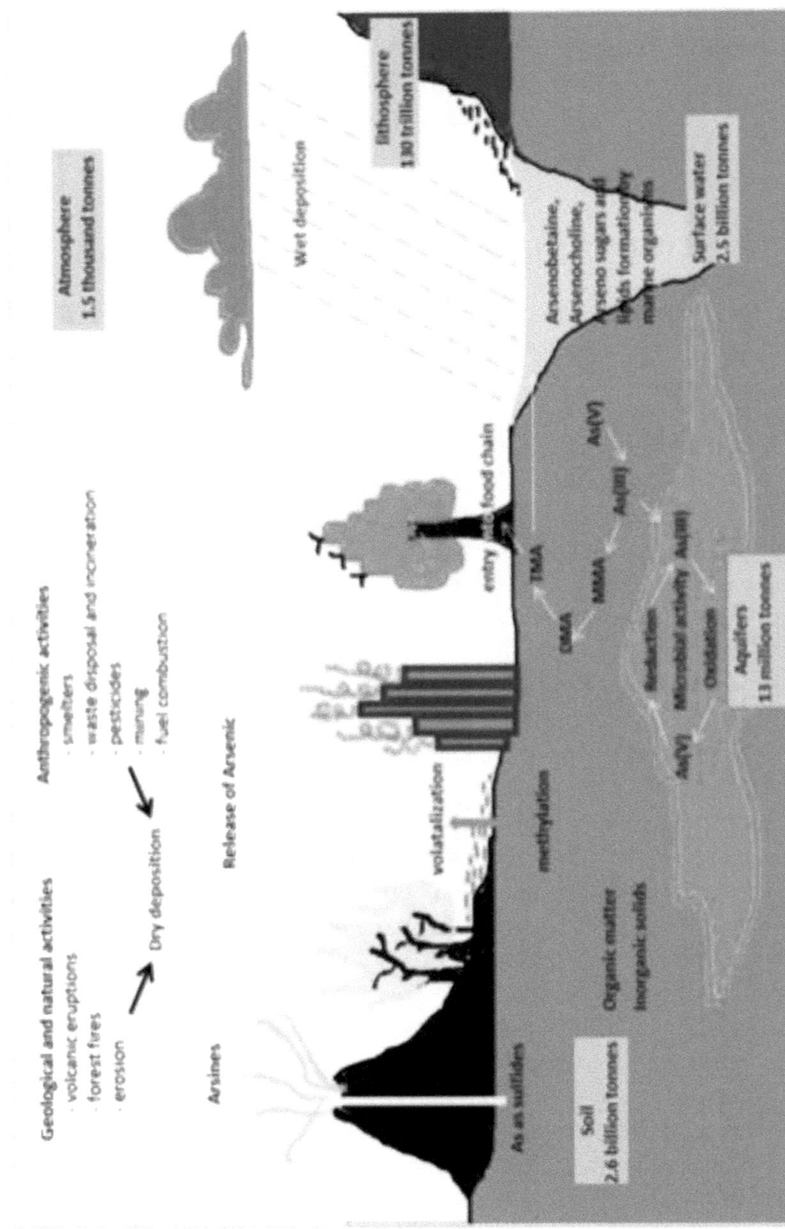

FIGURE 7.2 (See color insert.) Biogeochemical cycle of As. Pink boxes indicate total amount of As present in particular environment.

Source: Data obtained from Sohn (2014).

Oremland and Stolz, 2003). As(III) present in the environment is taken up by the CAOs to be oxidized to obtain metabolic energy through the process or detoxified by HAOs. Microbes and marine organisms are also capable of methylating the As species producing MMA, DMA, and TMA, which may also be volatilized into the atmosphere (Qin et al., 2006). These species may thereby enter into the food chain as they are found to be uptaken into the flora and fauna. Sohn (2014) presented a brief estimation of the total amount of As stored in a particular environment. According to the report, As contained in the atmosphere, lithosphere, soil, aquifers, and ocean were 1.5 thousand tonnes, 130 trillion tonnes, 2.6 billion tonnes, 13 million tonnes, and 2.5 billion tonnes, respectively. Hence, in this study, we discuss the major bacterial groups predominating As-impacted different environments namely As-impacted agricultural fields, aquifers, surface waters, and acid mine drainages or mines and their roles in As mobilization.

7.5 MICROBIAL DIVERSITY IN DIFFERENT ARSENIC-IMPACTED ECOSYSTEMS

7.5.1 ARSENIC CONTAMINATED AQUIFERS

Microbes are considered to be the driving factors of many geochemical reactions in natural environments, where the conditions or stress prevailing in the environment, such as concentrations of As, SO_4^{2-}, NO_3^-, Fe, and pH value, are the shapers of microbial communities (Chen et al., 2017). Bengal Delta plain and its aquifers have been found to be the worst affected with As. Microbial diversity of northeastern states was found to be predominated by *Brevundimonas* and *Acidovorax* along with *Acinetobacter*, *Pseudomonas* followed by *Undibacterium*, *Herbaspirillum*, *Rhodococcus*, *Staphylococcus*, *Bosea*, *Bacillus*, *Ralstonia*, *Caulobacter*, and Rhizobiales members (Ghosh and Sar, 2013). Microbial community composition in such contaminated aquifers of West Bengal has been reported by various studies (Sarkar et al., 2013; Paul et al., 2015; Ghosh et al., 2014). The predominance of chemolithotrophic As(III) oxidizing *Agrobacterium/Rhizobium*, *Ochrobactrum,* and *Achromobacter* along with other genera *Acidovorax, Albidiferax, Hydrogenophaga, Bosea*, and *Polymorphum* were found. Diverse, yet near consistent community, composition was found to be mostly represented by genera *Pseudomonas*, *Flavobacterium*, *Brevundimonas*, *Polaromonas*,

Rhodococcus, Methylo-versatilis, and *Methylotenera* through culture-dependent and independent approaches. Highly diverse potentially As(III) oxidizing microorganisms and Fe(III) reducing *Desulfuromonadles* were detected. Less diverse As(V) reducing iron-oxidizing *Gallionellaceae*have also been detected (Hassan et al., 2015). Islam et al. (2005) reported the presence of *Pseudomonas* species, facultative anaerobes able to respire alternative electron acceptors including nitrate, but not As(V) or Fe(III), *Clostridium* species capable of fermenting organic matter and reduce a range of metals including As(V), *Geobacter*, a dissimilatory metal-reducing bacteria and *Nitrosolobus* species, which are aerobic ammonia-oxidizing bacteria of the β-proteobacteria in As-rich groundwaters of Bangladesh. Hery et al. (2010) studied the microbial diversity in As rich groundwaters of Bangladesh and found similar results with some new families and genera like *Bacteroidetes, Acidobacteria, Verrucomicrobiales, Pleomorphomonas, Geodermatophilaceae, Vogisella, Burkholderiales,* and *Methylophilaceae* predominating such environments.

Microbial diversity associated with As contaminated soils and aquifers of other parts of Bengal Delta Plain has been reported by many studies (Cavalca et al., 2013; Sarkar et al., 2013; Ghosh et al., 2014; Paul et al., 2015). *Acinetobacter, Pseudomonas, Psychrobacter, Alishewanella, Thiobacillus, Pseudomonas, Hydrogenophaga, Enterobacteriaceae, Sulfuricurvum, Massilia, Arthrobacter, Dietzia, Planococcus, Brevundimonas, Aquabacterium,* and *Geobacter* have been reported to be the associated microbial diversity in the As-rich groundwater and sediments, respectively, in Inner Mongolia, China (Li et al., 2013). Archaeal sequences found were mostly related to methanogens (Wang et al., 2015). *Agrobacterium tumefaciens, Pseudomonas fluorescens, Variovorax paradoxus, A. tumefaciens, Flavobacterium* sp., *Microbacterium* sp., and *Arthrobacter*were detected in the As-rich soil of Montana (Macur et al., 2004). *Pseudomonas, Bacillus, Psychrobacter, Vibrio, Citrobacter, Enterobacter,* and *Bosea* were found to be the As-resistant genera in the As-contaminated groundwater of Taiwan (Liao et al., 2011). *Sulfuro*spirillum*, Desulfotomaculum auripigmentum,* and *Geobacter uraniumreducens* were the associated bacterial genera in the As-rich Cambodian sediments (Lear et al., 2007). Arsenic metabolizing members of *Hydrogenobaculum, Metallosphaera, Acidimicrobium, Thiomonas,* and chemolithotrophic. Hyperthermophilic populations, including *Stygiolobus* and *Hydrogenobaculum,* were detected in the geothermal springs of Yellowstone National Park

(Macur et al., 2004). Dominant As-resistant isolates from contaminated soils were identified by their fatty acid methyl ester profiles as *Acinetobacter*, *Edwardsiella*, *Pseudomonas*, *Salmonella*, *Enterobacter*, and *Serratia* (Turpeinen et al., 2004).

Bacterial diversity in groundwater has been found to differ with increasing depth and decreasing redox potential. The upper layer of the groundwater is generally found to be predominated by As(III) oxidizing bacteria, such as *Achromobacter, Acidovorax, Sinorhizobium, Agrobacterium, Alkaligenes faecalis, Variovorax sp., Xanthobacter autotrophicus, Rhizobium* NT-26, etc. Evidently, shallow groundwater contains higher dissolved oxygen thus creating a mild oxic condition and higher redox potential prevails compared with that of deeper groundwater where oxidation of the reduced species is possible. On the contrary, at greater depths with almost negative redox potential As(V) reducing bacterial populations are more into play. Diversity is predominated by *Pseudomonas, Acinetobacter, Sphingomonas parapaucimobilis, Arthrobacter, Desulfomicrobium, Wolinella, Clostridium,* etc. Another study by Osborne et al., 2015 reported a dissimilatory As(V) reducing bacterium belonging to the genus *Desulfuromonas* isolated from the redox front of an aquifer of West Bengal. Under reducing conditions, microbial As(V) reduction is a major process that contributes largely to the groundwater pollution that too, by the more toxic species As(III).In agreement to this, Ghosh et al. (2015) found presence of As(III) oxidizing bacterial communities predominated by *Acinetobacter*, *Acidovorax,* and *Hydrogenophaga*in gray sand aquifers, that is, shallow aquifers with little iron oxyhydroxide (FeOOH) coating imparting a gray color to the sand), whereas such communities were absent in deeper brown sand aquifers (brown color imparted by heavy Fe coatings). An abundance and diversity of dissimilatory As(V) reductase (*arr*) genes at a depth of 190–230 m while no such observation was found for shallow groundwater within 5–185 m (Chen et al., 2017). On the contrary, several novel arsenite oxidase genes (*aoxB*s) have been reported from the upper layers of the aquifer sediments (0–7 m) and were found to be specific for arsenite-oxidizing bacteria (Fan et al., 2008). Microbes use As(V), available in the aquifer sediments, as terminal electron acceptor under reducing conditions resulting in their metabolic advantage.A novel arsenic (As)-resistant, arsenate-respiring, alkane-metabolizing bacterium, *Rhizobium arsenicireducens,* isolated from As-rich groundwater of West Bengal was characterized by physiological and genomic properties. The cells were found to reduce O_2, As^{5+}, NO^{3-}, SO_4^{2-}, and Fe^{3+} as alternate electron acceptors (Mohapatra et al., 2017). Moreover, the potential availability of the heterogenous dissolved organic carbon in the aquifers can influence the community composition of indigenous heterotrophic microbial flora, which in turn affects As biogeochemical cycle (Ghosh et al., 2015). Many iron-reducing bacteria have been found to be equally potent As(V) reducer. *Shewanella* sp. ANA-3, *Shewanella putrefaciens* strain CN-32 *Sulfurospirillum barnesii* strain SES-3, *Shewanella* alga strain BrY, etc., have been known to be potent Fe(III) and As(V) reducers and have shown to have implications in increased As mobilization in groundwater postreduction (Cummings et al., 1999; Kocar et al., 2006; Tufano et al., 2008). It seems a natural adaptation by the bacterial population indigenous to the aquifer sediments rich in FeOOH coatings. FeOOH coatings on mineral surfaces act as a very good adsorbent for As(V). Fe reducing bacteria in process of their Fe(III) reduction to Fe(II) may dislodge the adsorbed As(V) or it might undergo simultaneous dissimilatory reduction to As(III).

7.5.2 ARSENIC CONTAMINATED AGRICULTURAL PADDY FIELDS

Geogenic As contamination of groundwater and soil also poses a threat to the sustainable agricultural fields mainly in paddy cultivation, where cultivation of paddy plants is done in waterlogged condition (Brammer and Ravenscroft, 2009). Therefore, prolonged irrigation with these contaminated groundwaters deposit As in the topsoil (with mean As deposition on land per year of about 5.02 kg/h; Roychowdhury et al., 2005) of agricultural fields and even contribute in direct absorption of As by the plant system in As contaminated areas of South and southeast Asian countries Bangladesh, Cambodia, China, India, Laos, Myanmar, Nepal, Pakistan, Taiwan, and Vietnam. Roychowdhury et al. in 2005 reported considerable uptake of As by various parts of rice plants (approximately 3.1–13.1%, 0.54–4.08%, and 0.36–3.45% of As is taken up by the root, stem, and leaf, respectively) from the soil. Elevated concentration of As in agricultural soil may exert negative impacts such as land degradation in terms of loss of yield, that is, decline in crop production and disease like 'straight head disease' with empty panicle in rice (Rahman et al., 2008). A high concentration of this toxic metalloid definitely has an impact on the soil microbial community. The speciation of As in paddy soil and rice plants have indicated the presence of both inorganic (As(V) and As(III)) and organic species (MMA, DMA, and

TMA) of As (Zheng et al., 2013, Roychowdhury, 2008; Pal et al., 2009). It is well known that higher eukaryotes, that is, plants are incapable of methylating As on their own (Zheng et al., 2013) and methylated forms of As found in plants are generated by soil micro-organisms (Lomax et al., 2012); therefore, a definite role of micro-organisms in contaminated paddy fields exists. An important assistance imparted by PGPMs (plant growth promoting micro-organisms), which is a diverse group of microbes including bacteria, cyanobacteria, etc., is the altered speciation of As in the soil through methylation and subsequent change in the bioavailability of As to the plants. Further, a change in As speciation also affects As uptake and transport in plants, favorably reducing As concentrations in crop plants or increasing As accumulation in phytoremediator plants (Upadhyay et al., 2018). Apart from As bio transforming, that is, oxidizing, reducing, and methylating bacteria, Fe(III) reducing, nitrate-reducing, nitrate dependent Fe(II) oxidizing along with methanoarchaea and cyanobacteria have been reported to be presented and play a consorted role in biogeochemical cycles. Arsenic methylation and volatilization have been reported to be facilitated by organic matter content of the agricultural field soil and are performed by numerous anaerobic prokaryotes and abundant autotrophic sulfate-reducing bacteria as well as methanoarchaea in paddy field soils under anaerobic condition (McBride and Wolfe, 1971; Liu et al., 2009). A study on quantification of volatile As species released from *Escherichia coli* expressing the As(III) S-adenosyl methionine methyltransferase gene has been reported by Yuan et al., (2008). Moreover, Mestrot et al. (2009) demonstrated that loss of volatile As majorly in form of trimethyl arsines and less amounts of Me_2AsH were obtained in manured samples suggesting that the As concentrations in the paddy field soil can be indeed reduced by increasing biomethylation of As. Majumder et al. (2013) characterized few As tolerant, arsenite-oxidizing bacterial strains from contaminated soils of West Bengal. A report from Japanese paddy soil documented the role of a dissimilatory arsenate-reducing bacterium *Geobacter* sp. OR-1 in As dissolution (Ohtsuka et al., 2013).

Predominant microbial diversity from paddy fields under flooded and nonflooded conditions have been found to vary considerably, expectedly due to varying redox potential under the two conditions. The genera *Ohtae-kwangia, Geobacter, Anaeromyxobacter, Desulfuromonas, Desulfocapsa, Desulfobulbus,* and *Lacibacter* were found in relatively high abundance in the flooded soil, whereas the genera *Acinetobacter*, *Ignavibacterium*,

Thiobacillus, and *Lysobacter* were detected in relatively high-abundance in the nonflooded soil (Das et al., 2016). Arsenic- resistant strains belonging to *Arthrobacter, Alkaligenes, Oceanimonasdoudoroffi, Arsenicicoccus-bolidensis, Microbacteriumhydrocarbonoxydans, Kocuriaerythromyxa, Staphylococcus succinus, Variovoraxparadoxus*, and *Bacillus megaterium* have been reported from As-contaminated paddy field soils of Bangladesh (Bachate et al., 2008).

Dissimilatory Fe(III) reducing bacterial community are dominant in the rhizospheric region where their increased activity and abundance (about 26% of the total-microbial community)have been found to cause Fe(III) reduction and thereby As release retained in Fe plaque (Lovley et al., 2004; Cummings et al., 1999; Rowland et al., 2007). Significant values of microbially mediated Fe(III) reduction (i.e., up to 24% of total reduced Fe; Wang et al., 2009) in rice paddies have been documented (Ratering and Schnell, 2001; Hori et al., 2009). A relatively high abundance of iron-reducing bacteria under continuously flooded condition compared with the intermittent-flood treatment implied greater relative iron reduction and possibly As release from the iron oxides under the former condition (Somenahally et al., 2011). However, it is still under speculation whether the FeRB activity actually results in As release or would serve in potentially As absorption upon the formation of secondary iron-oxide phases as a result of Fe(III) reduction (Kocar et al., 2006; Tufano et al., 2008; Masue-Slowey et al., 2010).

The detection of SRBs, such as members of *Desulfovibrionaceae*, related to *Desulforhabdus, Desulfobulbus, Desulfobotulussapovorans,* and *Desulfosarcinavariabilis Syntrophobacterwolinii*, and members of genus *Desulfotomaculum* in rice field soil and rice roots suggests that sulfate reduction could also be actively occurring in the rice rhizosphere (Ouattara and Jacq, 1992; Wind et al., 1999). Other SRBs were found to belong to *Desulfobacteriaceae,* genera *Desulfonema* and *Desulfosarcina,* members of *Clostridia* and *Deltaproteobacteria,* (Scheid and Stubner, 2006; Liu et al., 2009). Sulfides so produced as a result of sulfate reduction by SRB may undergo reaction with As(III) to form soluble and insoluble thioarsenic compounds (Newman et al., 1997; Oremland et al., 2004). Sulfate reduction could also result in incorporation of As(III) in Fe(II)sulfide minerals (Bostick and Fendorf, 2003), and thus potentially reducing As bioavailability to the rice plant. These studies provide substantial evidence that FeRB and SRB are present in high abundances in the rice rhizosphere and suggest that these bacteria might be more active in continuously flooded plots.

7.5.3 ARSENIC-RESISTANT BACTERIAL DIVERSITY IN MARINE ENVIRONMENT

Arsenic in the marine environment has been reported long back in 1902 by Bertrand when analytical techniques were not well developed. Later, Jones in 1922 reported that marine organisms naturally contained As in a range of 5–94 mg/kg, but most of the As present are in their organic forms. Arsenic concentrations in between 5.8 and 121.0 mg/kg have been found in deep-sea sediments from West Pacific Ocean, the Sea of Japan, North Atlantic Ocean, Indian Ocean, Equatorial Pacific Ocean, South Pacific Ocean with an average of 40 mg/kg for world oceans. (Onishi and Sandell, 1955, Bostrom and Valdes, 1969). Rather sediments from coastal and estuarine regions showed much lower As concentrations of around 14 mg/kg (Onishi, 1969; Francesconi and Edmonds, 1993). Andreae, 1979 found four species of As in seawater samples namely, arsenate, arsenite, methyl arsonate, and dimethyl arsenate. A good correlation of methylated As and rate of photosynthesis could be observed at the photic zone suggesting biotransformation of As species. These varying concentrations of As at various depth levels of the sea certainly has an impact on the marine bacterial ecosystem. A study on As metabolism in marine bacterium by Vidal and Vidal, 1980 indicated As metabolism by *Serratia marinorubra* found to produce As(III) and methylarsonic acid [CH_3AsO(OH)] upon growth with As(V). Another study performed by Jeanthon and Prior, 1990 on characterization of heterotrophic bacteria from hydrothermal vent polychaete annelids showed the predominance of heavy metal tolerant members of *Alteromonas, Pseudomonas* spp., *Vibrio* spp., etc. A much greater metabolic versatile bacterial diversity was detected clustering the isolates in 18 different taxon with mixotrophic sulfur-oxidizing capabilities. Sabry et al. (1996) isolated 81 marine bacterial isolates and studied their metal tolerance and antibiotic resistance patterns. A correlation of 100% was found between lead and all antibiotics tolerance/resistance followed by 95% for copper and penicillin and 83.3% for nickel and ampicillin. Arsenic resistance, accumulation, and transformation have also been reported in few other marine isolates such as *Marinomona scommunis* and has been suggested as potential candidate for As bioremediation (Takeuschi et al., 2007). *Microbacterium esteraromaticum* has been found to be the dominant member in As(III) enriched communities isolated from deep-sea sediments of the Southwest Indian Ocean Ridge (Chen et al., 2009).

7.5.4 ARSENIC CONTAMINATED ACID MINE DRAINAGES

Acid mine drainages (AMDs) are naturally formed in the environment on oxidation of pyrite minerals to produce high iron and sulfate contents. These are enhanced upon mining activities that expose the pyrite-rich minerals to the infiltrating oxygen-rich rain water causing such oxidation processes. Acid mine drainages are highly acidic with pH prevailing in the range of 2–3 that enhance mobility of dissolved metals/metalloids (e.g., Cu, Pb, Zn, Hg, As). Higher heavy metals concentration in the AMDs has toxic ecological effects perturbing the natural microbiome and reducing species richness and total biomass (Kelly, 1988). Isolation and application of sulfate-reducing bacteria (SRB)for AMD bioremediation purposes have been widely proposed as microbial sulfate reduction causing metal precipitation with sulfides and formation of soluble sulfides of metalloids, like As, Sb, Mo, increases alkalinity (Garcia et al., 2001; Kappler and Dahl, 2001; Kaksonen et al., 2003; Bijmans et al., 2009, 2010) coupled with organic matter oxidation (Giloteaux et al., 2012). Therefore, various studies on bacterial communities of differently As-impacted acid mine drainages have shown varied phylogenetic members of bacterial species capable of tolerating low pH and heavy metal stress. Such an environment host micro-organisms capable of modifying the physicochemical conditions of the AMD water either by detoxification or by metabolic exploitation (Bruneel et al., 2005). An As(V) respiring *Desulfomicrobium* sp. isolated from an AMD was reported by Macy et al., 2000. A bacterial community analysis for As-resistant bacterial populations in mine tailings through denaturing gradient gel electrophoresis revealed that *Caulobacter, Sphingomonas, Pseudomonas,* and *Rhizobium* like populations are associated with As mobilization (Macur et al., 2001). Giloteaux et al. (2012) reported various phylotypes of SRBs among which *Desulfohalobiaceae* members were detected to dominate the acidic sample (pH 1.4) with high iron and As concentrations. Other SRB present in highly As contaminated Carnoule`s acid mine drainage (France) were found to be the members of *Desulfobulbaceae, Desulfomicrobiaceae, Syntrophobactereceae, Desulfotomaculum, Firmicutes,* and *Nitrospiraceae,* etc. High Fe concentrations in AMDs favor the development of iron-oxidizing bacteria, such as *Acidithiobacillus ferrooxidans* and *Leptospirillum ferrooxidans* (Baker and Banfield, 2003; Hallberg and Johnson, 2003). Bruneel et al. (2006) found a Fe oxidizing neutrophilic bacterium *Gallionella ferruginae* to be the dominant member in Carnoules AMD that could efficiently remove Fe(III), As(V), and As(III) from water. Among other bacterial genera found associated with Fe, As, and S cycling was *Actinobacteria, Desulfobacterium,* and *Thiomonas*, respectively. The diversity of culturable As-resistant bacteria across a soil As gradient at an abandoned tungsten–tin mining area was, as expected, inversely related to the soil As content. The predominating As-resistant bacteria were *Firmicutes* (mainly *Bacillus* spp.) followed by *γ-Proteobacteria* (mainly *Pseudomonas* spp.) and a strain of the species *Rahnella aquatilis* exhibiting strong endurance to As (Valverde et al., 2011).

7.6 PHYLOGENETIC ANALYSIS OF THE BACTERIAL DIVERSITY FOUND ACROSS THE CONTAMINATED SITES

An intensive and in-depth literature survey on bacterial populations associated with different As-contaminated environments suggest a diverse bacterial community composition across the sites considered in this study. A neighbor joining tree with bootstrap iterations of 500 has been constructed with the various bacterial genera detected in different As-impacted sites to establish their phylogenetic relatedness (Fig. 7.3). Various bacterial classes, such as *Alphaproteobacteria, Betaproteobacteria, Gammaproteobacteria, Deltaproteobacteria, Epsilonproteobacteria, Bacteroidetes, Firmicutes, Nitrospira, Aquificeae, Acidobacterium, Verrucomicrobium, Ignavibacterium,* etc., have been found to play important roles in biogeochemical cycling of As in various As-contaminated sites with varying dominance. The aquifers highly contaminated with As have been found to contain maximum As-resistant bacterial diversity with maximum no. of genera found in these environments. Figure 7.4 represents the abundance of various bacterial classes found in four different As-impacted ecosystems. *Proteobacterial* members are generally found to predominate all four As-contaminated ecosystems. Although contaminated aquifers are mainly predominated by *Alphaproteobacteria, Betaproteobacteria, Gammaproteobacteria,* and *Firmicutes*, agricultural fields have been found to be rich in *Deltaprobacteria, Bacteroidetes,* and *Firmicutes*. As-impacted acid mine drainages were found to show a homogenous distribution of various proteobacterial classes and others like *Nitrospira, Aquificeae,* and *Acidobacterium.* Very less As-resistant bacterial populations belonging

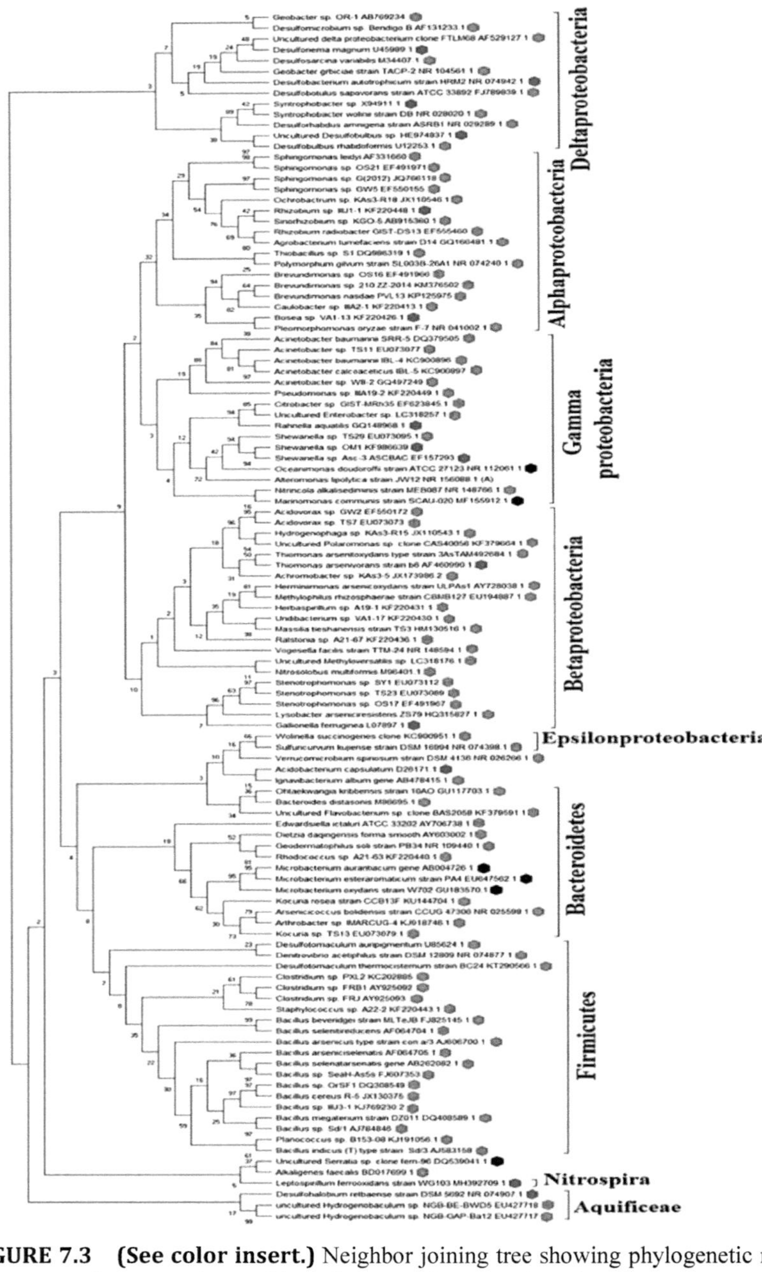

FIGURE 7.3 (See color insert.) Neighbor joining tree showing phylogenetic relatedness among the genera detected in various arsenic contaminated environments

to *Gammaproteobacteria* and *Bacteroidetes* were found to be associated with marine environments.

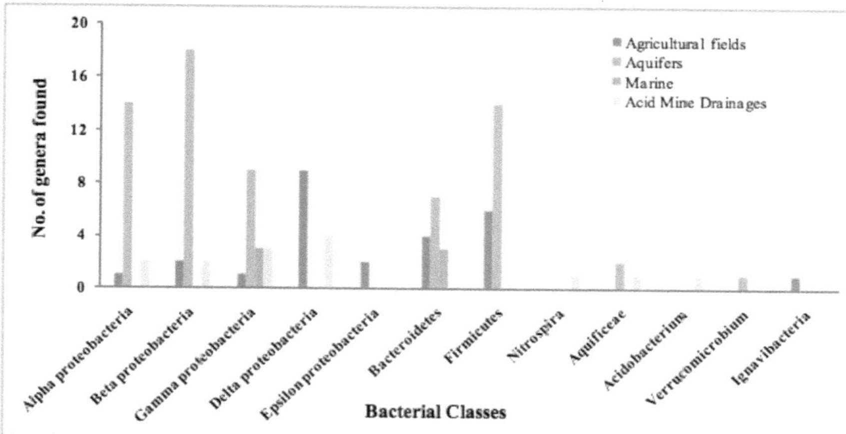

FIGURE 7.4 (See color insert.) Abundance of different bacterial classes in four As-impacted ecosystems.

7.7 METABOLIC DIVERSITY OF THE ARSENIC-RESISTANT BACTERIAL GENERA

The exhaustive analysis of bacterial populations associated with different As-contaminated ecosystems could detect a total of 120 genera whose metabolic potentials were extensively studied from the available literature. Metabolic capabilities, with respect to oxidation of As(III), Fe(II), NH_3; reduction of As(V), reported genera detected in As-impacted environments have been studied and represented in Figure 7.5. An unweighted paired group arithmetic mean analysis (UPGMA) performed through MVSP 32 shows a clear relatedness of As and Fe oxidation potentials of the bacterial diversity detected (Fig. 7.6a). The second broader clade relates chemo-organotrophy and/orchemoheterotrophy with various terminal electron acceptor reduction (e.g., As(V), Fe(III), SO_4^{2-}, NO_3^-) implying lithotrophy and methane production. As methanogenesis is one of the outcomes of anaerobic heteropolymer degradation, therefore, organotrophic or heterotrophic mode of nutrition is justified to show relatedness with methane production, Arsenate reduction is statistically found to be

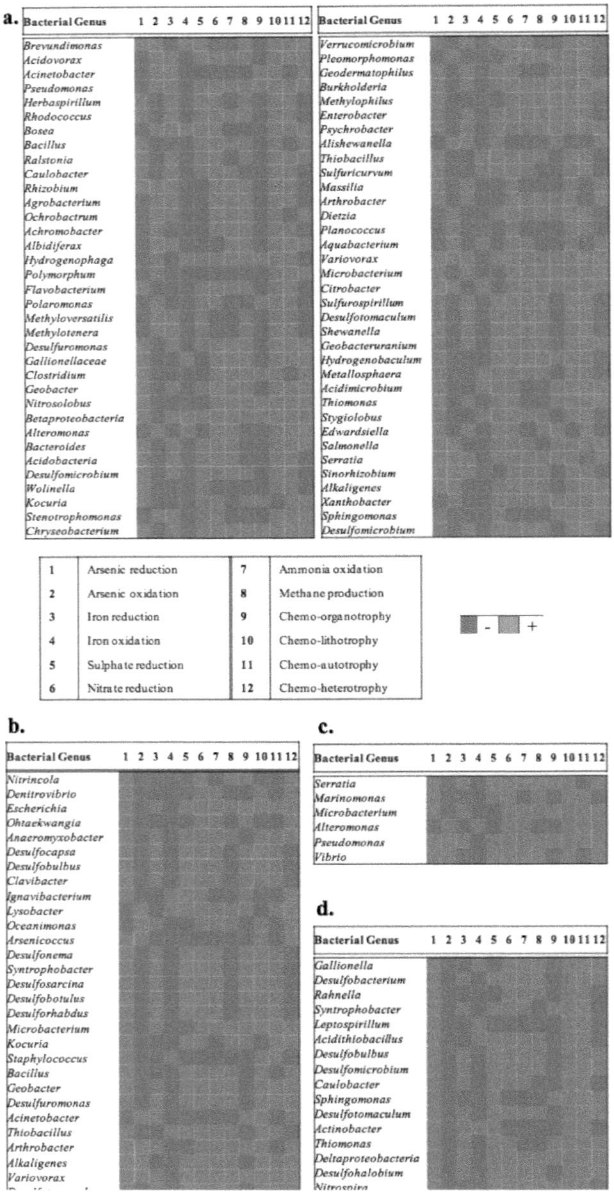

FIGURE 7.5 (See color insert.) Metabolic response of various bacterial genera to 12 parameters indicated in legend in four As contaminated ecosystems (a: aquifers, b: agricultural fields, c: marine environment, d: acid mine drainages)

better associated with potential chemoautolithotrophy, nitrate reduction, and ammonia oxidation. This observation is in well accordance with Guan et al. (2017) who proposed that As metabolism is associated with denitrification. Another UPGMA to analyze the metabolic relatedness of the bacterial genera was performed with the transposed data. The dendrogram suggests that the bacterial genera detected are highly metabolically versatile and clade depending on their metabolic diversity irrespective of their phylogenetic relatedness.

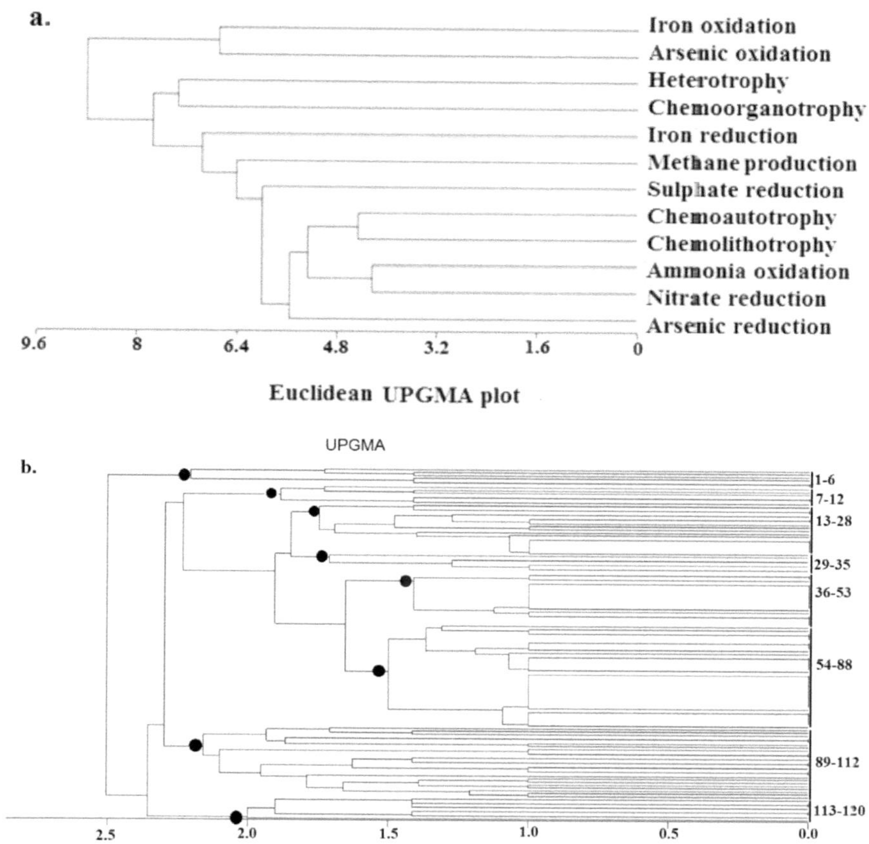

FIGURE 7.6 UPGMA dendrogram showing relatedness among metabolic characteristics of the dominant genera (a); UPGMA dendrogram showing cladding of dominant bacterial genera depending on their metabolic response (data according to Fig. 7.3) (b). The numbers representing the clades denote bacterial genera mentioned in Table 7.2. The clades that have been considered are marked in black circles.

TABLE 7.2 Bacterial Genera Corresponding the Clades in Figure 7.4.

CLADES	MICRO-ORGANISMS
1-6	*Arsenicoccus, Denitrovibrio, Massilia, Alishewanella, Acinetobacte*
7-12	*Ignavibacterium, Ohtaekwangia, Nitrincola, Kocuria, Edwardsiella*
13-28	*Citrobacter, Gallionella, Rahnella, Nitrospira, Hydrogenobaculum, Methylophilus, Nitrosolobus, Clostridium, Marinomonas, Sulfuricurvum, Acidithiobacillus, Thiobacillus, Stygiolobus, Metallosphaera*
29-35	*Flavobacterium, Salmonella, Alkaligenes, Vibrio, Bacillus*
36-53	*Thiomonas, Geodermatophilus, Syntrophobacter, Desulforhabdus, Desulfobotulus, Desulfosarcina, Desulfonema, Clavibacter, Desulfobulbus, Desulfocapsa, Enterobacter, Bacteroides, Caulobacter, Verrucomicrobium, Acidobacteria*
54-88	*Caulobacter, Arthrobacter, Serratia, Pseudomonas, Variovorax, Microbacterium, Deltaproteobacteria, Escherichia, Chryseobacterium, Psychrobacter, Burkholderia, Betaproteobacteria, Desulfohalobium, Desulfotomaculum, Desulfomicrobium, Desulfobulbus, Desulfobacterium, Desulfuromonas, Geobacter, Anaeromyxobacter, Geobacteruranium, Ralstonia, Shewanella, Sulfurospirillum, Rhodococcus*
89-112	*Herbaspirillum, Polaromonas, Planococcus, Polymorphum, Methylotonera, Methyloversatilis, Stenotrophomonas, Bosea, Aquabacterium, Alteromonas, Albidiferax, Oceanimonas, Ochrobactrum, Agrobacterium, Sinorhizobium, Rhizobium, Lysobacter, Dietzia, Hydrogenophaga, Pleomorphomonas, Achromobacter, Xanthobacter, Pseudomonas*
113-120	*Acidovorax, Actinobacter, Leptospirillum, Staphylococcus, Wolinella, Sphingomonas, Acidimicrobium, Brevundimonas*

7.8 CONCLUSIONS

This review provides a comprehensive study of the bacterial diversity associated with various As-contaminated ecosystems. Bacterial populations were found to be predominated by proteobacterial and low G+C members capable of As metabolism. Besides, these bacterial genera were also found to be metabolically versatile and diverse. Bacterial genera showing similarity in their metabolic activities irrespective of their phylogenetic lineage might act in consortia to elevate the rate of biotransformation processes. Therefore, this review shall act as a useful resource for designing sustainable bioremediation technologies exploiting the metabolic capabilities of these bacterial genera to bio transform As under varying conditions in extensively As-stressed environments.

ACKNOWLEDGMENTS

The authors thankfully acknowledge the School of Environmental Studies, Jadavpur University, West Bengal, India for providing a scientific platform. The fellowship provided by CSIR (09/096(0833)18-EMR-1) to Dr. Soma Ghosh is duly acknowledged.

KEYWORDS

- **arsenic**
- **agricultural fields**
- **aquifers**
- **mines**
- **marine ecosystem**
- **bacterial diversity**
- ***Proteobacteria***

REFERENCES

Alam, M. G. M.; Allinson, G.; Stagnitti, F.; Tanaka, A.; Westbrooke, M. Arsenic Contamination in Bangladesh Groundwater: A Major Environmental and Social Disaster. *Int. J. Environ. Health Res.* **2002,** *12* (3), 235–253.

Arao, T.; Ishikawa, S.; Murakami, M.; Abe, K.; Maejima, Y.; Makino, T. Heavy Metal Contamination of Agricultural Soil and Countermeasures in Japan. *Paddy Water Environ.* **2010,** *8* (3), 247–257.

Azcue, J. M.; Nriagu, J. O. Impact of Abandoned Mine Tailings on the Arsenic Concentrations in Moira Lake, Ontario. *J. Geochem. Explor.* **1995,** *52* (1–2), 81–89.

Bachate, S. P.; Cavalca, L.; Andreoni, V. Arsenic-resistant Bacteria Isolated from Agricultural Soils of Bangladesh and Characterization of Arsenate-reducing Strains. *J. Appl. Microb.* **2009,** *107* (1), 145–156.

Baker, B. J.; Banfield, J. F. Microbial Communities in Acid Mine Drainage. *FEMS Microb. Ecol.* **2003,** *44* (2), 139–152.

Benson, L.; Spencer, R. J. A Hydrochemical Reconnaissance Study of the Walker River Basin, California and Nevada, 1983.

Bhattacharya, P.; Samal, A. C.; Majumdar, J.; Santra, S. C. Arsenic Contamination in Rice, Wheat, Pulses, and Vegetables: A Study in an Arsenic Affected Area of West Bengal, India. *Water Air Soil Pollution* **2010,** *213* (1–4), 3–13.

Bijmans, M. F.; van Helvoort, P. J.; Dar, S. A.; Dopson, M.; Lens, P. N.; Buisman, C. J. Selective Recovery of Nickel Over Iron from a Nickel–Iron Solution Using Microbial Sulfate Reduction in a Gas-lift Bioreactor. *Water Res.* **2009,** *43* (3), 853–861.
Bostick, B. C.; Fendorf, S. Arsenite Sorption on Troilite (FeS) and Pyrite (FeS2). *Geochimica et Cosmochimica Acta* **2003,** *67* (5), 909–921.
Boström, K.; Valdes, S. Arsenic in Ocean Floors. *Lithos* **1969,** *2* (2), 351–360.
Brammer, H.; Ravenscroft, P. Arsenic in Groundwater: A Threat to Sustainable Agriculture in South and South-east Asia. *Environ. Int.* **2009,** *35* (3), 647–654.
Bright, D. A.; Dodd, M.; Reimer, K. J. Arsenic in sub-Arctic lakes Influenced by Gold Mine Effluent: The Occurrence of Organoarsenicals and 'Hidden' Arsenic. *Sci. Total Environ.* **1996,** *180* (2), 165–182.
Bruneel, O.; Duran, R.; Casiot, C.; Elbaz-Poulichet, F.; Personné, J. C. Diversity of Micro-organisms in Fe-As-rich Acid Mine Drainage Waters of Carnoules, France. *Appl. Environ. Microb.* **2006,** *72* (1), 551–556.
Casiot, C.; Lebrun, S.; Morin, G.; Bruneel, O.; Personné, J. C.; Elbaz-Poulichet, F. Sorption and Redox Processes Controlling Arsenic Fate and Transport in a Stream Impacted by Acid Mine Drainage. *Sci. Total Environ.* **2005,** *347* (1–3), 122–130.
Cavalca, L.; Corsini, A.; Zaccheo, P.; Andreoni, V.; Muyzer, G. Microbial Transformations of Arsenic: Perspectives for Biological Removal of Arsenic From Water. *Fut. Microb.* **2013,** *8* (6), 753–768.
Chen, S.; Shao, Z. Isolation and Diversity Analysis of Arsenite-resistant Bacteria in Communities Enriched from Deep-sea Sediments of the Southwest Indian Ocean Ridge. *Extremophiles* **2009,** *13* (1), 39–48.
Chen, X.; Zeng, X. C.; Wang, J.; Deng, Y.; Ma, T.; Guoji, E.;. ... Wang, Y. Microbial Communities Involved in Arsenic Mobilization and Release from the Deep Sediments into Groundwater in Jianghan Plain, Central China. *Sci. Total Enviro.* **2017,** *579*, 989–999.
Cummings, D. E.; Caccavo, F.; Fendorf, S.; Rosenzweig, R. F. Arsenic Mobilization by the Dissimilatory Fe (III)-Reducing Bacterium *Shewanella alga* BrY. *Environ. Sci. Technol.* **1999,** *33* (5), 723–729.
Dahal, B. M.; Fuerhacker, M.; Mentler, A.; Karki, K. B.; Shrestha, R. R.; Blum, W. E. H. Arsenic Contamination of Soils and Agricultural Plants Through Irrigation Water in Nepal. *Environ. Pollut.* **2008,** *155* (1), 157–163.
Das, S.; Chou, M. L.; Jean, J. S.; Liu, C. C.; Yang, H. J. Water Management Impacts on Arsenic Behavior and Rhizosphere Bacterial Communities and Activities in a Rice Agro-ecosystem. *Sci. Total Environ.* **2016,** *542*, 642–652.
Dixit, R.; Malaviya, D.; Pandiyan, K.; Singh, U. B.; Sahu, A.; Shukla, R.; Paul, D. Bioremediation of Heavy Metals from Soil and Aquatic Environment: An Overview of Principles and Criteria of Fundamental Processes. *Sustainability* **2015,** *7* (2), 2189–2212.
Dixit, S.; Hering, J. G. Comparison of Arsenic (V) and Arsenic (III) Sorption onto Iron Oxide Minerals: Implications for Arsenic Mobility. *Environ. Sci. Technol.* **2003,** *37* (18), 4182–4189.
Dopp, E.; Hartmann, L. M.; Florea, A. M.; Von Recklinghausen, U.; Pieper, R.; Shokouhi, B.; ... Obe, G. Uptake of Inorganic and Organic Derivatives of Arsenic Associated with Induced Cytotoxic and Genotoxic Effects in Chinese Hamster Ovary (CHO) Cells. *Toxicol. Appl. Pharmacol.* **2004,** *201* (2), 156–165.

Edmunds, W. M.; Ahmed, K. M.; Whitehead, P. G. A Review of Arsenic and its Impacts in Groundwater of the Ganges–Brahmaputra–Meghna Delta, Bangladesh. *Environ. Sci. Processes Impacts* **2015,** *17* (6), 1032–1046.

Fan, H.; Su, C.; Wang, Y.; Yao, J.; Zhao, K.; Wang, Y.; Wang, G. Sedimentary Arsenite-oxidizing and Arsenate-reducing Bacteria Associated with High Arsenic Groundwater from Shanyin, Northwestern China. *J. Appl. Microbiol.* **2008,** *105* (2), 529–539.

Francesconi, K. A.; Edmonds, J. S. Arsenic in the sea. *Oceanogr Mar Biol.* **1993,** *31*, 111–151.

Garcia, C.; Moreno, D. A.; Ballester, A.; Blazquez, M. L.; Gonzalez, F. Bioremediation of an Industrial Acid Mine Water by Metal-tolerant Sulphate-reducing Bacteria. *Miner. Eng.* **2001,** *14* (9), 997–1008.

Gault, A. G.; Islam, F. S.; Polya, D. A.; Charnock, J. M.; Boothman, C.; Chatterjee, D.; Lloyd, J. R. Microcosm Depth Profiles of Arsenic Release in a Shallow Aquifer, West Bengal. *Miner. Magazine* **2005,** *69* (5), 855–863.

Gelova, G. A. *Hydrogeochemistry of Ore Elements;* Nedra: Moscow, 1997.

Ghosh, D.; Bhadury, P.; Routh, J. Diversity of Arsenite Oxidizing Bacterial Communities in Arsenic-rich Deltaic Aquifers in West Bengal, India. *Front. Microb.* **2014,** *5*, 602.

Ghosh, D.; Routh, J.; Bhadury, P. Characterization and Microbial Utilization of Dissolved Lipid Organic Fraction in Arsenic Impacted Aquifers (India). *J. Hydrol.* **2015,** *527*, 221–233.

Ghosh, S.; Sar, P. Identification and Characterization of Metabolic Properties of Bacterial Populations Recovered from Arsenic Contaminated Ground Water of North East India (Assam). *Water Res.* **2013,** *47* (19), 6992–7005.

Guan, X.; Yan, X.; Li, Y.; Jiang, B.; Luo, X.; Chi, X. Diversity and Arsenic-tolerance Potential of Bacterial Communities from Soil and Sediments Along a Gold Tailing Contamination Gradient. *Canadian J. Microb.* **2017,** *63* (9), 788–805.

Hassan, Z.; Sultana, M.; van Breukelen, B. M.; Khan, S. I.; Röling, W. F. Diverse Arsenic-and Iron-cycling Microbial Communities in Arsenic-contaminated Aquifers used for Drinking Water in Bangladesh. *FEMS Microb. Ecol.* **2015,** *91* (4). DOI: https://doi.org/10.1093/femsec/fiv026

Héry, M.; Van Dongen, B. E.; Gill, F.; Mondal, D.; Vaughan, D. J.; Pancost, R. D.; Polya, D. A.; Lloyd, J. R. Arsenic Release and Attenuation in Low Organic Carbon Aquifer Sediments from West Bengal. *Geobiology* **2010,** *8* (2), 155–168.

Holmes, D. E.; Giloteaux, L.; Barlett, M.; Chavan, M. A.; Smith, J. A.; Risso, C. Molecular Analysis of the Growth Rate of Subsurface Geobacter Species During in Situ Uranium Bioremediation. *Appl. Environ. Microb.* **2012,** *79* (5),1646–1653.

Hori, T.; Müller, A.; Igarashi, Y.; Conrad, R.; Friedrich, M. W. Identification of Iron-reducing Microorganisms in Anoxic Rice Paddy Soil by 13 C-acetate Probing. *ISME J.* **2010,** *4* (2), 267.

Inskeep, W. P.; Macur, R. E.; Harrison, G.; Bostick, B. C.; Fendorf, S. Biomineralization of As (V)-hydrous Ferric Oxyhydroxide in Microbial Mats of an Acid-sulfate-chloride Geothermal Spring, Yellowstone National Park1. *Geochimica et Cosmochimica Acta* **2004,** *68* (15), 3141–3155.

Islam, F. S.; Gault, A. G.; Boothman, C.; Polya, D. A.; Charnock, J. M.; Chatterjee, D.; Lloyd, J. R. Role of Metal-reducing Bacteria in Arsenic Release from Bengal Delta Sediments. *Nature* **2004,** *430* (6995), 68.

Jeanthon, C.; Prieur, D. Susceptibility to Heavy Metals and Characterization of Heterotrophic Bacteria Isolated from Two Hydrothermal Vent Polychaete Annelids, Alvinellapompejana and Alvinellacaudata. *Appl. Environ. Microb.* **1990,** *56* (11), 3308–3314.

Johnson, D. B.; Hallberg, K. B. Acid Mine Drainage Remediation Options: A Review. *Sci. Total Environ.* **2005,** *338* (1–2), 3–14.

Kaksonen, A. H.; Franzmann, P. D.; Puhakka, J. A. Performance and Ethanol Oxidation Kinetics of a Sulfate-reducing Fluidized-bed Reactor Treating Acidic Metal-containing Wastewater. *Biodegradation* **2003,** *14* (3), 207–217.

Kappler, U.; Dahl, C. Enzymology and Molecular Biology of Prokaryotic Sulfite Oxidation. *FEMS Microb. Lett.* **2001,** *203* (1), 1–9.

Kim, M. J.; Nriagu, J.; Haack, S. Carbonate Ions and Arsenic Dissolution by Groundwater. *Environ. Sci. Technol.* **2000,** *34* (15), 3094–3100.

Kirk, J. L.; Beaudette, L. A.; Hart, M.; Moutoglis, P.; Klironomos, J. N.; Lee, H.; Trevors, J. T. Methods of Studying Soil Microbial Diversity. *J. Microb. Methods* **2004,** *58* (2), 169–188.

Kocar, B. D.; Herbel, M. J.; Tufano, K. J.; Fendorf, S. Contrasting Effects of Dissimilatory Iron (III) and Arsenic (V) Reduction on Arsenic Retention and Transport. *Environ. Sci. Technol.* **2006,** *40* (21), 6715–6721.

Koschinsky, A.; Hein, J. R. Uptake of Elements from Seawater by Ferromanganese Crusts: Solid-Phase Associations and Seawater Speciation. *Mar. Geol.* **2003,** *198* (3–4), 331–351.

Kudo, K.; Yamaguchi, N.; Makino, T.; Ohtsuka, T.; Kimura, K.; Dong, D. T.; Amachi, S. Release of Arsenic from Soil by a Novel Dissimilatory Arsenate-Reducing Bacterium, Anaeromyxobacter sp. Strain PSR-1. *Appl. Environ. Microb.* **2013,** *79* (15), 4635–4642.

Lear, G.; Song, B.; Gault, A. G.; Polya, D. A.; Lloyd, J. R. Molecular Analysis of Arsenate-reducing Bacteria within Cambodian Sediments Following Amendment with Acetate. *Appl. Environ. Microb.* **2007,** *73* (4), 1041–1048.

Lerda, D. E.; Prosperi, C. H. Water Mutagenicity and Toxicology in Riotercero (Cordoba, Argentina). *Water Res.* **1996,** *30* (4), 819–824.

Li, P.; Wang, Y.; Dai, X.; Zhang, R.; Jiang, Z.; Jiang, D.; ... Dong, H. Microbial Community in High Arsenic Shallow Groundwater Aquifers in Hetao Basin of Inner Mongolia, China. *PloS One* **2015,** *10* (5), e0125844.

Liao, V. H. C.; Chu, Y. J.; Su, Y. C.; Hsiao, S. Y.; Wei, C. C.; Liu, C. W.; ... Chang, F. J. Arsenite-oxidizing and Arsenate-reducing Bacteria Associated with Arsenic-rich Groundwater in Taiwan. *J. Contaminant Hydrol.* **2011,** *123* (1–2), 20–29.

Liaw, J.; Marshall, G.; Yuan, Y.; Ferreccio, C.; Steinmaus, C.; Smith, A. H. Increased Childhood Liver Cancer Mortality and Arsenic in Drinking Water in Northern Chile. *Cancer Epidemiol. Prev. Biomarkers* **2008,** *17* (8), 1982–1987.

Liesack, W.; Schnell, S.; Revsbech, N. P. Microbiology of Flooded Rice Paddies. *FEMS Microbiol. Rev.* **2000,** *24* (5), 625–645.

Liu, J.; Li, J.; Chi, X.; Qu, J.; Hu, Z.; Fang, S.; Zhang, Z. A Late-Carboniferous to Early-Permian Subduction–accretion Complex in Daqing Pasture, southeastern Inner Mongolia: Evidence of Northward Subduction Beneath the Siberian Paleoplate Southern Margin. *Lithos* **2013,** *177*, 285–296.

Liu, X. Z.; Zhang, L. M.; Prosser, J. I.; He, J. Z. Abundance and Community Structure of Sulfate Reducing Prokaryotes in a Paddy Soil of Southern China under Different Fertilization Regimes. *Soil Biol. Biochem.* **2009,** *41* (4), 687–694.

Lomax, C.; Liu, W. J.; Wu, L.; Xue, K.; Xiong, J.; Zhou, J.; ... Zhao, F. J. Methylated Arsenic Species in Plants Originate from Soil Microorganisms. *New Phytologist* **2012,** *193* (3), 665–672.

Lovley, D. R.; Holmes, D. E.; Nevin, K. P. Dissimilatory Fe (iii) and Mn (iv) Reduction, *Advanc. Microb. Phys.* **2004,** *49,* 219–286.

Lu, Y.; Rosencrantz, D.; Liesack, W.; Conrad, R. Structure and Activity of Bacterial Community Inhabiting Rice Roots and the Rhizosphere. *Environ. Microb.* **2006,** *8* (8), 1351–1360.

Macur, R. E.; Jackson, C. R.; Botero, L. M.; Mcdermott, T. R.; Inskeep, W. P. Bacterial Populations Associated with the Oxidation and Reduction of Arsenic in an Unsaturated Soil. *Environ. Sci. Technol.* **2004,** *38* (1), 104–111.

Macur, R. E.; Wheeler, J. T.; McDermott, T. R.; Inskeep, W. P. Microbial Populations Associated with the Reduction and Enhanced Mobilization of Arsenic in Mine Tailings. *Environ. Sci. Technol.* **2001,** *35* (18), 3676–3682.

Masue-Slowey, Y.; Kocar, B. D.; Bea Jofré, S. A.; Mayer, K. U.; Fendorf, S. Transport Implications Resulting from Internal Redistribution of Arsenic and Iron Within Constructed Soil Aggregates. *Environ. Sci. Technol.* **2010,** *45* (2), 582–588.

McBride, B. C.; Wolfe, R. S. Biosynthesis of Dimethylarsine by Methanobacterium. *Biochemistry 10* (23), 4312–4317.

Meharg, A. A.; Rahman, M. M. Arsenic Contamination of Bangladesh Paddy Field Soils: Implications for Rice Contribution to Arsenic Consumption. *Environ. Sci. Technol.* **2003,** *37* (2), 229–234.

Mench, M.; Vangronsveld, J.; Beckx, C.; Ruttens, A. Progress in Assisted Natural Remediation of an Arsenic Contaminated Agricultural Soil. *Environ. Pollution* **2006,** *144* (1), 51–61.

Mestrot, A.; Uroic, M. K.; Plantevin, T.; Islam, M. R.; Krupp, E. M.; Feldmann, J.; Meharg, A. A. Quantitative and Qualitative Trapping of Arsines Deployed to Assess Loss of Volatile Arsenic from Paddy Soil. *Environ. Sci. Technol.* **2009,** *43* (21), 8270–8275.

Mohapatra, B.; Sarkar, A.; Joshi, S.; Chatterjee, A.; Kazy, S. K.; Maiti, M. K.; ... Sar, P. An Arsenate-reducing and Alkane-metabolizing Novel Bacterium, Rhizobium arsenicireducens sp. nov., Isolated from Arsenic-rich Groundwater. *Arch. Microbiol.* **2017,** *199* (2), 191–201.

Mukhopadhyay, R.; Rosen, B. P.; Phung, L. T.; Silver, S. Microbial Arsenic: from Geocycles to Genes and Enzymes. *FEMS Microbiol. Rev.* **2002,** *26* (3), 311–325.

Newman, D. K.; Kennedy, E. K.; Coates, J. D.; Ahmann, D.; Ellis, D. J.; Lovley, D. R.; Morel, F. M. Dissimilatory Arsenate and Sulfate Reduction in *Desulfotomaculum auripigmentum* sp. nov. *Arch. Microbiol.* **1997,** *168* (5), 380–388.

Nickson, R.; McArthur, J.; Burgess, W.; Ahmed, K. M.; Ravenscroft, P.; Rahmanñ, M. Arsenic Poisoning of Bangladesh Groundwater. *Nature* **1998,** *395* (6700), 338.

Nimick, D. A.; Moore, J. N.; Dalby, C. E.; Savka, M. W. The Fate of Geothermal Arsenic in the Madison and Missouri Rivers, Montana and Wyoming. *Water Resources Res.* **1998,** *34* (11), 3051–3067.

Nordstrom, D. K. Worldwide Occurrences of Arsenic in Ground Water. *Science* **2002,** *296,* 2143–2145.

Onishi, H.; Sandell, E. B. Geochemistry of Arsenic. *Geochimica et Cosmochimica Acta,* **1995,** *7* (1–2), 1–33.

Oremland, R. S.; Stolz, J. F. The Ecology of Arsenic. *Sci.* **2003,** *300* (5621), 939–944.

Oremland, R. S.; Stolz, J. F. Arsenic, Microbes and Contaminated Aquifers. *Trends Microbiol.* **2005,** *13* (2), 45–49.
Oremland, R. S.; Stolz, J. F.; Hollibaugh, J. T. The Microbial Arsenic Cycle in Mono Lake, California. *FEMS Microb. Ecol.* **2004,** *48* (1), 15–27.
Osborne, T. H.; McArthur, J. M.; Sikdar, P. K.; Santini, J. M. Isolation of an Arsenate-respiring Bacterium from a Redox Front in an Arsenic-polluted Aquifer in West Bengal, Bengal Basin. *Environ. Sci. Technol.* **2015,** *49* (7), 4193–4199.
Ouattara, A. S.; Jacq, V. A. Characterization of Sulfate-reducing Bacteria Isolated from Senegal Ricefields. *FEMS Microbiol. Lett.* **1992,** *101* (3), 217–228.
Overesch, M.; Rinklebe, J.; Broll, G.; Neue, H. U. Metals and Arsenic in Soils and Corresponding Vegetation at Central Elbe River Floodplains (Germany). *Environ. Pollution* **2007,** *145* (3), 800–812.
Pal, A.; Chowdhury, U. K.; Mondal, D.; Das, B.; Nayak, B.; Ghosh, A.; Chakraborti, D. Arsenic Burden from Cooked Rice in the Populations of Arsenic Affected and Nonaffected Areas and Kolkata City in West-Bengal, India. *Environ. Sci. Technol.* **2009,** *43* (9), 3349–3355. DOI: 10.1021/es803414j
Paul, D.; Kazy, S. K.; Gupta, A. K.; Pal, T.; Sar, P. Diversity, Metabolic Properties and Arsenic Mobilization Potential of Indigenous Bacteria in Arsenic Contaminated Groundwater of West Bengal, India. *PloS One* **2015,** *10* (3), e0118735.
Plumlee, G. S.; Smith, K. S.; Montour, M. R.; Ficklin, W. H.; Mosier, E. L. Geologic Controls on the Composition of Natural Waters and Mine Waters Draining Diverse Mineral-deposit Types. *The Environmental Geochemistry of Mineral Deposits. Part B: Case Studies and Research Topics*, **1999,** *6*, 373–432.
Rahman, M. A., Hasegawa, H., Rahman, M. M., Miah, M., Tasmin, A. Straighthead Disease of Rice (Oryza sativa L.) Induced by Arsenic Toxicity. *Environ. Exp. Bot.* **2008,** *62*, 54–59.
Ratering, S.; Schnell, S. Nitrate-dependent Iron (II) Oxidation in Paddy Soil. *Environ. Microbiol.* **2001,** *3* (2), 100–109.
Rodríguez-Lado, L.; Sun, G.; Berg, M.; Zhang, Q.; Xue, H., Zheng, Q.; Johnson, C. A. Groundwater Arsenic Contamination Throughout China. *Science* **2013,** *341* (6148), 866–868.
Rosen, B. P. Biochemistry of Arsenic Detoxification. *FEBS Lett.* **2002,** *529* (1), 86–92.
Rowland, H. A. L.; Pederick, R. L.; Polya, D. A.; Pancost, R. D.; Van Dongen, B. E.; Gault, A. G.; ... Lloyd, J. R. The Control of Organic Matter on Microbially Mediated Iron Reduction and Arsenic Release in Shallow Alluvial Aquifers, Cambodia. *Geobiology* **2007,** *5* (3), 281–292.
Roychowdhury, T. Impact of Sedimentary Arsenic Through Irrigated Groundwater on Soil, Plant, Crops and Human Continuum from Bengal Delta : Special Reference to Raw and Cooked Rice. *Food Chemical Toxicol.* **2008,** *46*, 2856–2864. https://doi.org/10.1016/j.fct.2008.05.019
Roychowdhury, T.; Tokunaga, H.; Uchino, T.; Ando, M. Effect of Arsenic-Contaminated Irrigation Water on Agricultural Land Soil and Plants in West Bengal, India. *Chemosphere* **2005,** *58* (6), 799–810.
Sabry, S. A.; Ghozlan, H. A.; Abou-Zeid, D. M. Metal Tolerance and Antibiotic Resistance Patterns of a Bacterial Population Isolated from Sea Water. *J. Appl. Microbiol.* **1997,** *82* (2), 245–252.

Saha, G. C.; Ali, M. A. Dynamics of Arsenic in Agricultural Soils Irrigated with Arsenic Contaminated Groundwater in Bangladesh. *Sci. Total Environ.* **2007,** *379* (2–3), 180–189.

Sancha, A. M. Full-scale Application of Coagulation Processes for Arsenic Removal in Chile: A Successful Case Study. In *Arsenic Exposure and Health Effects III;* Elsevier Science Ltd.: 1999; pp 373–378.

Santini, J. M.; Sly, L. I.; Schnagl, R. D.; Macy, J. M. A New Chemolithoautotrophic Arsenite-oxidizing Bacterium Isolated from a Gold Mine: Phylogenetic, Physiological, and Preliminary Biochemical Studies. *Appl. Environ. Microbiol.* **2000,** *66* (1), 92–97.

Santra, S. C.; Samal, A. C.; Bhattacharya, P.; Banerjee, S.; Biswas, A.; Majumdar, J. Arsenic in Foodchain and Community Health Risk: A Study in Gangetic West Bengal. *Procedia Environ. Sci.* **2013,** *18*, 2–13.

Sarkar, A.; Kazy, S. K.; Sar, P. Characterization of Arsenic Resistant Bacteria from Arsenic Rich Groundwater of West Bengal, India. *Ecotoxicology* **2013,** *22* (2), 363–376.

Silver, S.; Phung, L. T. Genes and Enzymes Involved in Bacterial Oxidation and Reduction of Inorganic Arsenic. *Appl. Environ. Microbiol.* **2005,** *71* (2), 599–608.

Smedley, P. L. Arsenic in Rural Groundwater in Ghana: Part Special Issue: Hydrogeochemical Studies in sub-Saharan Africa. *J. Afr. Earth Sci.* **1996,** *22* (4), 459–470.

Sohn, E. The Toxic Side of Rice. *Nature* **2014,** *514* (7524), S62–S63.

Somenahally, A. C.; Hollister, E. B.; Yan, W.; Gentry, T. J.; Loeppert, R. H. Water Management Impacts on Arsenic Speciation and Iron-reducing Bacteria in Contrasting Rice-rhizosphere Compartments. *Environ. Sci. Technol.* **2011,** *45* (19), 8328–8335.

Su, C.; Puls, R. W. Arsenate and Arsenite Removal by Zerovalent Iron: Effects of Phosphate, Silicate, Carbonate, Borate, Sulfate, Chromate, Molybdate, and Nitrate, Relative to Chloride. *Environ. Sci. Technol.* **2001,** *35* (22), 4562–4568.

Takeuchi, M.; Kawahata, H.; Gupta, L. P.; Kita, N.; Morishita, Y.; Ono, Y.; Komai, T. Arsenic Resistance and Removal by Marine and Non-marine Bacteria. *J. Biotechnol.* **2007,** *127* (3), 434–442.

Tufano, K. J.; Fendorf, S. Confounding Impacts of Iron Reduction on Arsenic Retention. *Environ Sci. Technol.* **2008,** *42* (13), 4777–4783.

Turpeinen, R.; Kairesalo, T.; Häggblom, M. M. Microbial Community Structure and Activity in Arsenic-, Chromium-and Copper-contaminated Soils. *FEMS Microbiol. Ecol.* **2004,** *47* (1), 39–50.

Upadhyay, M. K.; Yadav, P.; Shukla, A.; Srivastava, S. Utilizing the Potential of Microorganisms for Managing Arsenic Contamination: A Feasible and Sustainable Approach. *Front. Environ. Sci.* **2018,** *6*, 24.

Valverde, A.; González-Tirante, M.; Medina-Sierra, M.; Santa-Regina, I.; García-Sánchez, A.; Igual, J. M. Diversity and Community Structure of Culturable Arsenic-resistant Bacteria Across a Soil Arsenic Gradient at an Abandoned Tungsten–tin Mining Area. *Chemosphere* **2011,** *85* (1), 129–134.

Vidal, F. V.; Vidal, V. M. V. Arsenic Metabolism in Marine Bacteria and Yeast. *Mar Biol.* **1980,** *60* (1), 1–7.

Wang, S.; Mulligan, C. N. Occurrence of Arsenic Contamination in Canada: Sources, Behavior and Distribution. *Sci. Total Environment.* **2006,** *366* (2–3), 701–721.

Wang, X. J.; Yang, J.; Chen, X. P.; Sun, G. X.; Zhu, Y. G. Phylogenetic Diversity of Dissimilatory Ferric Iron Reducers in Paddy Soil of Hunan, South China. *J. Soils Sediments* **2009,** *9* (6), 568–577.

Welch, A. H.; Westjohn, D. B.; Helsel, D. R.; Wanty, R. B. Arsenic in Ground Water of the United States: Occurrence and Geochemistry. *Groundwater* **2000,** *38* (4), 589–604.
Williams, T. M.; Rawlins, B. G.; Smith, B.; Breward, M. In-vitro Determination of Arsenic Bioavailability in Contaminated Soil and Mineral Beneficiation Waste from Ron Phibun, Southern Thailand: A Basis for Improved Human Risk Assessment. *Environ. Geochem. Health* **1998,** *20* (4), 169–177.
Wind, T.; Stubner, S.; Conrad, R. Sulfate-reducing Bacteria in Rice Field Soil and on Rice Roots. *System. Appl. Microbiol.* **1999,** *22* (2), 269–279.
Yin, X.; Chen, J.; Qin, J.; Sun, G.; Rosen, B.; Zhu, Y. Biotransformation and Volatilization of Arsenic by Three Photosynthetic Cyanobacteria. *Plant Physiol.* **2011,** pp 111.

CHAPTER 8

Sea Buckthorn: Its Sustainable Utilization for Biodiversity Conservation

N. MALIK[1*], S. SAXENA[1], H. NANGYAL[2], T. ZOHRA[3], and H. IKRAM[4]

[1]*Department of Botany and Microbiology, S.G.R.R. (P.G.) College, Constituent of S.G.R.R Education Mission, Dehradun 248001, Uttarakhand, India*

[2]*Department of Botany, Hazara University, Khyber Pakhtunkhwa, Pakistan*

[3]*Department of Biotechnology, Quaid-i-Azam University, Islamabad, Pakistan*

[4]*Department of Biochemistry, University of Karachi, Karachi 75270, Pakistan*

[*]*Corresponding author. E-mail: malinivedita8730@gmail.com*

ABSTRACT

Sea buckthorn (Hippophae Linn.) is a unique and valuable multipurpose plant species. It is one of the most magical plant resources for firewood, fodder, and serves as a soil binder species. It plays an important role in soil erosion control, reclamation of degraded and wastelands, wildlife habitat enhancement, and farmstead protection. However, despite the rich and exploited nutritional and medicinal value, sea buckthorn requires more research regarding its environmental aspect especially in fragile mountainous regions which are most susceptible to human impact. Therefore, present review article is an attempt to open up a new horizon for researchers in the unexplored environmental aspect of this wonder plant and its role in biodiversity conservation.

8.1 INTRODUCTION

Genus *Hippophae* Linn. (2n=24), commonly known as sea buckthorn or Leh Berry is a shrub/small tree of the family Elaeagnaceae. The name is derived from its habit of growing near the sea, and from the possession of many spines or thorns that are reminiscent of some buckthorn species (of the genus *Rhamnus*). In ancient Greece, leaves of sea buckthorn added to horse fodder was well reputed to result in weight gain and shiny hair. This, in fact, gave the generic name to the plant, in Latin "Hippo" means horse and "phaos" means to shine thus, the name "*Hippophae*" meaning shining horse (Rongsen, 1992).

Sea buckthorn is a fast-growing, frost- and wind-resistant plant. It is highly adaptable to extreme climatic conditions including temperature ranges of –43°–+40°C, draught, high altitude, salinity, alkalinity, and inundation (Ruan and Li, 2002). The genus Hippophae is distributed between 27° and 69°N latitude and 7°W and 122°E longitude in the world (Rousi, 1971). Sea buckthorn is a very rich source of vitamins and important bioactive compounds (Xurong et al., 2001) (Fig. 8.1). It has an extensive subterranean rooting system with strong soil binding ability useful for soil stabilization, river bank control, and water retention (TISC, 2001) (Fig. 8.1). For these reasons, it is also called a wonderful plant (Lu, 1992). Recently it has attracted considerable attention from researchers around the world mainly for its nutritional, medicinal, and environmental value.

FIGURE 8.1 Ripe sea buckthorn berries rich source of various vitamins.

8.1.1 DISTRIBUTION OF SEA BUCKTHORN

According to the new taxonomic system, sea buckthorn includes seven species and eleven subspecies that were also given in Species Records of *Hippophae*, Germplasm Resource Information Network (Grin, 2007) (Table 8.1).

8.1.2 ROLE OF SEA BUCKTHORN IN BIODIVERSITY CONSERVATION

8.1.2.1 SEA BUCKTHORN AS AN EFFECTIVE NITROGEN FIXER

Sea buckthorn has a mighty and well-developed tap root system, having primary, secondary and tertiary roots covered with root hairs, found more prominently in the apical portion (Dwivedi et al., 2006). Some 80% of its feeding roots are in the topsoil (0.2–0.8 m) helping to prevent erosion (Lu, 1992). The sea buckthorn root system is so extensive that its roots can branch many times in a growing season and form a complex network of roots. Horizontal roots also have root turions (underground buds) that sprout and give rise to another plant. In this way, sea buckthorn bushes play an important role in protecting riverbanks, preventing floods and clogging mud, which would otherwise be washed away in floodwaters. A symbiotic nitrogen-fixing mycorrhizal bacteria *Frankia* (Actinomycetes), has been found which makes root nodules in the roots of sea buckthorn (Gatner and Gardener, 1970). This symbiosis between the fungus and sea buckthorn can fix the maximum amount of atmospheric nitrogen. It is estimated that the capacity of sea buckthorn roots to fix nitrogen is twice that of soybean. An 8–10-year-old sea buckthorn forest can fix 180 kg of nitrogen/ha/y (Lu, 1992). Besides fixing nitrogen, the perennial root nodule has the function of transforming difficult to dissolve organic and mineral matter into an absorbable state.

8.1.2.2 WATER AND SOIL LOSS CONTROLLING ABILITY

Compared with wasteland, 7-year-old sea buckthorn forest can reduce 99% of runoff and 96% of soil loss. Its root system makes it suitable to be planted even in the fragile slopes. An observation made in most of the cultivated areas showed that a 5-year-old plant has a taproot of about 5-m

TABLE 8.1 Distribution and the Status of Utilization of *Hippophae* sp. Linn.

The distribution and status of utilization of *Hippophae* taxons	The areas of distribution	The status of utilization
1. *H. rhamnoides.* Subsp. *Rhamnoides*	Scandinavian countries, Baltic Sea countries, Germany, Belgium, Netherlands, Ireland, Poland, U.K., France, Russia	Many varieties are cultivated in some European countries and Canada
2. *H. rham.* Subsp. *Sinensis*	The North, Northwest, Southwest of China	Wild resources are used for ecological restoration and berries are processed for products. Some new varieties are in tests.
3. *H. rham.* Subsp. *Yunnanensis*	Sichuan, Yunnan, Tibet of China	Wild resources are used for ecological restoration only.
4. *H. rham.* Subsp. *Mongolica*	Siberia of Russia, Mongolia, Xinjiang of China	More than 60 varieties are cultivated in Russia, Mongolia, many East European counties. Many West European counties, Canada and China introduced the varieties for test.
5. *H. rham.* Subsp. *Turkestanica*	India, Pakistan, Afghanistan, Turkmenistan, Kyrgyzstan, Uzbekistan, Kazakhstan, Iran, Turkey, Xinjiang, Tibet of China	Wild resources are used for ecological restoration and berries are processed for various products on commercial level in India for the production of food, medicine, and cosmetics.
6. *H. rham.* Subsp. *fluviatilis*	Around Alps Mountains: Germany, France, Switzerland, Austria, Czech, Slovakia, Italy.	Most of wild resources are protected as forest species. Some berries are collected for processing products
7. *H. rham.* Subsp. *carpatica*	The Capathinan Mountains, Transsylvanian Alps.	Most of the wild resources are protected as forest species.
8. *H. rham.* Subsp. *Caucasica*	The Caucasus Mountains, Georgia, Azerbaijan, Armenia, Ukraine, Romania, Turkey, Bulgaria, Iran, Russia.	Most of the wild resources are protected as forest species. Some selected varieties are cultivated for test.

TABLE 8.1 *(Continued)*

The distribution and status of utilization of *Hippophae* taxons	The areas of distribution	The status of utilization
9. *H. goniocarpa*	Sichuan, Qinghai of China	Most of wild resources are protected as forest species. Very few studies have been done.
10. *H. goniocarpa* Subsp. *Litangensis*	Sichuan, Qinghai of China	Most of the wild resources are protected as forest species. Very few studies have been done.
11. *H. neurocarpa*	Sichuan, Qinghai, Gansu of China	Most of wild resources are protected as forest species. Very few studies have been done.
12. *H. neurocarpa* Subsp. *Stellatopilosa*	Sichuan, Qinghai, Tibet of China	Most of the wild resources are protected as forest species. Very few studies have been done.
13. *H. tibetana*	Sichuan, Qinghai, Gansu, Tibet of China, Nepal, India	Most of the wild resources are protected as grassland species. Very few studies have been done.
14. *H. gyantsensis*	Tibet of China	Most of the wild resources are protected as forest species. Some berries are collected for producing Tibetan medicine.
15. *H. salicifolia*	The southern slope of Himalayan Mt. Tibet of China, Bhutan, Nepal, India	Most of the wild resources are protected as forest species. Some berries are collected for producing products.

deep. The plant is highly appropriate for soil and water conservation. Few years back it has been proved to have high economic advantages of the plant. Generally, these plants grow wild and get rotten in mountain areas. As the plant increases the fertility of the soil, farmers destroy these plants and normally cultivate potatoes and other food crops. Genetic diversity is the basis for plant adaptation, evolution, and breeding. Sea buckthorn is extremely variable in height, from a small bush less than 50 cm to a shrub tree more than 20 m high (Rousi, 1971).

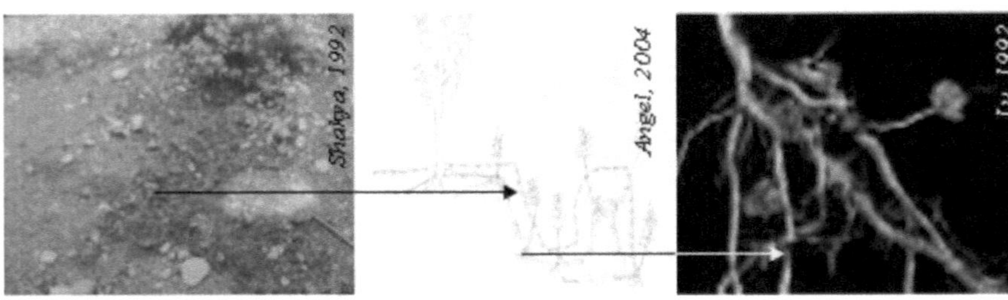

FIGURE 8.2 Subterrain rooting system of sea buckthorn.

8.1.2.3 *ABILITY TO PROMOTE THE GROWTH OF TREE SPECIES*

When pine and poplar were planted with sea buckthorn in mixed forests, it was found that sea buckthorn greatly promoted better growth of pine and poplar (Jasra, 1998). Studies have shown that sea buckthorn promotes the growth of other plants like poplars, pines, and other tree species in mixed stands (Lei et al., 1983; Shi et al., 1987).

8.1.2.4 *SEA BUCKTHORN IN IMPROVING SOIL FERTILITY*

Sea buckthorn is pioneer plant within a fragile and marginal context and forest can greatly increase the contents of nitrogen, phosphorus, and organic matter in the soil. A study conducted in China shows that the contents of nutrient elements in the woodland of sea-buckthorn plantations are clearly higher than that of farmland and barren hill, which show that sea-buckthorn plantations are a good kind of trees for soil improvement because of its strong roots. The total amounts of roots reach 7770 kg/hm^2 at a stand in 3 years, and biological energy and chemical energy produced by the roots affect the physical and chemical nature in the soil. The root's death would increase the contents of organic matters in the soil. In addition, the sea-buckthorn is the no-bean-family plant with the ability to fix Nitrogen; therefore, it will increase the contents of nutrient elements (Xiaoning et al., 2002) (Table 8.2).

TABLE 8.2 Soil Layers for Qing'an (8 age) and Lingyou (11 age).

Soil layers (cm)	Qing'an (8 age)			Lingyou (11 age)		
	Woodland	Farmland	Grassland	Woodland	Farmland	Grassland
0–10	3.12	1.01	1.34	4.45	0.79	4.13
20–30	1.73	0.88	0.53	1.40	0.38	1.64
40–50	1.10	0.62	0.47	0.93	0.36	0.81

8.1.2.5 LANDSCAPE MANAGEMENT TOOL

One of the most promising tools to control land degradation is revegetation and sea buckthorn is one of the species successfully used on a large scale. It can help to control desertification, conserve land and water resources, and integrate economic exploitation with ecological rehabilitation. A living windbreak is a linear arrangement of plants, primarily trees, and shrubs, established to reduce harmful effects of strong winds, such as soil erosion (Fig. 8.3). It also helps protect crops, manage snow accumulation and create wildlife habitat. Plants that serve as windbreaks must be resistant to the drying effects and physical injuries caused by wind and sea buckthorn is well suited to this task.

8.1.2.6 AS MAINTAINING ECOLOGICAL BALANCE

It has been observed that a number of wildlife species depend on sea buckthorn stems, leaves, flowers, roots, fruit, and seed. In the Loess Plateau of China, 51 bird species are entirely dependent and 80 bird species are relatively dependent on sea buckthorn for their food (ICIMOD, 2006). It plays a significant role in conserving wildlife by providing an excellent habitat for a number of wildlife species. Sea buckthorn provides long-term benefits in terms of maintaining the ecological equilibrium and improving the environment. In the Canadian prairies, sea buckthorn is considered to be a valuable habitat for the sharptail grouse, Hungarian Patridge, and pheasant (Shroeder, 1995) Under sea buckthorn forests (e.g., Jianping County, China) several species of birds (*Crosoptilon manchurian, Turdus ruficollis*, etc.) thrive on its fruits; many carnivorous animals (*Canis lupus, Vulpes vulpes, Felis bengalensis,* etc.) catch prey and take shelter in its forests. Edible fungi and rabbits can be trapped for food from these forests (Rongsen, 1992; Dharmananda, 2004).

FIGURE 8.3 Sea buckthorn plantations along degraded river beds.

8.1.2.7 USE OF SEA BUCKTHORN FOR MARGINAL LAND RECLAMATION

Sea buckthorn forest in the mountain areas plays a vital role in the amelioration of different forms of life found on the mountain ecosystem (Acharya et al., 2010). Certain characters of sea buckthorn, such as wide ecological adaptation, fast growth, strong coppicing, and suckering habit coupled with efficient nitrogen fixation make this plant well suited in soil and water conservation, soil improvement, and marginal land reclamation (Rongsen, 1992; Khosla et al.,1994).

8.1.2.8 SEA BUCKTHORN FOR SLOPE STABILIZATION AND EROSION CONTROL

Sea buckthorn can stabilize slopes reduce surface runoff and streamflow, formed by the rainstorm. Therefore, it can change the runoff and sediment transportation characteristics. Together with the change of slope, land tillage method, and engineering works in slope and gully, soil erosion may be controlled. The developed root system of sea buckthorn has formed underground big biomass. The root of a 3-year-old sea buckthorn can expand about 6 m in horizontal level direction and over 2 m in the vertical direction (Acharya et al., 2010). From the various experiments, it was found that the surface runoff coefficient in sea buckthorn forestland is only 0–3% at 50 mm daily rainfall (Cifen and Zhao, 2003). Various studies clearly proved the function of sediment retention by sea buckthorn stand (Table 8.3).

TABLE 8.3 Accumulated Infiltration Within 30 min in a Rainfall.

Landform unit	Sea buckthorn forest (mm)	Waster grassland (mm)	Farmland (mm)
Positive slope	708.323	77.160	94.107
Negative slope	179.231	94.767	28.401
Slope bottom	131.788	59.890	64.487
Slope top	205.425	62.640	75.266
average	306.192	73.615	65.566

Source: Cifen and Zhao (2003).

8.1.2.9 CONTROL OF LANDSLIDES

The frequent occurrence of landslides in the fragile mountainous regions of Trans-Himalayan range which is widely distributed by sea buckthorn species is mainly attributed to climatological, geographical, and geological conditions. Besides denudation of hill slopes, loose and unconsolidated periglacial material constitute the majority of mountainous slopes. As a sequel, the slightest precipitation causes destabilization of this type of loose material that results in landslides. To overcome this problem sea buckthorn has been developed in various states of India like Himachal Pradesh. The shrub has been found very effective in checking landslides (Gupta, 2008).

8.1.2.10 ROLE IN REDUCING POLLUTION

Sea buckthorn is useful in lessening pollution resulting from erosion of contaminated mine waste since it can be used to revegetate a variety of mine spoils. Because sea buckthorn is naturally resistant to pests, it has limited need for pesticides that are potentially damaging to the environment. In parts of North America, it has been planted as cover along highways where deicing salt prevents the growth of many other woody plants. Thus, sea buckthorn helps to prevent erosion and release of pollutants from roadsides (Rajchal, 2009).

8.2 CONCLUSIONS

The present study provides comprehensive information on the role of sea buckthorn in biodiversity conservation. The information will be helpful for researchers, local people, policymakers to develop an effective action plan for sustainable use, conservation, and management of sea buckthorn in cold desert region in particular and the Indian Himalayan region in general. Despite the large populations of this species in our mountains, it remains under-utilized due to the lack of further research and development. This wonder plant can become the most promising tool for biodiversity conservation and can be successfully used on a large scale. Sea buckthorn has proved to be an important species for soil and water conservation, windbreak, and sand fixation and eco-environment due to its strong adaptability, fast growth, the capability of root shoot propagation, and high density of the canopy. Even though this plant grows naturally in the Trans-Himalayan region, it has not been given sufficient emphasis for its role in environmental conservation. Sea buckthorn has many medicinal as well as economic and socio-cultural values if its conservation and development efforts are not practiced; it is likely to be extinct in years to come. Therefore, in-situ and ex-situ conservation approaches must be applied to guard against its removal from our mountain ecosystem.

KEYWORDS

- **environmental value**
- **farmstead protection**
- **medicinal aspect**
- **reclamation**
- **soil erosion**
- **wonder plant**

REFERENCES

Acharya, S.; Stobdan, T.; Singh, S. B. Sea buckthorn (*Hippophae* sp. L.): New Crop Opportunity for Biodiversity Conservation in Cold Arid and Trans- Himalayas. *J. Soil Water Conserv.* **2010,** *9* (3), 201–204.

Ansari, A. H. Sea buckthorn (*Hippophae* Linn. ssp) – A Potential Resource for Biodiversity Conservation in Nepal Himalayas. International Workshop on Underutilized Plant Species. 06–08 May 2003, Leipzig, Germany.

Cifen, B.; Zhao, G. Stabilization of Slopes and Watersheds by Seabuckthorn (*Hippophae rhamnoides* L.) in Arid and Semi-arid Areas of China. In *Seabuckthorn: A Multipurpose Wonder Plant;* Daya Publishing House: India 2003; pp 409–416.

Dharmananda, S. *Sea Buckthorn,* 2004. http://www.Itmonline.org/Arts/Seabuckthorn. Html. (accessed Oct 12, 2017)

Dwivedi, S. K.; Singh, R,; Ahmed, Z. The Seabuckthorn. Field Research Laboratory. Defence Research and Development Organization (DRDO). Leh (Ladakh)-194101 (J&K), India, 2006.

Garner, E. M. S; Gardener, I. C. Observations on the Fine Structure of the Rootnodule Endophyte of *Hippophae rhamnoides* L. *Arch. Microbiol.* **1970,** *70*, 183–96.

Grin. Species Records of *Hippophae.* Germplasm Resource Information Network. United States Department of Agriculture, Agricultural Research Service, Beltsville Area, Maryland, USA, 2007. Accessed on http://www.ars-grin.gov/cgi-bin/npgs/html/splist. pl?5698. 22 March, 2007.

Gupta, R. D. Grow Seabuckthorn (A wonder medicinal plant) for Greening Cold Desert Himalayas: In Wastelands News, 2008, pp 40–42 (May–July 2008).

Icimod, 2006. Seabuckthorn: A Multipurpose Plant Species for Fragile Mountains. Sustainable Options for the Mountains Best Practices and Appropriate Technologies. Accessed on http://www.icimod.org/sus_options/bp_soil.htm. Feb 25, 2006.

Jasra A. W. Seabuckthorn: A Medicinal Plant for High Arid-Regions. Asia Pacific Mountain Network National Arid Land Development and Research Institute (NADRI), Pakistan 1998, Vol. *3* (1).

Khosla, P. K.; Sehgal, R. N.; Sharma, V.; Bhatt, A. K.; Singh V. Genetic Resources. Of Sea Buckthorn in Himachal Pradesh. A report of Sea Buckthorn Task Force, 1994.

Lei, Q. Zhong, Y.; Yang, H. Effects of Nitrogen Supply from Seabuckthorn Inartificial Woods of Simon Poplar. *Forest Sci. Technol.* **1983,** *4*, 21–24.

Lu, R. Sea Buckthorn–A Multipurpose Plant for Fragile Mountains. International Center for Integrated Mountain Development ICIMOD. Occasional Paper No. 20. Kathmandu, Nepal, 1992.

Rajchal, R. Sea Buckthorn (*Hippophae salicifolia*) Management Guide: Submitted to The Rufford Small Grants for Nature Conservation, 2009.

Rongsen, L. Seabuckthorn: A Multipurpose Plant Species for Fragile Mountains. ICIMOD Occasional Paper No.20, Kathmandu, Nepal, 1992, pp 5–20.

Rousi, A. The Genus Hippophae L.: A Taxonomic Study. *Annales Botanic Fennica.* **1971,** *8*, 177–227.

Shi, Z.; Su, B.; Guo, Y.; Mixed Afforestation with Chinese Pine and Sea Buckthorn. *Ningxia Agri. Forest. Sci. Technol.* **1987,** *5*, 54.

Shroeder, W. R. Improvement of Conservation Trees and Shrubs. PFRA Shelterbelt Centre Supp. Rpt. 95–1, 1995, pp 42.

TISC. Proceeding of Workshop on Ecology and Distribution of Seabuckthorn (Hippopha-espp) Resource in Northwest Mountains of Nepal.HMG/DANIDA NARMSAP, Tree Improvement and Silviculture Component, Hattisar, Kathmandu, 2001.

Xiaoning, T.; Baoli, A.; Xiufeng, W.; Jing, N. Soil Improvement of Sea buckthorn Plantations and Its Characteristics of the Roots in Loess Plateau: In 12th ISCO Conference, Beijing, 2002.

Xurong, T.; Peter, M.; Tigerstedt, A. Variations of Physical and Chemical Characters Within and Elite Sea Buckthorn (*Hippophae rhamnoides* L.) Breeding Population. *Scientia Horticulturae* **2001,** 88, 203–214.

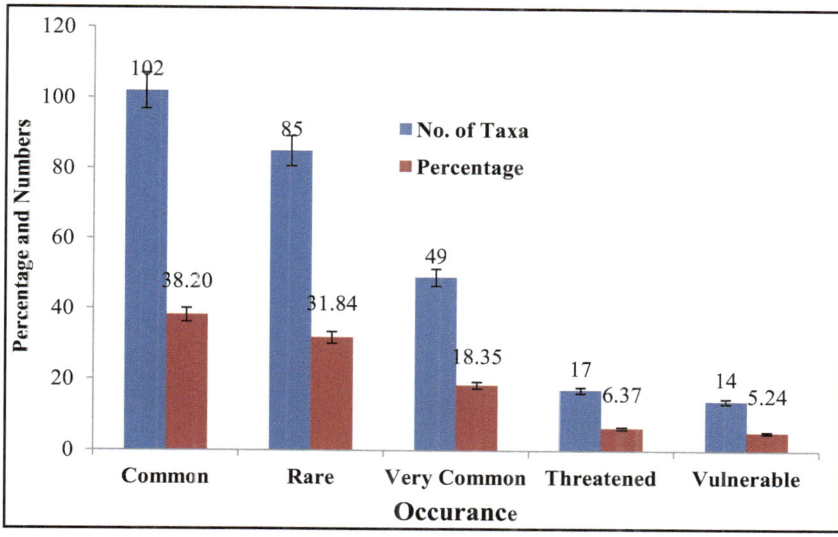

FIGURE 3.2 Summary of conservation status of the Taxa.

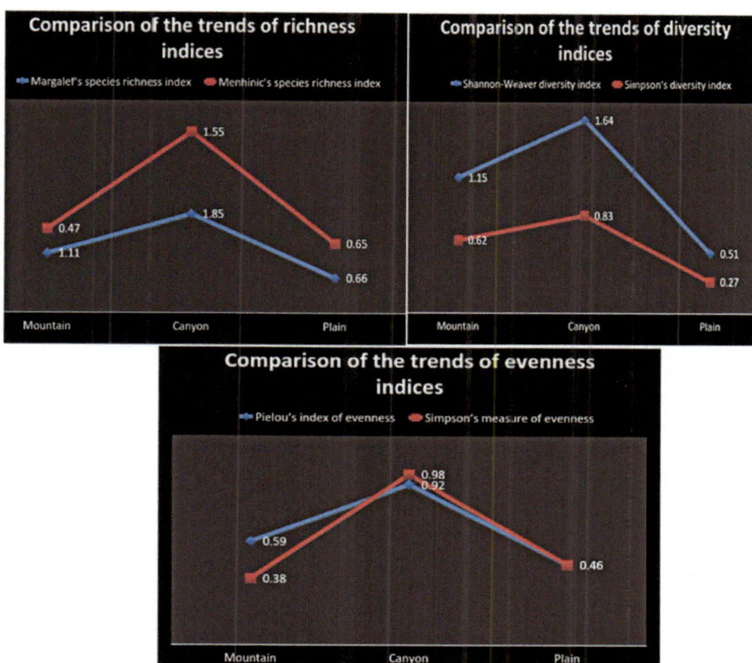

FIGURE 5.1 Comparison of the trends of richness, diversity, and evenness indices among three different habitat types of the study area in Iran.

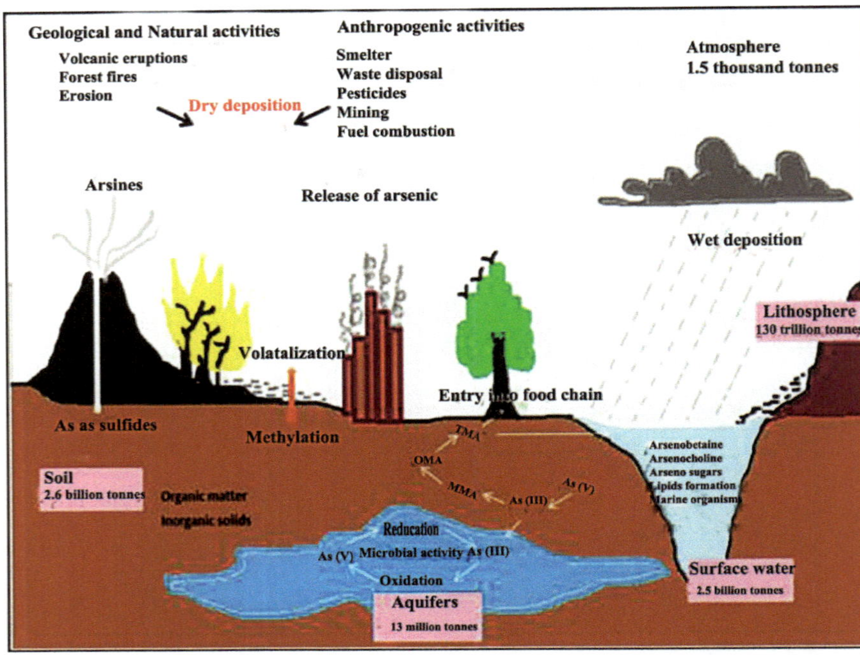

FIGURE 7.2 Biogeochemical cycle of As. Pink boxes indicate total amount of As present particular environment.

Source: data obtained from Sohn (2014).

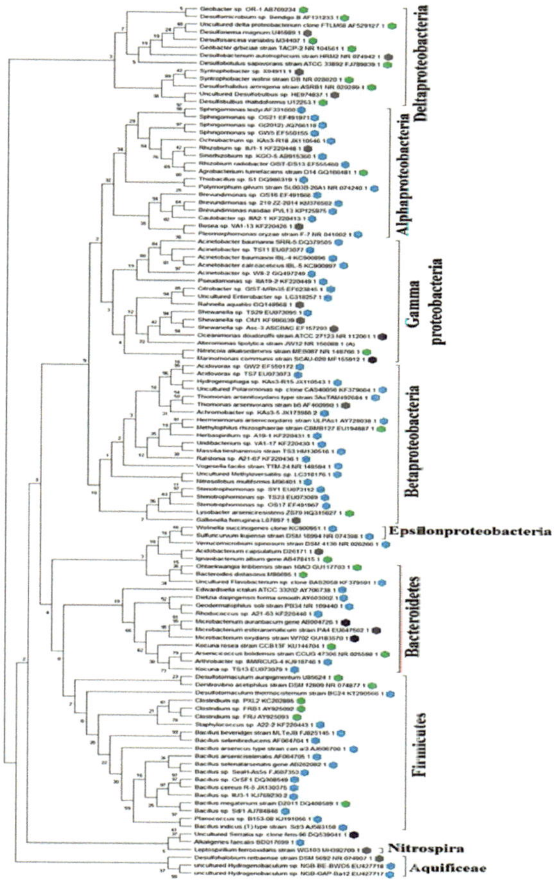

FIGURE 7.3 Neighbor joining tree showing phylogenetic relatedness among the genera detected in various arsenic contaminated environments

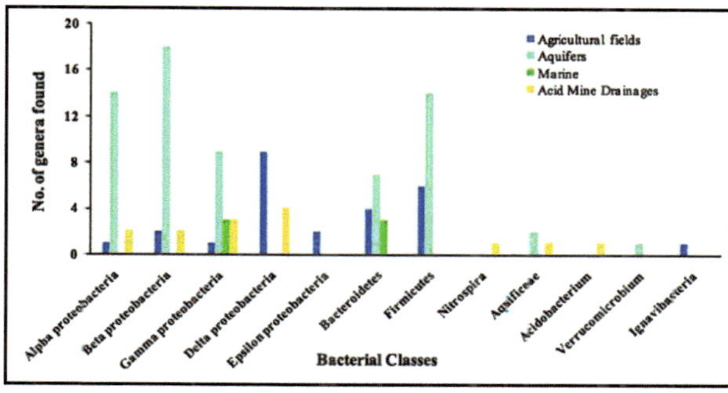

FIGURE 7.4 Abundance of different bacterial classes in four As-impacted ecosystems.

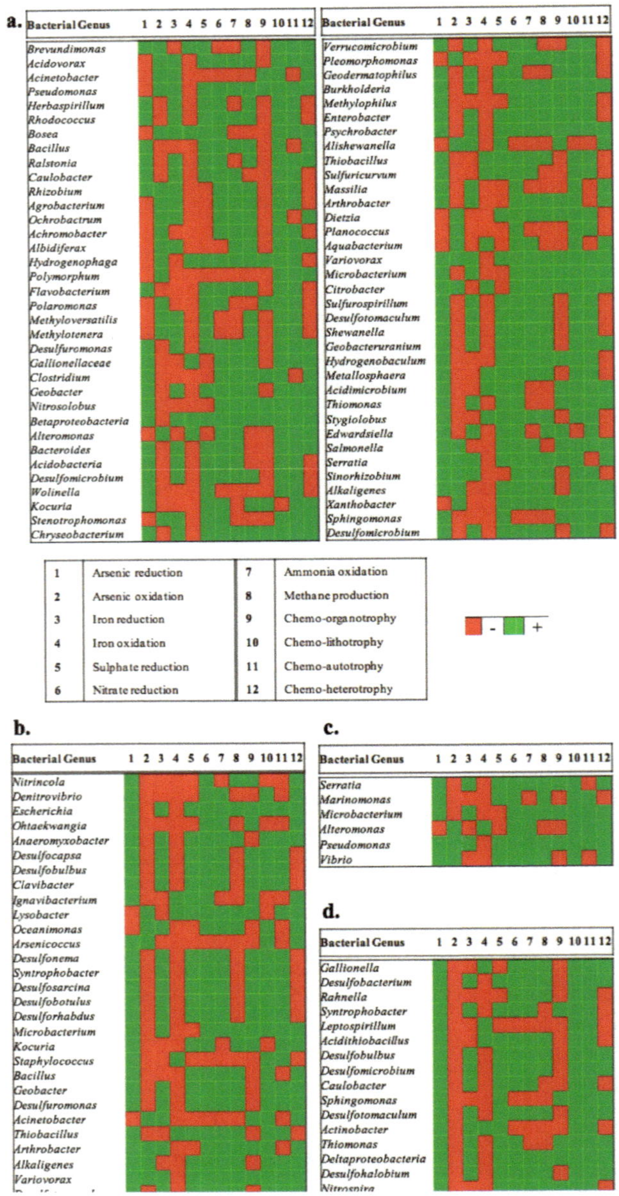

FIGURE 7.5 Metabolic response of various bacterial genera to 12 parameters indicated in legend in four As contaminated ecosystems (a: aquifers, b: agricultural fields, c: marine environment, d: acid mine drainages)

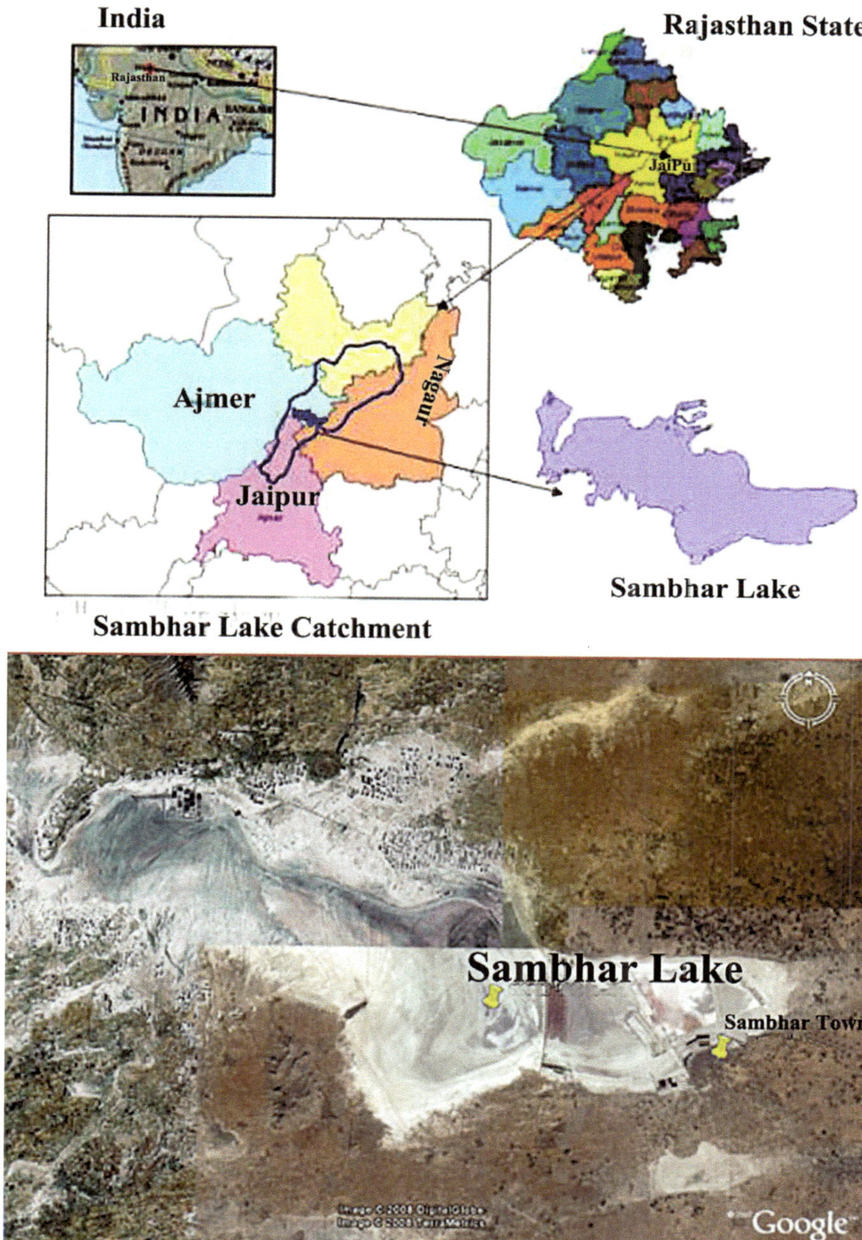

FIGURE 10.1 Location map and a satellite picture of Sambhar Salt Lake.

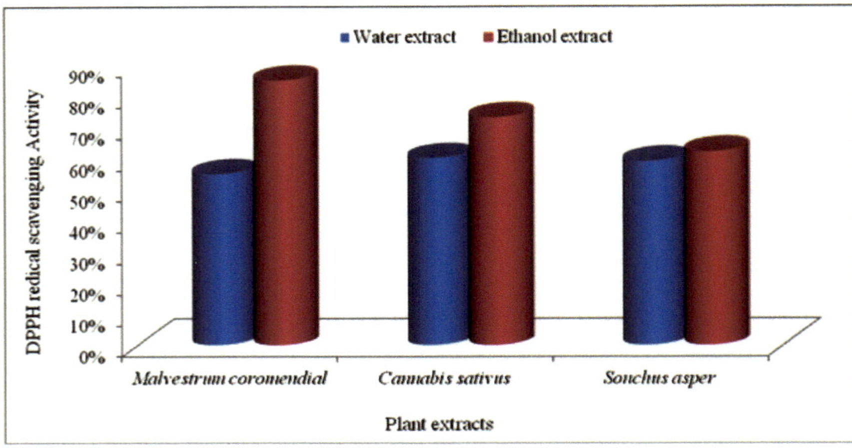

FIGURE 12.1 DPPH radical scavenging activity of the three selected medicinal plants.

FIGURE 13.1 A basic tree of life shows the relations between three eukaryotic groups: fungi and animals are sister groups, with plants as their next closest relative.
Source: Adopted from http://tolweb.org/tree/.

FIGURE 14.3 Investigation Locations for Cyanobacteria and Diatoms in Egypt.

Source: Hamed (2008).

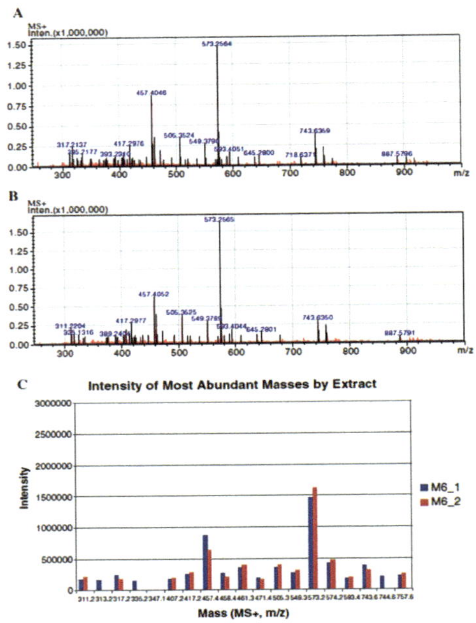

FIGURE 15.4 Profiling of different batches of methanol extract *G. changii* (Chu et al., 2013). (A) batch 1; (B) batch 2; (C) comparison of the intensity of the most abundant masses of the two methanol extracts *G. changii.*

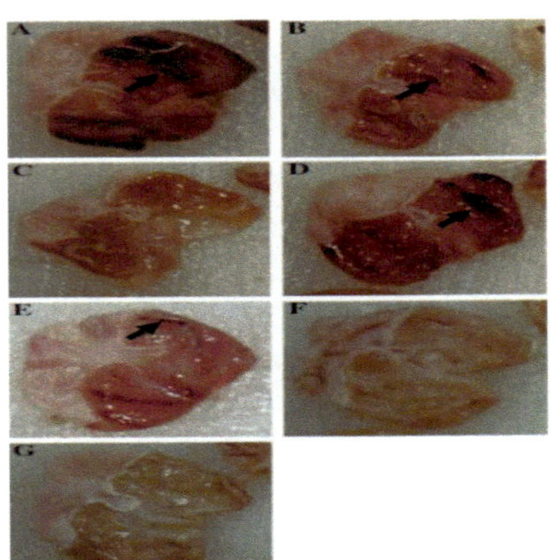

FIGURE 15.5 Influence of methanol extract *G.changii*on gastric lesions following induction of gastric mucosal injury (Chu et al., 2013). (A) Diluent. (B) Non-pecific plant extract (NSPE). (C) Positive control from chemical company of Malaysia (OMP). (D) 2.5 mg/kg b.w. MeOHGCM6. (E) 5 mg/kg b.w. 6. (F) 250 mg/kg b.w. (G) 500 mg/kg b.w.

CHAPTER 9

Environmental Pollution: Increasing Risk of Neurodegenerative Diseases

H. IKRAM[1*] and H. NANGYAL[2]

[1]*Department of Biochemistry, University of Karachi, Karachi 75270, Pakistan*

[2]*Department of Botany, Hazara University, Mansehra, Khyber Pakhtunkhwa, Pakistan*

[*]*Corresponding author. E-mail: hikram@uok.edu.pk*

ABSTRACT

Prevalence of neurodegenerative disorders is related to environmental pollutants and factors. Along with environmental factors, there is genetic vulnerability of predisposition of these. Pre/postnatal exposure to environmental cues such as pollutants, organic solvents, pesticides, herbicide, insecticides and heavy metal ions predisposes the individual to neurodegenerative disorders due to defective growth and development of brain. Proper policy making and safety measures are needed in this regard, to decrease the risk of neurodegenerative disorders.

9.1 INTRODUCTION

Advancing loss of neuronal tissues in the central nervous system results in neurodegenerative disorders. Neurons start functioning abnormally and then decease gradually, as they lack the ability to regenerate on their own (Agrawal and Biswas, 2015). Several different factors contribute to the predisposition and precipitation of neurodegenerative disorders. This may also involve interaction between genetic predisposition and environmental factors. As a result of industrialization and bad waste management strategies, people are becoming more exposed to the toxic effects of environmental pollutants. Globally, neurodegenerative disorders are becoming a major threat to mental health with more cases reported each year. Between 1990 and 2010, neurological, mental, and substance abuse disorders were increased by 41% (Patel et al., 2016). Protein misfolding and aggregation use to be the underlying cause for neurodegenerative diseases, such as Parkinson's, prion diseases, and others (Bose and Cho, 2016). Inhaled nanosized particles in the airborne particulate matter have been linked to several health effects including oxidative stress (OS) and neuroinflammation that may ultimately result in neurodegeneration and cognitive impairment (Heusinkveld et al., 2016). Heavy metal ions or other inhaled pollutants could precipitate neurodegenerative disorders by triggering protein misfolding (Tamás et al., 2014). Mitochondrial dysfunction, OS, and impaired enzymatic activities could be the resultant of excessive metal accumulation toxicity in the CNS (Farina et al., 2013). Environmental pollutants (serving as environmental cues) affect mental and physical wellness in the elderly as epigenetics suggests that environmental factors affect genetic imprinting (Modgil et al., 2014). Pre/postnatal exposure to these environmental cues predisposes the individual to neurodegenerative disorders due to defective growth and development of the brain.

9.2 POLLUTANTS PRECIPITATING NEURODEGENERATIVE DISORDERS

Various factors can contribute to progressive neurodegeneration and have put mental health at risk, globally. These pollutants and mechanism of their action are discussed in this chapter.

9.2.1 *HEAVY METAL IONS*

Two crucial metabolic processes: protein aggregation and OS, involve metal ions (Gaeta and Hider, 2005). Various metal ions can trigger reactions like protein misfolding, altered molecular pathways, and OS. All these would be affecting the cellular functions directly or indirectly, resulting in progressive neuronal death, leading to neurodegenerative disorders. These metal ions are present in industrial waste and could easily be disseminated into the air due to their nanosize. Through the respiratory tract they can easily reach brain and cause irreversible neuronal damage by targeting particular pathway/mechanism.

Monnet-Tschudi et al. (2006) have reported increased expression of amyloid precursor proteins, due to lead and mercury exposure. Both mercury (Hg) and lead (Pb) stimulate beta-amyloid synthesis, which is a triggering factor in Alzheimer's disease (AD) and OS. Glial cells mediate inflammatory reactions in the brain and reactivity of glial cell is also triggered by pesticides, lead, and mercury. Amyotrophic lateral sclerosis (ALS) or motor neuron disease develops due to pesticides, lead, and mercury (Cannon and Greenamyre, 2011). Characteristic of Parkinson's and Alzheimer's diseaes, that is, amyloid plaques, neurofibrillary tangles, phosphorylation of Tau protein (P-Tau) and beta-amyloid (Aβ) peptide are observed due to the exposure to neurotoxic metal ions, such as cadmium, lead, aluminum, mercury, and arsenic (Chin-Chan et al., 2015).

Metal ions like cadmium (Cd), not only increase oxidative damage, interact with other metals and exert estrogen-like effects but also increase the permeability of blood–brain barrier (BBB), thereby predispose the brain to neurotoxicity of other substances (Tobwala et al., 2014). Cd reaches central and peripheral neurons via olfactory or nasal route and zinc transporters facilitate its transport across the BBB (King et al., 2000). Cerebral edema and cellular dysfunction may result due to intracellular accumulation of Cd in the neurons and multifactorial oxidative damage (Patra et al., 2011).

In Earth's crust, Aluminum (Al), is the third most abundant element. Aluminum oxide is formed by the combination of aluminum with oxygen in the moist air. Al is reported to be involved in the pathophysiology of parkinsonism, dementia, ALS, and dialysis encephalopathy (Kawahara, 2016). Immunoexcitotoxicity induced by Al, results in neurodevelopmental- and neurodegenerative disorders. Upon activation by Al, microglia in the brain release neurotoxic concentrations of immune mediators, such as chemokines, cytokines proinflammatory, and excitotoxins. Glutamate receptors also play a very important role in this process (Blaylock, 2012).

Dietary essential metal ions, such as iron (Fe) and manganese (Mn), are required in the physiological system for normal enzyme activity. However, pathological conditions, such as neurodegeneration, may result due to excessive exposure to these metals. Fe, Mn, and Hg toxicity share a common feature of OS, particularly in mitochondria (Farina et al., 2013).

Despite its requirement for cellular homeostasis, development, and normal growth, Mn at high levels may lead to idiopathic Parkinson's disease with extrapyramidal symptoms including muscle and joint pain, apathy, and anorexia (Bowman et al., 2011). Concentrations of γ-aminobutyric acid, glutamate, and dopamine are also altered due to elevated Mn levels (Aschner et al., 2002). Just like Mn, iron is required in the central nervous system for several essential processes, such as synthesis and metabolism of neurotransmitters, oxidative phosphorylation, myelin production, and oxygen transportation. Elevated levels of iron can trigger hydroxyl radical production resulting in cellular damage due to oxidative modification of DNA, carbohydrates, proteins, and lipids (Ward et al., 2014). Including Parkinson's- and Alzheimer disease, brain iron accumulation leads to other neurodegenerative disorders such as hereditary ferritinopathies affecting the basal ganglia, pantothenate kinase-associated neurodegeneration type 2, and aceruloplasminemia. Complex interaction of iron with genetic variations, is one of the contributing factors in the pathophysiology of Friedreich ataxia and Huntington disease (Berg and Youdim, 2006).

9.2.2 PESTICIDES

The trend of using pesticides, to increase harvest yield, is increasing day by day. A pesticide should be used in amounts lethal to the targeted pests but not to other species. However, most of the time, pesticides are used in amounts more than recommended. This 'abuse' of pesticides is affecting human and other life forms (Aktar et al., 2009). Low dose/long term use of pesticides, such as pyrethroids, organophosphates, paraquat, dieldrin, and maneb are considered to be crucial etiological factors for main neurodegenerative diseases, such as Alzheimer's, Parkinson's, and ALS. These pesticides have been reported to mediate neurodegeneration by inducing mitochondrial dysfunction, OS, and α-synuclein fibrillization, ultimately resulting in neuronal cell loss (Baltazar et al., 2014).

Occupational pesticide exposure is linked to ALS, deficits in cognitive function, dementia and AD. A comparative study had shown that higher prevalence and risk rates of AD, Parkinson's disease, multiple sclerosis and suicide in districts with high pesticide use as compared to others with low use of pesticides. Population living in areas with high pesticide use had an increased risk for suicide attempts and AD (Parrón et al., 2011).

9.2.3 HERBICIDES

One of the most widely used herbicides in the world is Paraquat (1,1′-dimethyl-4,4′-bipyridinium dichloride). It is preferred over other herbicides due to its rapid action and low cost. It is highly toxic, nitrogen herbicide, slowly absorbed across the skin but it is not readily absorbed from the gastrointestinal tract, and is even more. Major acute toxicological effects are generated due to its accumulation in the kidneys, and lungs (Liou et al., 2001). Animal studies have demonstrated that paraquat can cause neurodegeneration of dopaminergic neurons. Alteration of ASK-1/SAPK, p53, and specific signal transduction pathways by paraquat results in the proapoptotic protein expression. The neurotoxicity mediated by paraquat, involves stress pathways of both ER and mitochondria. ROS formation is induced by paraquat through inhibition of mitochondrial ETC, activation of NADPH oxidases have and redox cycling (Franco et al., 2010). Pathophysiological effects of herbicides paraquat and rotenone-induced Parkinson's disorder are mediated through OS by damaging nigral dopaminergic neurons (Cannon and Greenamyre, 2011). Besides Parkinson's, these herbicides may cause other neurodegenerative disorders too, depending upon the neuronal cells damaged/degraded by ROS. Involvement of p53 gene indicates that it can intervene with gene expression and may affect the neurological disorders at a molecular level too.

9.2.4 NANOPARTICLES

Nanoparticles measure about 100 nm in diameter and are very tiny molecules. Nanoparticles are extremely efficient for cancer therapy, as well as drug delivery. Their small size makes penetration in cells, easier for them. Nanoparticles also reduce toxic- and side effects of drugs. However, despite all these beneficial effects, nanoparticles could impose a risk to the patient. They could cross the blood-brain barrier after penetrating the human body through different pathways. Once they reach the central nervous system, they cause neurodegeneration, neurotoxicity, and neuro-inflammation (Mushtaq et al., 2015). Requirement of synthetic bio persistent fibers in physiological systems is fulfilled by the carbon nanotubes. Bronchial epithelial cells and keratinocytes, upon in vitro incubation with high doses of single walled carbon nanotubes, showed changes in cell morphology, mitochondrial dysfunction, OS, and lipid peroxidation (Sayes et al., 2006). Quantum dots are nanoparticles which range in size from 2.5 up to 100 nm. These are not toxic themselves but produce cytotoxicity due to the release of free Cd^{2+} ions that serve to be highly toxic (Kirchner et al., 2005). Gold nanoparticles are potential drug carriers. Cytotoxic effects of these gold nanoparticles are not due to gold itself, but it is attributed to the stabilizer CTAB (Cysteine and cetylteimethyl ammonium bromide). Even though it is removed during washing but residual amounts are left even after washing, resulting in considerable cytotoxicity (De Jong and Borm, 2008).

For drug delivery, brain is the most challenging and ultimate target. With the aging population, incidences of degenerative diseases in the brain is increasing globally. As nanoparticles have more efficiency to cross BBB, targeting them to brain for therapeutic purposes, would increase predisposition to neurodegenerative disorders.

9.2.5 INSECTICIDES

Insecticides, ovicides, and larvicides are used not only in industry, medicine, and agriculture at large scale but also for domestic purposes at a very small scale. This attributes to direct exposure to these insecticides, even though in very much small quantity. Pyrethroids (allethrin, deltamethrin, and permethrin), are used as domestic pesticides. Expression of Nurr1 genes is reduced in striatum nuclei, due to exposure to permethrin in the prenatal period. This would cause a decline in the Nurr1 inhibitory effects causing the elevated generation of proinflammatory mediators (Carloni et al., 2012). Studying the effects of allethrin exposure during prenatal, postnatal, and perinatal periods showed significantly increased BBB permeability (Sinha et al., 2004) that can predispose the individuals to other neurotoxins and facilitate neurodegeneration. Both in vitro and in vivo studies have shown that exposure to deltamethrin causes apoptosis. Deltamethrin activates endoplasmic reticulum stress and causes calcium overload by interacting with sodium ion channels, leading to the unfolding of proteins (Hossain and Richardson, 2011). Protein unfolding is predisposing and precipitating factor in the pathophysiology of several neurodegenerative disorders. Therefore, early life exposure would affect brain development and a predispose individual to neurological disorders in the later age.

9.2.6 SOLVENTS

Industries, such as degreasants, paints, pharmaceuticals, printing inks, adhesives, pesticides, as well as household uses, like cosmetics, and household cleaners employ organic solvents. Exposure to carbon disulfide, methanol, benzene, toluene, ketones, n-hexane, and other solvents, is reported to produce Parkinson's disorder. Occupational exposures to these organic solvents occur by inhalation of solvent vapors as these are volatile in nature. Solvent exposure increases the risk of Parkinson's disease, essential tremor, AD, ALS, multiple sclerosis, and other neurodegenerative disorders. Solvent exposure/inhalation does not only affect central nervous system but it also causes peripheral neuropathy (Dick, 2006). Solvents are lipophilic and hence, can penetrate BBB easily. Effects of solvent trichloroethylene (TCE), are widely studied. Dermal exposure and chronic inhalation of TCE increases the risk of Parkinson's disease (Gash et al., 2008). Solvents not only penetrate the BBB but can also increase the permeability of membrane by affecting their fluidity. Therefore, exposure to solvents could be further complicated by the synergistic action of other environmental pollutants/ metal ions/nanoparticles.

9.3 RECOMMENDATIONS

Exposure to environmental pollutants is increasing day by day due to the increased industrialization and development. Keeping this increased predisposition to neurotoxins in view, we recommend following for minimizing the risk of neurodegeneration due to environmental pollutants:

- Policies should be made for proper industrial waste management to decrease heavy metal ion contaminants in water/food sources.
- Pesticides should be used with appropriate work practices, spills, and personal protective equipments.
- To function as its own "carrier," the drug should be formulated at a nanoscale (Baran, 2002). This would exclude the use of toxic metal ions like Cd for nanoparticle synthesis.
- In nanoparticles, cadmium selenium/zinc sulfate quantum dots should be preferred over cadmiun/telluride quntam dots. As selenium/zinc sulfate quantum dots show no cellular toxicity (Cho et al., 2007).

KEYWORDS

- **pollution**
- **environment**
- **diseases**
- **neurodegenerative**
- **pesticides**
- **metals**
- **solvent**

REFERENCES

Agrawal, M.; Biswas, A.; Molecular Diagnostics of Neurodegenerative Disorders. *Front. Mol. Biosci.* **2015,** *2*, 54.

Aktar, M. W.; Sengupta D.; Chowdhury, A. Impact of Pesticides Use in Agriculture: Their Benefits and Hazards. *Interdiscip. Toxicol.* **2009,** *2* (1), 1–12.

Aschner, M. Shanke, G.; Erikson, K.; Yang, J.; Mutkus, L. A. The Uptake of Manganese in Brain Endothelial Cultures. *Neurotoxicology* **2002,** *23*, 165–168.

Baltazar, M. T.; Dinis-Oliveira, R. J.; de Lourdes Bastos, M.; Tsatsakis, A. M. Duarte, J. A.; Carvalho, F Pesticides Exposure as Etiological Factors of Parkinson's Disease and Other Neurodegenerative Diseases—a Mechanistic Approach. *Toxicol. Lett.* **2014,** *230* (2), 85–103.

Baran, E. T.; Özer, N.; Hasirci, V. In Vivo Half-life of Nanoencapsulated L-asparaginase. *J. Mat. Sc.: Mat. Med.* **2002,** *13*, 1113–1121.

Berg, D.; Youdim, M. B. Role of Iron in Neurodegenerative Disorders. *Top Magn. Reson. Imag.* **2006,** *17* (1): 5–17.

Blaylock, R. L. Aluminum Induced Immunoexcitotoxicity in Neurodevelopmental and Neurodegenerative Disorders. *Curr. Inorg. Chem.* **2012,** *2* (1), 46–53.

Bose, S.; Cho, J.; Targeting Chaperones, Heat Shock Factor-1, and Unfolded Protein Response: Promising Therapeutic Approaches for Neurodegenerative Disorders. *Ageing Res. Rev*. **2016**. DOI: http://dx.doi.org/10.1016/j.arr.2016.09.004.

Bowman, A. B.; Kwakye, G. F.; Herrero Hernández, E.; Aschner, M. Role of Manganese in Neurodegenerative Diseases. *J. Trace. Elem. Med. Biol.* **2011,** *25* (4), 191–203.

Cannon, J. R.; Greenamyre. T The Role of Environmental Exposures in Neurodegeneration and Neurodegenerative Diseases. *Tox. Sci.* **2011,** *124* (2), 225–250.

Carloni, M.; Nasuti C.; Fedeli, D.; Montani, M.; Amici, A.; Vadhana, M. S.; Gabbianelli, R. The Impact of Early Life Permethrin Exposure on Development of Neurodegeneration in Adulthood. *Exp. Gerontol.* **2012,** *47,* 60–66.

Chin-Chan, M.; Navarro-Yepes, J.; Quintanilla-Vega, B Environmental Pollutants as Risk Factors for Neurodegenerative Disorders: Alzheimer and Parkinson diseases. *Front. Cell. Neurosci.* **2015,** *9*, 124.

Cho, S. J.; Maysinger, D.; Jain, M.; Röder, B.; Hackbarth, S.; Winnik, F. M. Long-term Exposure to CdTe Quantum Dots Causes Functional Impairments in Live Cells. *Langmuir* **2007,** *23* (4), 1974–1980.

De Jong, W. H.; Borm, P. J. A. Drug Delivery and Nanoparticles: Applications and Hazards. *Int J Nanomed.* **2008,** *3* (2), 133–149.

Dick, F. D. Solvent Neurotoxicity. *Occup Environ Med* **2006,** *63* (3), 221–226.

Farina, M.; Avila, D. S.; Rocha, J. B. T.; Aschner, M. Metals, Oxidative Stress, and Neurodegeneration: A focus on Iron, Manganese and Mercury. *Neurochem. Int.* **2013,** *62* (5), 575–594.

Franco, R.; Li, S.; Rodriguez-Rocha, H.; Burns, M.; Panayiotidis, I. M. Molecular Mechanisms of Pesticide-induced Neurotoxicity: Relevance to Parkinson's Disease. *Chem. Biol. Interact.* **2010,** *188* (2), 289–300.

Gaeta, A.; Hider, R. C. The Crucial Role of Metal Ions in Neurodegeneration: The Basis for a Promising Therapeutic Strategy. *Br. J. Pharmacol.* **2005,** *146* (8), 1041–1059.

Heusinkveld, H. J.; Wahle, T.; Campbell, A.; Westerink, R. H. S.; Tran, L.; Johnston, H.; Stone, V.; Cassee, F. R.; Schins, R. P. F. Neurodegenerative and Neurological Disorders by Small Inhaled Particles. *Neuro Toxicol.* **2016,** *56*, 94–106.

Hossain, M. M.; Richardson, J. R. Mechanism of Pyrethroid Pesticide–Induced Apoptosis: Role of Calpain and the ER Stress Pathway. *Toxicol. Sci. 122* (2), 512–525.

Kamel, F, Umbach, D. M.; Bedlack, R. S.; Richards, M.; Watson, M.; Alavanja, M. C. R.; Blair, A.; Hoppin, J. A.; Schmidt, S.; Sandler, D. P. Pesticide Exposure and Amyotrophic Lateral Sclerosis. *Neurotoxicology* **2012,** *33*, 457–462.

Kawahara, M Link Between Aluminum Neurotoxicity and Neurodegenerative Disorders. *Nihon Rinsho.* **2016,** *74* (7), 1176–1185.

King, L. M.; Banks, W. A.; George, W. J. Differential Zinc Transport into Testis and Brain of Cadmium-sensitive and -Resistant Murine Strains. *J. Androl.* **2000,** *21*, 656–663.

Kirchner, C.; Liedl, T.; Kudera, S.; Pellegrino, T.; Muñoz Javier, A.; Gaub, H. E.; Stölzle, S.; Fertig, N.; Parak, W. J. Cytotoxicity of Colloidal CdSe and CdSe/ZnS Nanoparticles. *Nano Lett* **2005,** *5* (2), 331–338.

Liou, H. H.; Chen, R. C.; Chen, T. H.; Tsai, Y. F.; Tsai, M. C. Attenuation of Paraquatinduced Dopaminergic Toxicity on the Substantia Nigra by (-)-Deprenyl in Vivo. *Toxicol. Appl. Pharmacol.* **2001,** *172*, 37–43.

Modgil, S.; Lahiri, D. K.; Sharma, V. L.; Anand, A. Role of Early Life Exposure and Environment on Neurodegeneration: Implications on Brain Disorders. *Translat. Neurodeg.* **2014,** *3*, 9.

Monnet-Tschudi, F.; Zurich, M. G.; Boschat, C.; Corbaz, A.; Honegger, P Involvement of Environmental Mercury and Lead in the Etiology of Neurodegenerative Diseases. *Rev. Environ. Health* **2006,** *21* (2), 105–117.

Mushtaq, G.; Khan, J. A.; Joseph, E.; Kamal, M. A. Nanoparticles, Neurotoxicity, and Neurodegenerative Diseases. *Curr. Drug. Metab.* **2015,** 16 (8), 676–684.

Parrón, T.; Requena, M.; Hernández, A. F.; Alarcón, R. Association Between Environmental Exposure to Pesticides and Neurodegenerative Diseases. *Toxicol. App. Pharmacol.* **2011,** *256* (3), 379–385.

Patel, V.; Chisholm, D.; Parikh, R.; Charlson, F. J.; Degenhardt, L.; Dua, T.; Ferrari, A. J.; Hyman, S.; Laxminarayan, R.; Levin, C.; Lund, C.; Medina Mora, M. E.; Petersen, I.; Scott, J.; Shidhaye, R.; Vijayakumar, L.; Thornicroft, G.; Whiteford, H.; DCP MNS

Author Group. Addressing the Burden of Mental, Neurological, and Substance use Disorders: Key Messages from Disease Control Priorities, 3rd ed; *Lancet* **2016,** *387* (10028),1672–1685.

Patra, R. C.; Rautray, A. K.; Swarup, D. Oxidative Stress in Lead and Cadmium Toxicity and its Amelioration. *Vet. Med. Int.* **2011,** *2011*, 457327.

Sayes, C. M.; Liang, F.; Hudson, J. L.; Mendez, J.; Guo, W.; Beach, J. M.; Moore, V. C.; Doyle, C. D.; West, J. L.; Billups, W. E.; Ausman, K. D.; Colvin, V. L. Functionalization Density Dependence of Single-walled Carbon Nanotubes Cytotoxicity In Vitro. *Toxicol. Lett.* **2006,** *161* (2), 135–142.

Sinha, C.; Agrawal, A. K.; Islam, F.; Seth, K.; Chaturvedi, R. K.; Shukla, S.; Seth, P. K. Mosquito Repellent (pyrethroid-based) Induced Dysfunction of Blood-brain Barrier Permeability in Developing Brain. *Int. J. Dev. Neurosci.* **2004,** *22* (1), 31–37.

Tamás, M. J.; Sharma, S. K.; Sebastian Ibstedt, S.; Jacobson, T.; Christen, P. Heavy Metals and Metalloids As a Cause for Protein Misfolding and Aggregation. *Biomolecules 4* (1), 252–267.

Tobwala, S.; Wang, H.-J.; Carey, J. W.; Banks, W. A.; Ercal, N. Effects of Lead and Cadmium on Brain Endothelial Cell Survival, Monolayer Permeability, and Crucial Oxidative Stress Markers in an in Vitro Model of the Blood-Brain Barrier. *Toxics* **2014,** *2*, 258–275.

Wang, B.; Yanli Du, Y. Cadmium and Its Neurotoxic Effects. *Oxidative Medicine and Cellular Longevity* **2013,** *2013*, 898034

Ward, R. J.; Zucca, F. A.; Duyn, J. H.; Crichton, R. R.; Zecca, L.; The Role of Iron in Brain Ageing and Neurodegenerative Disorders. *Lancet Neurol* **2014,** 1.

CHAPTER 10

Community-Based Conservation of the Sambhar Salt Lake, Jaipur, India

S. KULSHRESHTHA[1*], G. V. REDDY[2], and H. NANGYAL[3]

[1]*Department of Zoology, Government Shakambhar P. G. College, Sambhar Lake, Jaipur, Rajasthan, India*

[2]*Department of Forests, Government of Rajasthan, Jaipur, India*

[3]*Department of Botany, Hazara University, Mansehra, Khyber Pakhtunkhwa, Pakistan*

[*]*Corresponding author. E-mail: drseemakulshreshtha@gmail.com*

ABSTRACT

Saline lakes of arid zone are of considerable economic, ecological, and scientific values besides being a fragile ecosystem. Sambhar Lake, Rajasthan, a "Wetland of International Importance" and a "Ramsar Site" is struggling with an array of conservation threats like siltation and disturbances in the lake bed, excessive resource-utilization and unsustainable development causing a massive shrinkage of the water sheet. Maximizing economic benefit by the market force is proving death knell for the survival of this Ramsar Site. An initiative taken for saving this ancient flamingo habitat through the community conservation approach, by forming biodiversity management committees (BMCs) and maintaining people's biodiversity registers (PBRs) in accordance to Biodiversity Act, 2002 (India) and the community mobilization by Communication, Education, Public Awareness (CEPA) CBD toolkit shall prove to be the best alternative in finding out a sustainable solution.

10.1 INTRODUCTION

Biodiversity of the nation is closely linked to its ecological security and the prosperity. Continued deterioration of the biodiversity across the globe resulted in the Convention on Biological Diversity (CBD) of the nations at Reo de Genero in 1992. For the first time, biological resources of the country were considered to be brought under legal framework. India, being a party to CBD and a mega-biodiversity country, enacted the Biological Diversity Act, 2002 (BDA), which was notified in 2004.

The BDA, 2002 deals with the sustainable use of biological resources, conservation issues and their apt and equal benefit-sharing. The beauty of the BD Act is its decentralized functional approach and an amazing capacity to link at the grass-root level with the formation of National Biodiversity Authority (NBA), State Biodiversity Boards (SBBs) and Biodiversity Management Committees (BMCs) at the national, state, and local levels respectively.

10.1.1 BIODIVERSITY HERITAGE SITES

Section 37 of the Biological Diversity Act, 2002 has following provision: "The state government in consultation with local bodies may notify in the official gazetteers of biodiversity importance as 'Biodiversity Heritage Sites' (BHS) and with consultation of central government may frame rules for the management and conservation of BHS". The Rajasthan Biological Diversity Rules, 2010 have following provisions for the notification and management of the BHSs: Rule 22(1) says that the board shall, in consultation with the local bodies and other key stakeholders, take necessary steps to notify areas of significant biodiversity values as "Heritage Sites." On recommendation of the board after consultation with the central government, the state government may issue notification to this effect. Rule 22(2) says that the board shall frame guidelines on the selection, management and conservation and other aspects of heritage sites, ensuring that these provide decision-making role for relevant BMC.

10.1.2 SIGNIFICANCE AND OBJECTIVES OF BIODIVERSITY HERITAGE SITES

Biodiversity heritage site is a biodiversity-rich area and an important component of local ecosystem, being conserved and managed by the society.

1. Unwise anthropological activities are causing the great loss of the biological resources. Traditional conservational management approach is the most suitable method for dealing with such challenges. Special attention must be paid to identify such traditionally managed areas.
2. These traditionally managed areas exhibit a deep relationship between the nature and human beings. The dependency of communities on the biodiversity for the secure livelihood, economy, and cultural framework is a universal truth.
3. Being designated as a "Bio heritage Site" means ensuring the conservation of the biological resource of the particular area. Sacred grooves and "Orans" also can be proposed as BHS.
4. It may also instil the respect for conservational values in the local communities, who are supposed to be the watchdogs of the local biodiversity which in turn, may put stop to harmful conservational practices.

10.1.2.1 CONSERVING THE SAMBHAR SALT LAKE AS A "BIO-HERITAGE SITE" THROUGH LOCAL COMMUNITY AND SHAKAMBHAR GOVT P. G. COLLEGE INITIATIVES

10.1.2.1.1 Overall Mission of the Project

To work with all the stakeholders and increase the awareness about the ecological significance and engage them in conservation and protection of the Sambhar Lake leading to notify as "Bio-heritage Site." Saline lakes, commonly found in the semiarid and arid regions, are of considerable economic, ecological, and scientific values besides being a unique ecosystem. A vast array of lakes from ephemeral playas to ancient lakes has endorheic drainage basins which are requirements for the retention of water in the lake basin (Hammer and Williams, 1996).

10.1.3 STUDY AREA

The arid zone of Rajasthan has five salt lakes, that is, Deedwana, Pachpadra, Kuchaman, Phulera, and Sambhar Lakes. Out of these saline lakes, Deedwana Lake does not exist now and Fulera Lake has been dried.

The Sambhar Salt Lake is a "Wetlands of International Importance" and a "Ramsar Site." It is also an Important Bird Area (IBA) (INRJ-16 and A_4, A_{4i}, and A_{4iii} codes), being a wintering ground of thousands of migratory water birds, specially the flamingos. In the current decade, the lake is struggling with an array serious conservation challenges along with harsh climatic conditions like high temperature and scanty rainfall, causing a massive shrinkage of the water spread, that is, 192 km^2-7 km^2. If the situation continues, this ancient lake may slide into oblivion in the next two decades.

Economically speaking, this Salt Lake is an important source of salt, besides holding high aesthetic and socio-cultural values. Moreover, the massive salt production plant at the lake and its wonderful biodiversity have a great influence on people's economic condition, employment, and business.

10.1.3.1 GEOGRAPHICAL FEATURES

The elliptical-shaped Sambhar Lake falls in in Jaipur, Ajmer, and Nagour districts of Rajasthan (26° 52' N-27°02' N, 74°54'E-75°14'E). Having a size of 190 km^2 (22.5 × 3.2 × 0.003 km) when full, lake bed is almost flat with a slope of less than 10 cm/km. To the northwest and the west of the lake, Aravalli ranges are seen with the highest peak of 500 m called "Patiala Dungar" while northern and southern periphery of the lake has scattered hillocks.

10.1.3.2 LAKE BASIN

The lake is divided in two unequal parts by a 5.16 km long Jhapok–Gudhadam in the south–north direction. The eastern part (76.8 km) has two large reservoirs having concentrated brine and many salt kyars (pans) and canals and the western part holds saline water. The sluice gates are used for transferring the water from western to eastern part.

10.1.3.3 SALT KYARS AND CANALS

The oldest mainline kyars run parallel to the west side of broad gauge railway line. New kyars and Gudhakyars situated in southwest, Deodhanikyars

FIGURE 10.1 (See color insert.) Location map and a satellite picture of Sambhar Salt Lake.

situated on the eastern extremity, Jhapokkyars situated in west of the lake and Nawakyars have two railway lines on either side at the northwestern extremity of the lake. Sambhar canal is situated between the Deodhanikyar and the shores, sweeping out in a curve into center of eastern part of the lake. Nawa canal is situated in the north of Nawakyar. Gudha canal is in the west of Gudhakyar. Percolation to canals is situated in these kyars to extract subsoil brine of 15°Be–22°B density. If there is a failure of monsoon, sub-soil brine gets transferred to small pans and the salt obtained is, thus, rich in salt contents and free from algae hence, whiter in color. While, from the lake brine, salt is almost pink in color.

TABLE 10.1 Catchment Area of the Sambhar Salt Lake.

Districts	**Area (km^2)**	**%**
Ajmer	1032.50	18.08
Jaipur	1967.83	34.46
Nagaur	1042.68	18.26
Sikar	1667.37	29.20
Total	5710.38	100.00

10.1.3.4 WATER SOURCES TO THE LAKE

The lake is fed with four epimeral streams; Rupnagar enters the lake from the south, at a distance of 13 km, the north-east of Jhapokkyar. It rises in Ajmer district and blows toward lake through Kishangarh. The escape flow runs into a natural depression of "Julga Jheel" and its overflow reaches the lake. It has 77 ppm salinity in the water. Mendha enters the lake from the north about 3 km southeast of Nawa. It is the largest stream which originating in Sikar and runs south-west and the west before entering the lake. It covers 3600 km^2 area. It has got 1395 ppm salinity. Kharian enters the lake from north-west at a distance of about 6.5 km. from Nawa Kyar and its water bear salinity of 665 ppm. Khandel enters the lake at the eastern end of Deodhani. 11 ppm sodium chloride contents are found in its water.

10.1.3.5 SETTLEMENTS

Toward the eastern side of the lake is Sambhar town, while 38 small clusters of settlements, main being–Jhapok, Korsina, and Khakhardi situated in south and Gudha and Nawa in the north of the catchment.Western side of the lake is surrounded by several hillocks of Aravalli ranges.

Maximum households follow agriculture as their primary occupation, while 9% work as daily wageworkers. 4% households are directly involved in salt work, though only 1% are occupied in any other kind of factory labor. 8.7% of the total village population employed in salt work earn between Rs. 20,000–30,000 per annum through the saltpans, indicating that income generated through the work contributes substantially to the village families. 78% households in the villages around the lake own livestock, mostly goats. Some families engage in poultry while fewer own cows (10%).The proportion of women and men involved in agriculture or allied activities is almost the same. The overall rate of unemployment is low in the villages.

10.1.3.6 CLIMATE

The catchment of Sambhar is called the gateway of the arid eastern climatic zone, During the summer season, the climate is hot and dry with winds (30–36 km/h) and dust storms. Atmospheric temperature ranges from 38 to 44°C in summers and at its peak rises to even 47–48°C, causing the evaporation of shallow lake waters. In winters, the temperature declines and at peaks reaches even 4°C otherwise, the minimum average temperature is 8–19°C. However, the climate is flexible and greatly depends on the rain fall conditions which are quite low (100–500 mm) even in rainy season (July–October) in general. A total annual rainfall and its period of occurrence as well as intensity during the season exhibit wide variations resulting in drought with rare episodes of floods (Kulshreshtha et al., 2011).

10.1.4 CONSERVATION VALUES OF THE SAMBHAR SALT LAKE

Sambhar Lake is not merely a salt-producing factory but a socio-cultural, archaeological, geological, and ecological asset to the nation.

10.1.4.1 MYTHOLOGICAL AND RELIGIOUS VALUES

As far back as 5000 BC, Adiparva of Mahabharat and 18th–19th chapters of *Bhagavadpurana* (9th vol) describes it as a kingdom of the demons, Brishparva. Devyani, the daughter of Shukracharya, the chief priest married to Yayati, the emperor of *BharatVarsha*, 10th generation of Lord Brahma (Aggarwal, 1951). Sambhar Lake is mythologically considered a blessing of Shakambari Devi about 2500 years ago, which is also the goddess of the Chouhan Rajputs in 551 A.D. Apart from this, "Dadupalka," a religious center of Dadupanth, a sect made by Guru Dadudayal is also situated here.

10.1.4.2 HISTORICAL AND ARCHAEOLOGICAL VALUES

Sambhar was mentioned as the first capital of Chauhan Kings in 11th century in an interesting article by Diwan Bahadur Harvilas Sarda published in the Journal of Royal Asiatic Society, 1913 (Kumar, 2005). Archeological excavations also proved the origin of the Sambhar town in Kushan and Gupta period in 551 B.C. (Sharma, 1988a, b). During the times of Emperor Akbar from 1192 B.C.–1708 B.C., the earning from the lake was Rs. 2.5 lakhs per month (Aggarwal, 1951) which increased to Rs. 15 lakhs during Aurangzeb's period. Later after the year 1844, the lake was under *Shamlat*, the joint government of Jaipur and Jodhpur. About 60 settlements were situated on the periphery of the lake. In 1870, the salt production was taken over by the British government till independence of India in 1947. The central government renamed it as Hindustan Salt Ltd. Currently, the Sambhar Salts Ltd.—a joint venture of Hindustan Salts Ltd. and the state Government of Rajasthan is managing it. Sambhar Salts Ltd. is also the first iodization plant in the country (Gopal and Sharma, 1994).

10.1.4.3 GEOLOGICAL VALUE

Chemical weathering of Aravalli rocks and Felspar gypsum buried underneath the land plays an important role in the accumulation of high sodium salts on the surface (16% NaCl and 5.2% Na_2SO_4) decreasing to 5% at a depth of 30–60 cm (Hackett, 1980; Sharma, 1988a, b). A thick layer of 20 m saliferous clay lying above micaceous schists covered by lime stone nodules after drying gives cracks to the bed. Apart from this, some heavy metals and radioactive uranium has also been found.

10.1.4.4 ECONOMIC VALUES

10.1.4.4.1 Salt Production

Agarwal (1951, 1977) has described the history of salt production, process, cost revenues, and so on. According to the Official Report (2005) of Sambhar Salts Ltd., 5500 million tons of salt is still lying beneath the lake surface. Salt from the margins of the lake is called "Reshta" Salt which is, marketed separately. Sodium Sulfate, Bromine and Soda Ash are the by-products of the procedure.

10.1.4.4.2 Biodiversity/Ecological Values

Despite of the fact that the Sambhar Lake is a hypersaline lake, the biodiversity of this Salt Lake is of utmost importance, hence, shall be conserved.

10.1.4.4.3 Flora

The saline water changes colors in different seasons due to the algae *Arthrospira, Spirulina, Anabaenopsis* and *Dunaliella. D. salina* and *Arthrospiraplatensis* causes bluish-green color of the water from September to November and dark brick-red color in summers. 13 species of alga are reported from the lake water (Dayma, 2003). The catchment of the Sambhar Lake has dry thorny scrub forests, typical of arid and semi-arid areas. The Aravalis are covered with *Anogeissuspendula* and *Boswellia, Euphorbia. Prosopisspicigera, Acacia milotica, Acacia senegal, Capparis decidua, Salvadora persica, S. oleoides*, and so on are thedominanttrees. *Saccharumspontaneum, Cenchrus spp.*, and *S. benghalense* are the common coarse grasses. *Salsola* and *Suaeda* species are the common succulent halophytes seen on the moist margins of the lake.

10.1.4.4.4 Fauna

10.1.4.4.4.1 Invertebrates

Prof. M. L. Roonwal (1969) from Zoological Survey of India had carried out an intensive survey of the fauna of the Sambhar Lake after Adam

(1873 and 1874). A few zooplankton like *Cladocera* and *Copepods* such as *Branchianus, Miona, Cyclops*, and *Diaptomus* (Gopal and Sharma, 1994; Baid, 1952), *Berosusindicus, Ephydramalellia*, and *Eristalis* (Baid 1958, 1959, and 1969), shells of *Indoplanorbisexystus* and *Gabbiaorula* (Gopal and Sharma, 1994) were seen during rainy seasons when the salinity was low which disappeared due to hypersalinity later. A brine Shrimp *Artemiasalina* was reported by Baid (1958) which disappeared during floods of 1970s. The catchment has some common insect species like *Schistocerca* (locust), *Agrian* and *Petalura* (dragon flies), *Vespa* and *Polistis* sp (wasp). Sambhar Lake watershed area has first record of 35 species of aquatic beetles (Roonwal, 1969). *Cybistertripunctatusasiaticus* (sharp), *Eretessticticus* (Linn), and *Hyphoporousseverni* (Baid, 1959) and 22 species of family Dytiscidae (Vazirani, 1964) around Nariana, Phulera, Bhadurpur, and Koni villages were reported. However, we observed 16 species in 2006 and six species in 2007 and 2008.

10.1.4.4.4.2 Vertebrates

During heavy rains, sometimes, freshwater fishes enter from the stream water during which die later due to hypersalinity of the lake (Gopal and Sharma, 1994). However, no fish is ever seen during the present study. Frogs and other amphibians, variety of snakes *Eryx, Naja, Bungarus, and Ptyas*, lizards like *Hemidactylus, Gecko, Mabouia, Varanus*, and land tortoise *Testudo* are seen in catchment area. The flourishing ecology of the Sambhar Lake and a sizable number of avifauna including 20,000–25,000 Greater Flamingo *Phoenicopterusruber* and Near-threatened Lesser Flamingo *Phoeniconaias minor*, the regular visitors of the lake had been documented a century back (Adam, 1873, 1874; Hume, 1878; McCann, 1939). Across the century, specifically in the last two decades, the continuous decline in the avian diversity of the lake has led to the occurrence of 53 terrestrial and 42 aquatic bird species now. Blue Antelope *Boselaphustragocamelus* are commonly seen in the fallow lands on the periphery of the lake. Domestic animals are 7.5 million, having cattle, buffaloes, goats, camels, horses, and donkeys. Grazing is a major problem in the area for the forests. Rodents like gerbils are seen near the Deodhanikyars. Other mammals like *Mus, Talpa, Rattus, Funambulus, Gerbil*, Hedgehog *Hemiechinus* and mongoose *Herpestes*, monkey such as *Macaca* and *Presbytis* are observed A golden jackal *Canisaureus* with two pups was also seen roaming around the Deodhani Kyar of the lake in the year 2005.

A large colony of around 1000–1500 Flying foxes *Pteropus* is seen on a banyan tree near Deodhani kyars of the lake. Fruits of the tree make the favorite food of these bats.

10.1.5 CONSERVATION THREATS TO THE SAMBHAR LAKE

For want of serious protection measures the ecological health of the lake continuously deteriorating (WWF and AWB 1993; Kulshreshtha and Sharma 2009; Kulshreshtha *et al.*, 2009). The major problems are:

1. *Shrinkage of water spread*: A considerable reduction of the water sheet of the lake from more than 200 km^2 to only 7 km^2, has been caused by constructions of innumerable anicuts on inflowing rivers by the state irrigation department.
2. *Siltation of the lake bed*: Sandpits are being made on the dried lake bed causing siltation and reduction of the lake capacity and size due to the deposition of the silt at the lake bottom. Trespassing of vehicles adds to this exploitation.
3. *Excessive resource utilization*: Over-exploitation of the lake water for the salt-production by illegal salt-manufacturers who also steal brine using underground pipes, has led to the desiccation of the lake.
4. *Altered land-use pattern*: The eroded area of the catchment has changed from 24 to 33% during the last decade (Jain, 2005). It is unfortunate that Sambhar Salts Ltd. is using only 3.04% of the total lake area for salt extraction while 3.98% agricultural land of Sambhar catchment is being used illegally for the purpose, given the fact that fallow lands are limited due to the scarcity of water and dependency on rainfall. These anthropogenic pressures have increased the desertification (Dhabariya, 1984).
5. *Deforestation:* The forest covers are already sporadic. Large scale grazing and chopping of trees of *Acacia* sp. and *Prosopis* sp. in the watershed has led to the loss of biodiversity of the lake catchment. The rapid urbanization and development also taxing heavily on the arid forest cover. The construction of a 12.4 km long road from Nawa to Khakardi village, in turn, dividing the lake bed in to two parts has been the most unwisely anthropogenic action at the southwest of the lake.

6. *Unregulated tourism*: Tourism has no particular set up, nor any government care-taking, hence, economy on this approach is almost nil. The local hunters, tribes, and villagers continue to poach birds and flamingos as a common and age-old practice.
7. *Reduction in biodiversity:* Since, the Sambhar Salt Lake is a rain-fed playa, its physical, biological, and ecological characteristics greatly depend on the seasonal variations. The current decade is a witness to the decline in the population of a few plant species around the lake, such as, *Suaedafruticosa, Salsolabaryosoma, Salsolafoetida, Salvadorapersica, Portulacaoleracea, Eragrost-isciliaris, Trianthematriquetra, Saccharummunja, and Citrul-luscolocynthis.* In addition to these flora, some faunal species like, aquatic beetles from 36 to 7–8 species, aquatic birds from 97 to 52 species, populations of flamingos from 20,000–25,000 to 2000–2500 (Kulshreshtha et al., 2011) in winters months and disappearance of White-napped Tit *Parusnuchulis* due to the chopping of trees and subsequent loss of dry forest around the lake (Tiwari and Bhardwaj, 2013) are notable losses of faunal wealth of the lake.

10.2 OBJECTIVES

The objectives of the study were:

1. To work with the select local communities in documenting the biodiversity of Sambar Lake area, constitute BMCs and compile "Biodiversity Registers."
2. To monitor the ecological health of the lake by the changing bio-indicator faunal and floral species through BMC.
3. To identify prevailing anthropological threats responsible for the biodiversity loss.
4. To compile the existing scientific information on the Sambhar Lake and make it accessible to the common man through a database/website.
5. To identify the best practices on conservation and management of saline lakes going on in other parts of India and make it available for incorporation in the Sambhar Lake management plan.

6. To mobilize the community support to get declared the lake as a "Bio heritage Site"
7. To draw the attention of stakeholders to the extensive continuing anthropological desiccation while realizing the urgency of action needed at national, regional and global levels.

10.3 METHODOLOGY

Clusters of villages around the Sambhar Salt Lake targeted:

1. Korsina, Kaliyan, Habaspura, Jhapok in the south
2. Sinodhiya, Khakardi in the north
3. Mohanpura, Khardia, Kaseda, Ulhana, Luniawas
4. Gudha, Jabdinagar, Khandel

TABLE 10.2 Demographical Statistics of Target Villages.

S. No.	Names	Houses	Total population	Males	Females	Male child	Female child
1.	Sambhar	3865	22,327	11,517	10,810	2010	1865
2.	Korsina	335	1930	991	939	114	93
2.	Habaspura	208	1422	738	684	233	114
4	Jhapok	74	484	240	244	82	36
5.	Kakardi	191	1178	616	562	193	101
6.	Sinodiya	229	1841	966	875	214	173
7.	Mohanpura	163	1199	624	575	201	91
8.	Kaseda	133	988	508	480	101	49
9.	Luniawas	480	3053	1516	1537	451	224
10.	Khandel	325	2037	1029	1008	283	141
11.	Gudha	249	1372	748	622	212	185

Source: Census of Government of Rajasthan, 2011.

The planning of the methodology is based on Community, Education and Public Awareness (CEPA) approach as enunciated by the CBD toward the sustainable management of the lake which is as follows:

10.3.1 IMPLEMENTATION STRATEGY

10.3.1.1 BASIC APPROACH

Communication, Education and Public Awareness (CEPA) are important tools for the conservation and sustainable use of biodiversity. CEPA also forms an important link between society, ecological conservation, and ecosystem services and acts as a catalyst or motivational force to implement the policies of Conventions on Biological Diversity. Government of India, Biological Diversity Act, 2002 (BDA), notified in 2004 along with the Rajasthan Biodiversity Rules, 2010 will be followed for constituting BMCs and Biodiversity Registers.

10.3.2 PROPOSED ACTIVITIES TO ACHIEVE THE OBJECTIVES

1. Formations of BMCs in village; clusters,
2. Preparation and maintenance of Biodiversity Registers,
3. Visit Chilika Lake and the Kollaru Lake to understand the ongoing management and community participation,
4. Compile biodiversity information from the secondary sources and develop a website,
5. Organizing workshops for BMCs, stakeholders, and public representatives, and
6. Monitor population of key bio-indicator floral species specifically halophytes like *Salsola, Sueda,* and *Salvadora* and faunal species like Flamingos, winter birds, White-napped Tit, fruit bats, and aquatic beetles.

10.4 RESULTS

The proposed project by involving local governance and the awakened public would certainly have a significant impact on the restoration of the ecosystem and the catchment. We have been successful in achieving following outcomes:

1. Eight Biodiversity Committees in the four clusters of 14 villages have been formed;

2. One workshop organized for the training of Biodiversity Committee members;
3. Peoples Biodiversity Registers prepared for eight panchayats;
4. 108 BMC members trained in monitoring the biodiversity; and
5. The secondary literature collected and maintained in the Shakambhar Government P.G. College, Sambhar Lake.

10.5 DISCUSSION

Management decisions regarding saline lakes are often based solely on easily quantified direct economic benefits, while the considerable ecological values and diffuse economic impacts associated with their deterioration are rarely quantified.

The impact of Salt Lake sediments and salt-affected soils on global climate issues like ozone depletion, global warming and decreasing rainfall comes through volatile halogenated organic compounds (VOX) emission like monochlorinated, for example, chloromethane (CH_3Cl), and higher chlorinated VOX compounds such as trichloromethane ($CHCl_3$) (Ruecker et al., 2014).

With the increasing desertification, biological diversity and productivity of this unique ecosystem have also gone low. For the rehabilitation and restoration of the lake, a proper planning and strategy are required along with the intelligent lake basin management. Complex and expensive mitigation schemes for filling up the lake involves diversions of water from nearby rivers and redistributing drainage basin to the lake through pipelines, tunnels, deep navigable canals, and so on. This would still supply a small fraction of water to the lake besides changing the climate and the local fauna and flora. However, inviting useful and implementable proposals from the efficient MNCs/the World Bank by the state government on the priority basis could be a solution but such megaprojects need prior budgeting and a strong political will. Though, restoring the water volume to its initial state and maintaining it through these proposals is still technically difficult, unrealistic, and doubtful due to shallow water depth and compensation for the continued fast evaporation.

The concept of community conservation approach is the best which could prove to be fruitful in achieving the common goal of sustainable resource-utilization and restoring the lake involving society, economy, culture, and formation of the policy for the allocation of resources.

CEPA was adopted to provide for capacity-building. It teaches how to raise awareness by networking, engage stakeholders and mainstream biodiversity and plan communication strategically (Hasselink et al., 2007). It can also be an important tool/policy instrument for the development and implementation and management the National Biodiversity Strategy and Action Plans (NBSAP).

The 7th meeting of the State Board of Wildlife was held under the Chairmanship of Hon'ble Chief Minister, Government of Rajasthan on October 17, 2015 a decision was taken to discuss the proposal to form Sambhar Lake Development Authority on the lines of Chilka Lake Development Authority for its management and conservation. Irrigation department may be directed to look into the reduced level of water storage in the Sambhar Lake along with its reduction in water spread and suggest action for restoration of water in the lake. Illegal tube wells and pipelines laid should be acted upon. A dedicated team led by a suitable officer with members from revenue, WR/GW departments, PHED industries and district admin would oversee this task. In order to maintain nation's unique ecosystem and to allow wise-use of its resources, scientific management and action plans and recommendations of this project are absolutely necessary.

10.6 FUTURE PLANS AND PERSPECTIVES

1. The website enabling open access will be formed in the next phase;
2. One seminar will be organized for sharing the experiences of Biodiversity Committee members; and
3. A proposal to notify the Sambhar Salt Lake a "Biodiversity Heritage Site" will be moved by the BMCs formed in the area.

10.7 AN IMPACT ASSESSMENT

A mobilization and the awareness regarding the conservation of Sambhar Salt Lake has its wide impact on the psyche of the local communities as the continuous inquiries regarding the formations of the BMCs are pouring in. The success of any state biodiversity board is considered by the number of BMCs it forms. The more the number of BMCs, the more strength, and support for the greater cause of the local biodiversity conservation. The day is not far when like Kerala state, 100% BMCs will be formed at every panchayat level and work with a mission like a Telangana state board in a systematic and phased manner, yet in the most wisely manner to achieve the target.

ACKNOWLEDGMENTS

The authors are thankful to Rajasthan State Biodiversity Board (RSBB), Jaipur, Rajasthan for providing financial support for this Community conservation project.

KEYWORDS

- **saline lakes**
- **biodiversity management committees (BMCs)**
- **CEPA**
- **Ramsar site**
- **Flamingo habitat**

REFERENCES

Adam, R. M. *Notes on the Birds of Sambhar Lake and its Vicinity*; Stray Feath: Calcutta, 1873; Vol. 1, pp 361–404.

Adam, R. M. *Additional Notes on the Birds of Sambhar Lake and its Vicinity*; Stray Feath: Calcutta, 1874; Vol. 2, pp 65–466.

Agrawal, S. C. *Sambhar Lake Salt Source*; Government of India Publication: New Delhi, 1951; p 365.

Baid, I. C. On the Occurrence of *Artemiasalina* in Sambhar Lake. *Curr. Sci.*, **1958,** *27*, 58–59.

Baid, I. C. Some Observations on the Population of *Artemiasalina* in Sambhar Lake. *Proc. Natnl. Acad. Sci.* **1959a,** *28*, 416–423.

Baid, I. C. The Arthropod Fauna of Sambhar Salt lake, Rajasthan, India. *Oikoassay* **1969,** *19* (2), 292–230.

Dayama, G. P. *Algae of Sambhar Lake*. Ph.D. Thesis. University of Rajasthan, Jaipur, Rajasthan, 2001.

Dhabariya, S. S. In *Intensification of Desertification Hazard Through Sambhar Gap. Problem of Environmental Degradation in the Indian Arid Zone*. Proceedings of the 18th International Symposium on Remote Sensing of Environment, Paris, 1984; p 3.

Gopal, B.; Sharma, K. P. *Ramsar Site of India, Sambhar Lake*; World Wide Fund for Nature: India, 1994.

Hammer, U. T.; Williams, D. R. *Saline Lake Ecosystems of the World*; Junk Publishers, PO Box 163, Springer Publication: Dordrecht, Netherlands, 1996; Vol. 59, p 616.

Hume, A. O. The Birds of a Drought; Stray Feath, 1878; Vol. 7, pp 52–68.

IUCN,. The Ramsar Conference: Final art of the International Conference on the Conservation of Wetlands and Waterfowl. *Spec. Suppl. IUCN Bull.* **1971,** *2* (9), 1–4.

Jain, A. K. *Conservation Planning of Sambhar Lake, Rajasthan using Satellite Remote Sensing and GIS'*. A Thesis for M Tech submitted to Andhra University, India, 2005.

Kulshreshtha, S.; Sharma, B. K. In *Declining Avifaunal Diversity of Sambhar Lake is an Important Bio-indicator of the Ill Health of the Lake*. Proceedings of the International Congress for Conservation Biology and 22nd Annual Meeting, Chattanooga, Tennessee, USA, 13–17 July, 2008.

Kulshreshtha, S.; Sharma, B. K. Conservation Values and Conservation Management of Sambhar Salt Lake, Rajasthan (India): An Ignored Ramsar Site of Asia. *Asian J. Exp. Sci.* **2009,** *23* (2), 32.

Kulshreshtha Seema; Kulshreshtha, M.; Sharma, B. K. Ecology and Present Status of Flamingos at Sambhar Salt Lake, Rajasthan, India: A Critical Comparison with Past. Weltvogelpark Walsrode, Germany, *Flamingo Bull. Flamingo Spec. Group* **2011,** *18*, 24–27.

Kulshreshtha, Seema; Sharma, B. K. Ramsar Sites of Rajasthan–Ecology and Conservation of Sambhar Salt Lake and Keoladeo. In *Faunal Heritage of Rajasthan: Ecology and Conservation of Vertebrates*; Sharma, B. K., Kulshreshtha, S., Rahmani, A. R, Eds.; WWT Slimbridge Wetland Centre, UK, https://www.iucn.org/sites/dev/files/2016-2017_flamingo_sg_report.pdf, 2011.

Kumar, S. *Fauna of Sambhar lake (Rajasthan).* Wetland Ecosystem Series; Zoological Survey of India: Kolkata, India, 2005; p 200.

Kumar, S. Conservation of Sambhar Lake–An important Waterfowl Habitat and A Ramsar Site in India. In *Book of Abstracts of Taal, 12th World Lake Conference, Jaipur, Rajasthan, India*, Oct 28–Nov 2, 2007; Ministry of Environment and Forests, Govt of India, 2007; p 284.

McCann, C. The Flamingos (*Phoenicopterus rubber* antiguorum Temmi). *J. Bomb. Nat. His. Soc.* **1939,** *54*, 961–962

Jellison, R.; Egor, S. Z.; Sharma, P. A.D.; Melack, J. M.; Rosen, M.R.; Degermendzhy, A. G; Sharma, S. D., Sharma; Germàn, Zambrana *Conservation and Management Challenges of Saline Lakes: A Review of Five Experience Briefs† Lake Basin Management Initiative Briefs*, 2004. http://www.worldlakes.org/projects.asp?projectid=4&programid=2.

Roonwal, M. L. Fauna of Rajasthan India. Part I: General Introduction With a List of Collecting Localities and a Bibliography of Rajasthan. *Zool. Rec. Zool. Surv. Ind.* **1969,** *61* (3–4) (1963), 291–376.

Sangha, H. Flamingo Survey at Sambhar Lake (Rajasthan), India. W1/SSC. Flamingo Specialist Group. Newsletter, 1998; p 8.

Sharma, B. K.; Kulshreshtha, S. In *Diminishing Faunal Diversity of Sambhar Lake: An Ignored Ramsar Site of Asia*. Proceedings of the International Congress for Conservation Biology and 23rd Annual Meeting, Beijing, China, 11–16 July, 2009.

Tiwari, J. K.; Bhardwaj, D. B. White-naped Tit *Parusnuchalis*: An Endemic and Endangered Species. In *Faunal Heritage of Rajasthan: Ecology and Conservation of Vertebrate*; Sharma, B. K., Kulshreshtha, S., Rahmani, A. R., Eds.; 2011.

Vijayan, V. S.; Prasad, N.; Vijayan, L.; Murlidhar, S. *Inland Wetlands of India*; Salim Ali Center for Ornithology and Natural History (SACON), WWF, India and AWB (1993). Directory of Indian Wetlands WWF India: New Delhi and AWB: Kualalampur, 2004.

WWF India and AWB Directory of Indian Wetlands: Kuala Lumpur, 1993.

CHAPTER 11

Biodiversity and Conservation of the Orphans of Rio in Northern Africa–Egypt: A Case Study

A. M. ABDEL-AZEEM* and F. M. SALEM

Department of Science, University of Suez Canal, Ismailia 41522, Egypt

Corresponding author. E-mail: Zemo3000@yahoo.com

ABSTRACT

Ecologically and economically, the Fungi kingdom is of immense value and necessary for sustainable life on planet earth. There are six kingdom classifications of life, with Fungi classified as a kingdom of their own. Members of Kingdom Fungi are one of the most diverse and numerous groups of living organisms, making it the second-largest group of organisms after insects. Technological advances in molecular research have enabled mycologists to discover and identify fungal taxa. It is estimated that as many as 1.5 million fungal species exist. Fungi are ubiquitous, undertaking many roles both independently and in association with other organisms. Recently, the International Union for Conservation of Nature has recognized that fungal conservation is just as important as animal and plant conservation, and has called on governments worldwide to pay much more attention to fungal conservation. Fungal conservation in Northern Africa has not flourished and does not flourish in the shadow of botany. This chapter describes the diversity and current status of fungal conservation in Northern Africa and Egypt as a case study, threats, future perspectives and the role of The Arab Society for Fungal Conservation in Egypt for mycology.

11.1 INTRODUCTION

The Northern African countries (six countries and one disputed) are in an arid to semi-arid region with a Saharan climate in the south, an oceanic climate in the west, and a Mediterranean climate in the north (United Nations, 2016). These countries constitute an area of 8,258,701 km^2 and with a population of more than 210 million people.

Conservation of nature in Northern Africa (NA) increased in the last two decades, in which a total 599 protected areas were designated, proposed, not reported or adopted (United Nations list of protected areas 2014). Nature conservation and conservation training in Northern Africa are actively pursued and receive much international interest, but there is little awareness of fungi, of their importance, their uses, their unexplored biological diversity, and the need to protect them. The "Orphans of Rio" (Hawksworth, 1997), reflects how ineffective The Convention on Biological Diversity (Rio de Janeiro, 1992) has been in protecting fungi (Minter, 2010). The Convention of Biological Diversity is dedicated to promoting sustainable development. However, since many fungi are inconspicuous and have yet to be classified, the true biodiversity of Kingdom Fungi is unknown. Therefore, many fungi species are treated as Cinderella between both animal and plant kingdoms in which members are distributed across both kingdoms (Abdel-Azeem et al., 2012).

Egypt (1,000,000 Km^2), lies at the northeast corner of Africa at the junction of four biogeographical regions, Irano-Turanian, Mediterranean, Saharo-Sindian and Afrotropical. Egypt is bounded on the north and east by two largely enclosed seas, the Mediterranean Sea and the Red Sea. Egypt is divided by the River Nile into a western part including the Libyan Desert (681,000 km^2) and an eastern part comprising the Eastern Desert (223,000 km^2), and the Sinai Peninsula (61,000 km^2). The Nile basin, comprising the valley in the south (Upper Egypt) and Nile delta in the north (lower Egypt), forms a riparian oasis (40,000 km^2) that constitutes the densely inhabited farmlands of Egypt. This unique position is enhanced by the circumstance that it is divided by the Nile, the longest river in the world. Most of Egypt is either arid or hyper-arid; however, due to its very varied eco-zones, the country is home to a wide diversity of terrestrial, aquatic and marine habitats.

11.2 FACTORS LIMITING FUNGAL CONSERVATION IN NA

11.2.1 FUNGI OR PLANTS?

Fungi are different from animals and plants since at least 1970, scientists have agreed that fungi belong in their own separate biological kingdom which is likely to contain far more species than the plant kingdom. Where plants produce and animals consume, fungi are the recyclers. Fungi are just as much threatened as animals and plants by climate change, habitat destruction, invasive, pollution, over-exploitation and even, in some cases, persecution. The Nature Conservation Sector of the Egyptian Environmental Affairs Agency (EEAA) still considered fungi as plants (http://www.eeaa.gov.eg/English/main/protect_bio.asp) even after more than 40 years of emerging of the Fifth Kingdom system by Whittaker.

Estimates for the number of fungi in on Earth would have been considerably greater than the 1.5 M suggested by Hawksworth (1991), which is currently accepted as a working figure although recognized as conservative (Hawksworth, 2001). The 10th edition of Ainsworth and Bisby's Dictionary of the Fungi (Kirk et al., 2008) provided a total of 989.98 for the number of fungal species accepted to date (excluding taxa treated under Chromista and Protozoa). Kirk et al. (2008) reported 1039 species chromistan fungal analogs and 1165 as protozoan in which 1038 are regarded as protozoan fungal analogs: Percolozoa (Acrasida), Amoebozoa (Dictyostelia, Myxogastria, Protostelia), Cercozoa (Plasmodiophorida) which were previously treated as Myxomycota and Plasmodiophoromycota. Habitats important for threatened fungi may be different from habitats important for threatened animals and plants. Biodiversity can only be conserved if the well-being of fungi is given as much consideration as that of animals and plants: without fungi life on earth would be unsustainable.

Fungi provide enormously important ecosystem services (for example, soil fertility, mycorrhizas, crop protection, litter decomposition, checks, and balances). The economic value of such services has been estimated as running to trillions of US dollars. Fungi are also a very important source of unusual chemicals of great value in industry and medicine. The yeast used to make bread is a fungus, and many pharmaceuticals such as antibiotics, statins, and anticancer drugs are derived from fungi.

The International Union for Conservation of Nature has recognized that fungal conservation is just as important as animal and plant conservation, and has called on governments worldwide to pay much more attention to fungal conservation. Conservation of fungi in NA still overlooked and the challenges involved predictably are daunting in comparison with animals and plants. Only one culture collection was recorded by WFCC in NA (www.wfcc.info/ccinfo/collection/by_country) to conserve fungi and other microorganisms. According to The Micheli Guide to Fungal Conservation, all NA countries are totally deficient (http://www.fungal-conservation.org/micheli.htm), except Egypt, even though they signed The Rio Conventions' mandate (https://www.cbd.int/reports/). This fifth national report of Egypt recognized this scientific reality (it should particularly avoid the very common mistake of treating fungi as though they are plants as before).

11.2.2 ABSENCE OF FUNGAL DOCUMENTATION

Egypt and NA countries are not a sandbox and host to rich biological diversity of fungi. Conservation of nature by ancient Egyptians was recorded on walls of temples and papyrus sheets. Egypt considered as the cradle of fungi when ancient Egyptians produced a number of hieroglyphic depictions of plants (many of which are psychedelic) on walls and within texts throughout Egypt. Temples with countless pillars are shaped like huge mushrooms with tall stems, umbrella caps, and mushroom engravings distributed all over the country (Abdel-Azeem, 2010, 2014). Berlant (2005) theorized that the Egyptian White and Triple Crowns were originally primordia of the entheogenic *Psilocybe* (*Stropharia*) *cubensis*, and expressed in ancient Egyptian literature.

Egyptian iconography is therefore loaded with many other structures that were clearly designed to depict mushrooms. According to The Boston Museum of Fine Arts Catalogue (1982) depicting ancient Egyptian ear studs, and Aldred's (1971) analysis of similar studs, they obviously resemble mushrooms.

The Hearst (1550 B.C.) prescriptions 89–92 deal with the treatment of skin abrasions or contusions. They recommend the application of moldy bread crumbs, salt, and rags. Here one may assume an ancient observation on the possibility of molds being helpful in preventing skin infections in connection with abrasions. This is a long way from the modern antibiotics, but it suggests the same sort of observational skill (Leake, 1952).

11.2.3 UPDATED CHECKLISTS

Updated checklists are important tools in taxonomy, systematics, and conservation (Söderström et al., 2007, 2008) and this is a limiting factor in NA. In Egypt, only very few comprehensive assessment of local fungi has been published (e.g., El-Abyad and Abu-Taleb, 1993; El-Abyad, 1997; Moustafa, 2006; Abdel-Azeem, 2010; Moustafa and Abdel-Azeem, 2011; Abdel-Azeem and Salem 2013a, 2015).

11.2.4 RESPONSIBILITY FOR FUNGAL CONSERVATION

In the majority of NA countries, fungi still treated as plants and the responsibility of their conservation back to the Ministry of Environment. Since 1936, when Egypt became a party to the Convention Relative to the Preservation of Fauna and Flora in their Natural State, they have been among the pioneering countries in NA and world-wide taking an active interest in the conservation of biodiversity and the preservation of natural resources. In 1992, Egypt signed the Biodiversity Convention of Rio de Janeiro and ratification of this Convention was completed in 1994. In 1979, the Ministry of Agriculture established the Egyptian Wildlife Service as the first national institution concerned with the formulation and implementation of policies pertaining to the protection of wildlife. In 1982, it was replaced with the Egyptian Environmental Affairs Agency (http://www.eeaa.gov.eg), which has become recently a department of the State Ministry of Environmental Affairs. The National Biodiversity Unit, Academy of Science and Scientific Research (http://www.asrt.sci.eg), Universities, National Institutes, Zoological and Botanical Gardens, and NGOs are also charged with the task of protection of organism, including fungi (Abdel-Azeem and Soliman, 2011a).

11.2.5 EDUCATION AND FUNGAL CONSERVATION

Fungal conservation not integrated into the curricular and extracurricular activities in the majority of NA countries. In Egypt, Abdel-Azeem and Soliman (2011) started the integration of fungi in extracurricular activities of different high Schools. Later on, Soliman with Abdel-Azeem in 2011 run the activities of Science Across Egypt© Projectin the whole country through different activities to train science teachers, students and school managers on fungal conservation activities either directly or indirectly (http://blogs.nature.com/soapboxscience/2011/10/19/science-across-egypt-project).

11.2.6 BIOPROSPECTING AND FUNGAL CONSERVATION

Fungi (macro or micro) in NA countries suffered all the time from bioprospecting which generally described as the search for naturally occurring chemical compounds and biological material, especially in extreme or biodiversity-rich environments (Abdel-Azeem et al., 2012).

11.3 FUNGAL CONSERVATION EFFORTS IN EGYPT AS A CASE STUDY

Since 2007, Dr. Ahmed M. Abdel-Azeem, a mycologist at the University of Suez Canal, Egypt and his team of young, energetic mycologists, teachers and colleagues established substantial plans and long-sustained efforts to conserve the Egyptian fungi through different activities as follows:

- *First:* from 2007 up to 2010, Dr. Abdel-Azeem, (chairman of Community and Environmental Services Society in Saint Katherine-South Sinai,) organized many training courses to more than 200 Bedouin women in the Sinai Peninsula in Egypt on cultivating and marketing mushrooms for food (EuropeAid/123879/C/ACT/EG). This is regarded as another fixed income way to conserve the over-collection of medicinal plants which harbor many promising endophytic fungi.
- *Second:* in 2011, Dr. Abdel-Azeem won the prestigious Rubenstein Research Fellowship Award for his scientific research and commitment to online collaboration and outreach studies on the fungi of Egypt. Under supervision of Dr. David Minter (President of European Mycologist Association and International Society of Fungal Conservation), Abdel-Azeem also established a minimum of 400 EOL-compatible web pages, each relating to a different kind of fungi (non-lichen forming ascomycete) species known to occur in Egypt (http://eol.org/info/175). Since 2012, more than 260 pages have been published in EOL for the Egyptian taxa.

- *Third:* Dr. Abdel-Azeem and his team started campaigns to raise the awareness of fungal and environmental conservation among Egyptian schoolchildren. In 2011, Abdel-Azeem and Soliman surveyed 20 schools (400 students—online and onsite polls) to know if the students have any orientation on the topic of fungi. The results were frustrating; 86.4% of the surveyed students thought that fungi were micro-organisms and 0% correctly answered the question, "How many natural protectorates are there in Egypt?" Ironically, only 4.8% of students surveyed said they ever visited a protectorate in Egypt. The authors took the same survey to a sample of 40 journalists and the results were not much brighter (Abdel-Azeem and Soliman, 2011a).
- *Fourth:* Dr. Abdel-Azeem presented publications and provided translation of online articles, database websites on fungi, and fungal conservation publications (e.g., The Rio Convention and Fungi: a review of recent national biodiversity action plans and reports; (Abdel-Azeem and Minter, 2011); Saving the Forgotten Kingdom in Egypt (Abdel-Azeem, 2011); what will be the fate of the world's fungi? (Soliman and Abdel-Azeem, 2011); Arabic articles concerning conservation and translation into Arabic of "Changes to publication requirements made at the XVIII International Botanical Congress in Melbourne—what does e-publication mean for you?" (Abdel-Azeem and Soliman, 2011b, 2012); Biodiversity and Conservation of Fungi in Egypt (Soliman, 2011), botanists and zoologists fungal conservation needs you (Abdel-Azeem and Soliman, 2011c).

  As the Arabic editor of Cybertruffle's Robigalia website (http://www.cybertruffle.org.uk/robigalia/ara), Dr. Abdel-Azeem translated the database of the site and made it available for the Arabic speakers. The Cybertruffle's Robigalia provides a total number of 685,401 records of living organisms of all kinds. The conversion of the databases of fungi and associated organisms of Cybertruffle's Robigalia and Cybernome (http://www.cybertruffle.org.uk/cybernome/ara/index.htm) into Arabic language by Abdel-Azeem is a huge leap for Arab students, learners, researchers, educators, and amateurs mycologists (Abdel-Azeem et al., 2013).
- *Fifth:* Mycophobia ended when Dr. Abdel-Azeem introduced mycology in extracurricular activities at Port Said American

School, Egypt. High school students are totally engaged and have a fun time with science as they learn to research, isolate and identify indoor and outdoor fungi by deposition plate technique. Dr. Abdel-Azeem organizes school events to introduce leading fungi experts and brilliant international mycologists to school children. For Dr. Abdel-Azeem, blending the exciting world of mycology, fun and budding young scientists makes these events memorable (e.g., a visit by Professor Kari Steffen, Finland, to the Port Said American School, Egypt in 2011).

- *Sixth:* Dr. Abdel-Azeem raises national and international awareness on fungal conservation through organizing workshops in different Egyptian universities and research centers for scientists and for stakeholders and decision-makers. Training courses include more than 150 students on applying IUCN red-listing criteria for assessing and reporting on the conservation status of fungal species in Egypt. In addition, the recognized participation of undergraduate and postgraduate students in local international conference by their researches on Fungal conservation (16th Congress of European Mycologist, BioVision Alexandria 2012, 2014, TWAS-ARO 8th Annual Meeting 2012, Horizon 2020 Capacity building workshops (2011,2012), Eu-Jordan Networking in renewable Energy and Energy Efficiency (Jordan, 2013); third International Congress for Fungal Conservation (Turkey, 2013).
- *Seventh:* in 2013, Dr. Abdel-Azeem and his team at the University of Suez Canal, Botany Department, founded Arab Society for Fungal Conservation, the first society for fungal conservation in the Arab world and North of Africa (http://fungiofegypt.com/ASFC.html) (Abdel-Azeem and Salem, 2013b). It is as a private, nonprofit, voluntary, nongovernmental organization by the decree no. 699/2013 by the Ministry of Insurance and Social Affair in Egypt. In 2014, Arab Society for Fungal Conservation received a grant from http://www.speciesconservation.org/ to conserve and evaluate the current conservation status of 10 species of Egyptian Fungi.

In 2015, a protocol of cooperation between the Mycological Society of Romania and Arab Society for Fungal Conservation was signed in Iasi, Romania and in 2016 similar protocol signed with Korean Mycological Society and Arab Society for Fungal Conservation.

- *Eighth*: a good candidate for International Fungus Day published by Abdel-Azeem (2014) followed by the acceptation of the 20th of February 2016 as The Egyptian Fungus Day (http://www.fungiof-egypt.com/20Feb/).

11.4 STATUS OF FUNGAL CONSERVATION IN EGYPT

One of the most important fungal conservation steps is update your information concerning what are you going to conserve. How many species of fungi are there in Egypt? That is our important question! In the Fifth Egyptian National Report of CBD (https://www.cbd.int/doc/world/eg/eg-nr-05-en.pdf), Abdel-Azeem and Salem made an exhaustive revision of all the available literature and sources mentioned since 1813 and after the omission of duplicate names, name correction, allowance for synonyms and taxonomic assignments of all reported taxa from Egypt, the number of the Egyptian fungi recorded is 2422 taxa in comparison with El-Abyad (1997) and Abdel-Azeem (2010) who recorded 1246 and 2181 species respectively. On the kingdom level Fungi came first by 1964 species followed by Chromistan fungal analog (186) and Protozoan fungal analogs (61), respectively. The Ascomycota form the single largest group within this checklist, with about 1563 species, of which about 157 are lichen-forming, and about 1000 are known only from their conidial (asexual) states, 89 belonging to Chytridiomycota, 27 to Blastocladiomycota, 70 to Zygomycota, 47 to Glomeromycota, and 222 to Basidiomycota had been recorded up to the present time.

Threats to fungi throughout the globe are of concern since they are not only beautiful but also play a significant role in human welfare. Three steps were suggested by Moore et al. (2001) for fungal conservation: (1) conservation of habitats; (2) in situ conservation of non-mycological reserves/ecological niches; and (3) ex situ conservation especially for saprobic species growing in culture.

In Egypt, fungal diversity and impact of human activities are unknown and need of fungal conservation is ignored. Some threats can already clearly be identified: loss of habitat due to overgrazing, medicinal plants collection practices, climate change, human population growth, the vulnerability of Nile's river to pollution, bioprospecting, and the fragility of desert ecosystems. Other threats doubtless exist. Egypt is an important factor in explaining the wealth of fungi found there, but it is also an indication that mycology needs encouragement and funding so that systematic exploration can be carried out, inventories made, and checklists produced. In most cases, however, the scarcity of information about fungal populations makes conservation status evaluations beyond "data deficient" difficult or impossible.

Recently all the aforementioned steps were applied for conservation of fungi in Egypt through various activities they were: (1) by conserving the habitats of some national parks in Egypt (e.g., Saint Katherine Protectorate) through volunteers, (2) in situ conservation of desert plants reserves and ecological niches as microhabitats for fungi, (e.g., truffles awareness campaigns with local Bedouin tribes) and, (3) ex situ conservation for isolated saprobic taxa in a new Fungarium (culture collection to preserve different species of fungi) of Suez Canal University established by Abdel-Azeem et al.

One example of Egyptian fungi that needs conservation is desert truffles. Truffles undoubtedly, are of commercial importance, especially in North Africa (Trappe, 1971). The popularity of the highly-prized truffles is due to their constituents of essential nutrients. Bedouins usually use truffles as a substitute for meat. Due to Arab spring movements in three NA countries (Tunisia, Egypt, Libya) and their effects on the economic level of their populations, truffles were over-collected as a source of income (Abdel-Azeem and Salem, 2013b) and their threatened condition was studied (Abdel-Azeem and Salem 2013c). In order to maintain and improve their strategic importance, several conservation strategies, such as habitat preservation, are applied in situ (e.g. conservation of truffles though awareness campaigns with local Bedouin tribes have been carried by Abdel-Azeem's team in Sidi Barani desert).

11.5 CONCLUSIONS

Myco-entanglement and public perceptions of fungi in biodiversity conservation in NA especially in Egypt remains very low, and a lot of education is necessary. This work has to be considered as an important contribution to the nature conservation program in Egypt. One important theme is the sustainable use of fungal resources, enforcement of legislations and conventions concerning the fungal diversity in different habitats in Egypt. Environmental education is an important point in the activity concerning fungi conservation and application of appropriate protected area management systems. There is a need to integrate fungal diversity and conservation into the science curricula and extra-curricular activities of schools and colleges in Egypt. Egyptian mycologists have a responsibility to communicate these issues to public and politicians. This is difficult, as even the scientific community rarely acknowledges the true importance of fungi and their fundamental role in the conservation and protection of ecosystems. Obtaining the attention of politicians is even more difficult. Egyptian conservation legislation is strongly focused on protecting animals and plants and their habitats and fungi are still neglected. To solve such problems there should be collaboration between mycologists, amateur fungal groups, fungal conservation societies, regional natural parks, and environmental agencies.

KEYWORDS

- **climate changes**
- **desert**
- **fungal diversity**
- **fungarium**
- **truffles**

REFERENCES

Abdel-Azeem, A. M. The History, Fungal Biodiversity, Conservation, and Future Perspectives for Mycology in Egypt. *IMA Fungus* **2010,** *1* (2), 123–142.

Abdel-Azeem, A. M. Saving the Forgotten Kingdom in Egypt. *Fungal Conserv.* **2011,** *1,* 40–44.

Abdel-Azeem, A. M. A Good Candidate for International Fungus Day. *Fungal Conserv.* **2014,** *1* (4), 19–20.

Abdel-Azeem, A. M.; Minter, D. W. The Rio Convention and Fungi: A Review of Recent National Biodiversity Action Plans and Reports. *Fungal Conserv. 1,* 29–32.

Abdel-Azeem, A. M.; Salem, F. M. A Checklist of Egyptian Fungi: I. Protozoan Fungal Analogues. *Mycosphere*, **2013a,** *4* (4), 794–807.

Abdel-Azeem, A. M.; Salem, F. M. Arab Society for Fungal Conservation in Egypt. *Fungal Conserv.* **2013b,** *1* (3), 11–14.

Abdel-Azeem, A. M.; Salem, F. M. In *Fungal Conservation in Northern Africa and the Arab Society for Fungal Conservation.* Third International Congress on Fungal Conservation, Akyaka, Mugla, Turkey, 11–15 November 2013. Abstract book: 19, 2013c.

Abdel-Azeem, A. M.; Salem, F. M. Fungi Fimicola Aegyptiaci: I. Recent Investigations and Conservation in arid South Sinai. *Mycosphere* **2015,** *6* (2), 174–194.

Abdel-Azeem, A. M.; Soliman, G. S. Fungal Conservation and Encyclopedia of Life in Egypt. Sixteen Congress of European Mycologist, 19–23 September 2011, Porto Carras on the Sithonia peninsula, Greece. Abstract book: 64–66, 2011a.

Abdel-Azeem, A. M.; Soliman, G. S. Translation into Arabic of: Changes to Publication Requirements Made at the XVIII International Botanical Congress in Melbourne—What does e-publication Mean for You? *PhytoKeys* **2011b,** *7*, 63–68.

Abdel-Azeem, A. M.; Soliman, G. S. Translation intro Arabic: Botanists and Zoologists Fungal Conservation Needs You, 2011c. http://ahramag.com/modules/publisher/item.php?itemid=751 (accessed on February 5, 2016).

Abdel-Azeem, A. M.; Soliman, G. S. Translation into Arabic of: Changes to Publication Requirements Made at the XVIII International Botanical Congress in Melbourne – What Does e-publication Mean for You? *Mycokeys* **2012,** *2*, 45–50.

Abdel-Azeem, A. M.; Salem, F. M.; Mohamed, H. M.; Rashad, H. M.; Mohamed, R. M.; Khalil, W. F. Bioprospecting of Egyptian Fungi for Ecological and Medical Applications. TWAS-ARO 8th meeting, Bibilotheca Alexandrina, December 30–31, 2012.

Abdel-Azeem, A. M.; Mohamed, A. H.; Minter, D. W.; Salem, F. M.; Soliman, G. S. Fungi and Encyclopedia of Life: Abdel-Azeem's Team Contributions to English and Arabic EOL. First Young Reashersher Conference, Suez Canal University, Ismailia, 6–7 March 2013. Abstract book: 31, 2013.

Aldred, C. *Jewels of The Pharaoh: Egyptian Jewellery of the Dynastic Period*; Praeger Publishers: New York, 1971; PI. 68, p 210.

Boston Museum of Fine Arts *Egypt's Golden Age, the Art of Living in the New Kingdom, 1558–1085 B.C. Catalogue of the Exhibition*; Museum of Fine Arts: Boston, 1982.

Berlant, S. R. The Entheomycological Origin of Egyptian Crowns and the Esoteric Underpinnings of Egyptian religion. *J. Ethnopharmacol.* **2005,** *102* (2), 275–288. DOI: 10.1016/j.jep.2005.07.028.

El-Abyad, M. S. Biodiversity of Fungal Biota in Egypt. Up-dated Check-list. Publication of National Biodiversity Unit No. 7. Egyptian Environmental Affairs Agency, 1997.

El-Abyad, M. S.; Abu-Taleb, A. II-Soil Fungi. Ecology of Fungal Flora. In *Publication of National Biodiversity* Unit. No. 1. Habitat Diversity: 237–262; Egyptian Environmental Affairs Agency: Egypt, 1993.

Hawksworth, D. L. The Fungal Dimension of Biodiversity: Magnitude, Significance, and Conservation. *Mycol. Res.* **1991,** *95*, 641–655.

Hawksworth, D. L. Orphans in "botanical" Diversity. *Muelleria* **1997,** *10*, 111–123.

Hawksworth, D. L. The Magnitude of Fungal Diversity: The 1.5 Million Species Estimate Revisited. *Mycol. Res.* **2011,** *105*, 1422–1432.

http://blogs.nature.com/soapboxscience/2011/10/19/science-across-egypt-project (accessed on Feb 5, 2016).

http://eol.org/info/175 (accessed on Feb 5, 2016).

http://fungiofegypt.com/ASFC.html (accessed on Feb 5, 2016).
http://fungiofegypt.com/20Feb/ (accessed on Feb 5, 2016).
https://www.cbd.int/doc/world/eg/eg-nr-05-en.pdf (accessed on Feb 5, 2016).
http://www.cybertruffle.org.uk/robigalia/ara (accessed on Feb 5, 2016).
http://www.cybertruffle.org.uk/cybernome/ara/index.htm (accessed on Jan 15, 2016).
http://www.eeaa.gov.eg/English/main/protect_bio.asp (accessed on Nov 15, 2014).
http://www.fungal-conservation.org/micheli.htm (accessed on Feb 5, 2016).
http://www.speciesconservation.org (accessed on Feb 5, 2016).

Kirk, P.; Cannon, P. F.; Minter, D. W.; Stalpers, J. A. Ainsworth & Bisby's Dictionary of the Fungi, 10th ed.; CAB International: Wallingford, UK, 2008.

Leake, C. D. The Old Egyptian Medical Papyri; University of Kansas Press, 1952.

Minter, D. W. Conservation of Fungi: the Orphans of the Rio de Janeiro Convention. In: The First International Conference on Basic and applied Mycology, 9–11 March 2010, Assiut, Egypt, 2010; pp. 22–23.

Moore, D.; Nauta, M. M.; Evans, S. E.; Rotheroe, M. Fungal Conservation Issues: Recognizing the Problem, Finding Solutions. In: *Fungal Conservation: Issues and Solutions*; Moore, D., Nauta, M. M., Evans, S. E., Rotheroe, M. Eds.; Cambridge University Press: Cambridge, 2001; p 1–6.

Moustafa, A. F. Fungi of Egypt. Zygomycetes. AUMC Descriptions, No. 1. Assiut, 2006.

Moustafa, A. F.; Abdel-Azeem, A. M. An Annotated Check-list of Ascomycota Reported From Soil and Other Terricolous Substrates in Egypt. *J. Basic Appl. Mycol.* **2011,** *2*, 1–27.

Soliman, G. S. Biodiversity and Conservation of Fungi in Egypt (in Arabic), 2011. http://ahramag.com/modules/publisher/item.php?itemid=965 (accessed on Feb 5, 2016).

Soliman, G. S.; Abdel-Azeem, A. M. What Will be the Fate of the World's fungi? Earthsky (earthsky.org/biodiversity/what-will-be-the-fate-of-the-worlds-fungi), 2011 (accessed on Jan 31, 2016).

Trappe, J. M. A Synopsis of the Carbomycetaceae and Terfeziaceae (Tuberales). *Trans. Brit. Mycol. Soc.* **1971,** *57*, 85–92.

United Nations, 2016 http://unstats.un.org/unsd/methods/m49/m49regin.htm#africa (accessed on Feb 5, 2016).

United Nations List of Protected Areas, 2014. http://worldparkscongress.org/drupal/sites/default/files/documents/docs/2014%20UN%20List%20of%20Protected%20Areas%20EN%20web.pdf (accessed on Feb 5, 2016).

CHAPTER 12

Diversity of Functional Potential of Selected Medicinal Plants

TANZEEL ZOHRA[1], ALI AKBAR[2*], IMRAN ALI[3], HASNAIN NANGYAL[4], UZMA SITARA[5], AND MUHAMMAD ANWAR[3]

[1]*Department of Botany, PMAS- Arid Agriculture University Rawalpindi*

[2]*Department of Microbiology Faculty of Life Sciences University of Balochistan, Quetta*

[3]*Institute of Biochemistry, Faculty of Life Sciences University of Balochistan Quetta*

[4]*Department of Botany, Faculty of Life Sciences, Hazara University Mansehra Khyber Pakhtoonkhwa*

[5]*Food Quality Storage Research Institute, Southern-Zone Agricultural Research Centre, Pakistan Agricultural Research Council, Karachi-75270, Pakistan*

[*]*Corresponding author. E-mail: aliakbar.uob@gmail.com*

ABSTRACT

In this study, three plant extracts of medicinal importance (*Cannabis-sativa, Sonchusasper, Malvestrumcoromendalianum*) were analyzed for its phytochemical, antioxidant, and antibacterial potentials against multi-drug-resistant bacteria. The presence of bioactive compounds such as, saponins, tannins, alkaloids, and flavonoid was observed in its aqueous and ethanolic extracts. Tannins was absent in *Cannabis sativa*. Phenolic compounds and antioxidant compounds were confirmed using standard methods. The results showed higher antioxidant activity of the ethonolic extract compared to aqueous extract of all three plants. The same trend was noted for the total phenolic compounds investigated in all three extracts. Antibacterial potential of three plant extracts (aqueous and ethanolic) against multidrug-resistant bacteria (*Escherichia coli*, *Pseudomonas aeruginosa*, and *Staphylococcus aureus*) were evolved and were found active against different pathogens with different inhibition zones.

12.1 INTRODUCTION

Plant produces phytochemical compounds during its metabolic process, which are usually termed as secondary metabolites (Okwu, 2004; Mojab et al., 2003). Majority of such compounds are biologically active and exclusively present in medicinal plants. Such compounds are precursor for different valuable drugs used for the wellness of humanity (Sharanabasappa, 2007). It has been reported that more than 120 different bioactive compounds have been separated from plants. These compounds are widely used for medicinal purposes (Fabricant and Daniel, 2001).

The trace components present in phytochemicals and their interaction may change the drug response. The synergism of photochemical with drug has been studied by researchers that need clinical trials for in vivo investigation (Izhaki, 2002).

The synthetic pharmaceuticals have known toxicity compared to plants and herbs (Panchabhai, 2008). The research and development related to natural products and its chemotaxonomy are now making it more clear for common folks. The safe application of herbal chemical has increased the interest of biochemists and botanists in phytochemical investigations (Sharanabasappa, 2007). A large number and diversity of edible plants and vegetables are source of food and medicines worldwide (Suman, 2010; Sittiwet et al., 2008; Wahyuningrum et al., 2016).

Phytochemical components from plants are used for the treatment and risk reduction of different diseases such as gastric, systemic, and cardiovascular diseases and some other are best known for their anticancer potential (Sal and Hosseinzadeh, 2008). Scientific research and knowledge about the chemical compositions and mechanism of actions of plants phytochemicals has much more importance due to its emerging value as therapeutic agents and search for economically important chemicals. The phytochemicals such as flavonoids, saponins, tannins, gums, and oils are common precursors for the development of pharmaceutical substances used for the cure of different diseases. Investigation of the phytochemical components is helpful in determining the real medicinal value of folkloric remedies (Frankel, 1995). This research was performed to investigate some native plants for their phytochemical components, antioxidant and antimicrobial potential. In this study, we have analyzed the phytochemical, medicinal potential of three selected medicinal plants (*Cannabissativa, Sonchusasper*, and *Malvestrumcoromendalianum*).

12.2 MATERIALS AND METHODS

12.2.1 PLANT MATERIALS COLLECTION

Three different plants namely *M. coromendalianum*, *S. asper*, and *C. sativa* were collected for its phytochemicals antioxidant and antimicrobial potential studied because of their ethnobotanical importance.

The plants samples were collected from PMAS-Arid Agriculture University Rawalpindi and were identified by plant taxonomists from the department of botany, PMAS-Arid Agriculture University Rawalpindi.

12.2.1.1 EXTRACTION

The plant samples materials were cleaned with distilled water and dried in shade at room temperature for one week. Leaves of plant samples were carefully chopped after complete drying and grinded using electric grinder. The grinded powder of each sample was preserved in airtight glass bottles for phytochemical analysis.

12.2.1.2 PREPARATION OF EXTRACTS

A 5 g sample of each grinded plant was dissolved in 100 mL ethanol and distilled water, respectively. The solutions were mixed well and kept for 24 h. Aqueous solutions were boiled for 2 min and filtered through Whatman filter paper no.1. The extracts were then analyzed for the presence of different active group with the help of chemical analysis methods described by Ali (2009).

12.2.1.3 PHYTOCHEMICALS ANALYSIS

12.2.1.3.1 Test for Tannins

Few drops of ferric chloride (0.1%) was added in the plant filtrate (5 mL). The appearance of blue–black or brownish-green color was the indication of tannins presence in extracts.

12.2.1.3.2 Test for Saponins

An amount of 0.25 g crude plant materials was mixed well by vortex in 10 mL of distilled water in a clean test-tube. The test-tube was warmed in a water bath for 5 min and later shaken for observing the frothing. The appearance of the froth indicates saponins presence.

12.2.1.3.3 Test for Flavonoids

An amount of 2 mL plant samples extract was added to 4 mL of potassium hydroxide solution. Yellow coloration confirmed the flavonoids presence.

12.2.1.3.4 Test for Alkaloids

An amount of 2 mL concentrated hydrochloric acid was added to 2 mL of plant extracts, after mixing it, few drops of Mayer's reagent was added in the test tube. The appearance of white precipitate and green color confirmed the presence of alkaloids.

12.2.1.4 DETERMINATION OF THE TOTAL POLYPHENOLIC CONTENT

The presence of total polyphenols was determined by using Folin–Ciocalteu reagent following (Singleton et al., 1999). An amount of 1 mL plant extracts was transferred to a 25-mL volumetric flask containing 9 mL of ultra-pure water. Folin–Ciocalteu reagent (1 mL) was added and mixed properly with the help of vortex. Sodium carbonate (7%) 10 mL was added after 5 min incubation and the volume was completed with ultra-pure water. The mixture was incubated for 90 min at 23°C in dark, and absorbance was measured at 750 nm using spectrophotometer (UV–Vis lambda 40, Perkin Elmer, USA). The results were calculated by using a pre-prepared tannic acid calibration curve (0–100 mg/L). Blank was prepared using the same procedure with 1 mL of ultra-pure water replacing extract. The results are articulated comparing with the standard curve of tannic acid (mg TAN/g dry extract).

12.2.1.5 1, 1-DIPHENYL-2-PICRYLHYDRAZYL (DPPH) RADICAL SCAVENGING ASSAY

The antioxidant potential of the plants extract was evaluated using DPPH radical scavenging method. The DPPH solution (1 mL of 0.1 mM) was mixed with 0.5 mL of various concentrations (10–500 μg/mL) of plant extracts. The mixture was vortexed for a while and kept for 30 min at room temperature in dark place. The DPPH free radical reduction was measured with the help of spectrophotometer at OD value of 517 nm. The DPPH and methanol solution without extract was used as control. Ethanol or Methanol absolute will be taken as blank. The percentage of DPPH free radical activity was calculated by following formula (Hsu et al., 2007).

$$\% \text{ scavenging activity} = [(A_o - A_1) / A_o] \times 100$$

$$\text{Where } A_o = \text{control abs (DPPH sol)}$$

$$A_1 = \text{test sample}$$

12.2.1.6 ANTIBACTERIAL POTENTIAL DETERMINATION ASSAY

The ethanolic and water extract of three selected medicinal plants were evolved against different target bacteria for its antibacterial potential determination using agar well and disc diffusion methods following (Akbar et al., 2014). Fresh cultures of the target bacteria were prepared by inoculating bacteria to fresh nutrient broth and incubated for 24 h. In agar well method, the target bacteria lawn was prepared by spreading it over the surface of Mullar Hinton Agar (Himedia, India) and 6 mm well were made through sterile borer. The ethanolic extract (100 μL) priory dissolved in DMSO (2 mg/mL) were added in the well and incubated for

16–24 h at 37°C. Zone of inhibition was recorded in millimeter. In disc diffusion assay, the sterile filter paper disc (10 mm) was placed over the surface of target bacterial lawn on Muller Hinton Agar, 20 μL extract was poured on the dry filter paper and incubated for 16–24 h at 37°C. Zone of inhibition was recorded in millimeter.

12.3 RESULTS

The phytochemical screening of *M. coromandelianum, C. sativa*, and *S. asper* leaves verified the presence of different phytoconstituents in the plant materials. Major phytochemicals found were tannins, alkaloids, flavonoid, and saponins (Table 12.1). It was found, that flavonoids, alkaloids, saponins, and tannins are present in all selected plants, whereas presence of tannins was not confirmed in *C. satvia*.

TABLE 12.1 Preliminary Phytochemical Screening of Different Plants Species.

Chemical tests	Cannabis sativa	Sonchus asper	Malvestrum coromandelianum
Alkaloids	+	+++	++
Flavonoids	++	++	+
Saponins	+++	+	+
Tannins	–	++	++

Note: = + = Presence, – = Absence.

In this study, it was revealed that all three plants extract contained phenolic and antioxidant compounds. It was found, that the antioxidant activity of the ethanolic extracts was higher than that of aqueous extracts. The same trends were noted for total phenolic compounds in all three medicinal plants extracts. The total phenolic contents in *M. coromendial* was 0.75 g/g confirmed in ethanolic extract higher as compared to its aqueous extracts (0.35 g/g). Whereas, the polyphenolic compounds in *C. sativus* and *S. asper* were found as 0.70 g/g and 0.75 g/g in ethanolic extract respectively (Table 12.2).

The DPPH radical scavenging activity of *M. coromendial* was 85% in ethanolic extract higher (55%) than its aqueous extracts, whereas *C. sativus* and *S. asper* showed 73% and 62% radical scavenging activity higher compared to its aqueous extracts (60% and 59 %) respectively (Fig. 12.1).

TABLE 12.2 Polyphenolic Activity of the Extracts From Three Selected Medicinal Plants.

Name of plant	Total plolyphenolic content	
	Water extract	**Ethanolic extract**
Malvestrumcoromendial	0.35 g/g of extract	0.75 g/g of extract
Cannabis sativus	0.3 g/g of extract	0.70 g/g of extract
Sonchus asper	0.45 g/g of extract	0.75 g/g of extract

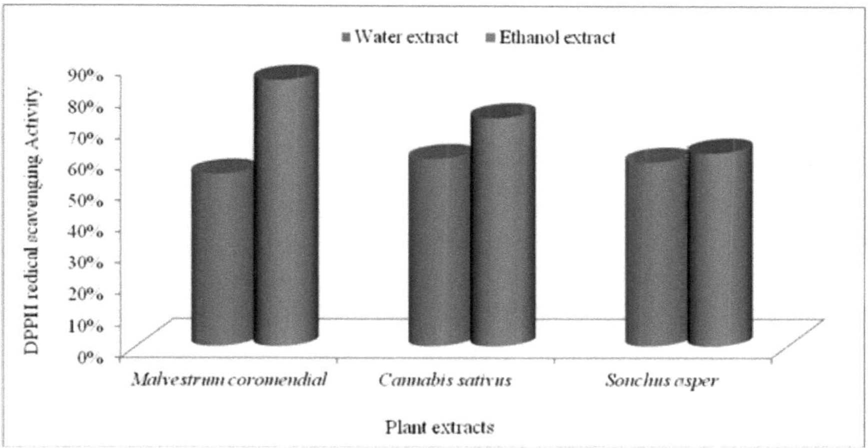

FIGURE 12.1 (See color insert.) DPPH radical scavenging activity of the three selected medicinal plants.

In this study, it was found that the extract from three different medicinal plants showed activities with different zone of inhibition against three different pathogens. The ethanolic extract from *M. coromendial* showed promising activity against all three target pathogenic bacteria, whereas its aqueous extract was found slightly active against *S. aureus* and with no activity against *E. coli* and *P. aerogenosa*. The ethanolic extract of *C. sativus* was found slightly active (3.3 ± 0.4 mm) against *P. aerogenosa* with no activity against *S. aureus* and *E. coli*. The aqueous extract of *C. sativus* showed good activity (22.5±2 mm) against *S. aureus* with no activity against *E. coli* and *P. aerogenosa*. The ethanolic and aqueous extracts of *S. asper* showed broad activity against all target pathogens (Table 12.3).

TABLE 12.3 Antibacterial Activity of the Extracts From Three Selected Medicinal Plants.

Target bacteria	Malvestrumcoromendial		Cannabissativus		Sonchusasper	
	EE	AE	EE	AE	EE	AE
S. aureus	24.3±1.3	2±1	0	22.5±2	33±2.5	35.5±1.5
E. coli	24.5±1.2	0	0	0	26±1.5	36±2.0
Pseudomonas aerogenosa	15.4±0.9	0	3.3±0.4	0	18±1.0	12.5±1.6

Note: The results are expressed in millimeter (mm) with Mean ± SD.

EE, ethanolic extracts; AE, aqueous extracts; *S. aureus*, *Staphylococcus aureus*; *E. coli*, *Escherichia coli*.

12.4 DISCUSSION

Alkaloids are the basic natural product present in plants that occurs primarily. It has an important position amongst the therapeutical plant substances. The pure alkaloids are useful medicinal agent as bactericidal, antispasmodic, and analgesic (Okwu, 2004). Alkaloids are antimicrobial in the form of salts with organic acids (Carr and Frei, 1999).

Flavonoids have known antioxidant potential with a radical scavenging properties having significant health value for human (Wahyuningrum et al., 2016). Some investigation suggested that it is biologically active due to its antioxidant potential (Gryglewski et al., 1987). It helps in the risk reduction of heart diseases and cancer activity and reduces the chances of tumor development (Okwu, 2004).

Saponins are group of glycosides with a soapy characteristic found in various plant species. It has also been reported with antifungal activities (Khan et al., 2011). Tannins are well-known antimicrobial agents, which are soluble in water. It is polyphenol in nature that precipitate the proteins found in many plant foods. Tannins have been reported with having various physiological and health effects such as antimicrobial, antiparasitic, and antiirritant effects. As *S. asper* is used in wounds healing due to high amounts of tannins in it. The plants containing tannin are useful in the treatment of intestinal disorder. It stops the increase of microbes by precipitating with microbial protein and making dietary proteins unavailable for them. It suppresses the growth of many fungi, yeasts, bacteria, and viruses (Chang et al., 2000).

Mavi et al. (2011) reported that the antioxidant activity of methanolic extract of *S. erzincanicus*. The same pattern of activity was confirmed in our study. Lone et al. (2012) reported antioxidant activity of *C. sativus*. Khan et al. (2012) reported antioxidant and polyphenolic activity of *S. asper* (L.) Hill.

Islam et al. (2010) reported good activity of ethanolic extracts from *M. coromendial* against different bacterial and fungal pathogens. Mushatq et al. (2012) reported antibacterial activity of chloroform extracts from *M. coromendial*. Lone et al. (2012) reported antibacterial activity of aqueous and acetone extract of *C. sativus* at concentrations (5 and 10 μg/mL) against *P. aeruginosa* and *Vibriocholera*. Xia et al. (2011) reported polyphenolic, antioxidant, and antibacterial activity of *S. asper*. They reported its antibacterial activity against Gram-negative (*E. coli*, *Salmonella enterica*, and *Vibrio parahaemolyticus*) and Gram-positive bacteria (*S. aureus*). There finding are in agreement to our study as in the current study *S. asper* showed good activity against three target bacterial pathogens in both aqueous and ethanolic extracts.

KEYWORDS

- ***Cannabissativa***
- ***Sonchusasper***
- ***Malvestrum coromendalianum***
- **multidrug-resistant bacteria**
- **biodiversity**

REFERENCES

Akbar, A.; Sitara, U.; Ali, I.; Muhammad, N.; Khan, S. A. Isolation and Characterization of Biotechnologically Potent *Micrococcus luteus* Strain From Environment. *Pak. J. Zool.* **2014,** *46* (4), 967–973.

Ali, R. A. Antimicrobial and Irritant Activities of the Extracts of Malvaparviflora, *Malvastrum coromandelianum* and *Amaranthus viridis*-A Preliminary Investigation. *Pak. J. Pharmacol.* **2009,** *1* (2), 20–23.

Carr, A. C.; Frei, B. Toward a New Recommended Dietary Allowance for Vitamin C Based on Antioxidant and Health Effects in Humans. *J. Clin. Nutr.* **1999,** *69* (6), 1086–1107.

Chang, Y. C.; Tai, K. W.; Huang, F. M. Cytotoxic and Nongenotoxic Effects of Phenolic Compounds in Human Pulp Cell Cultures. *J. Endodontics* **2000,** *26* (8), 440–443.

Fabricant, T.; Daniel, S. The Value of Plants Used in Traditional Medicines for Drug Discovery. *J. Pharm. Res.* **2001,** *109* (1), 69–75.

Frankel, E. Nutritional Benefits of Flavonoids. International Conference on Food Factors: Chemistry and Cancer Prevention, Hamamatsu, Japan. *Abstracts* **1995,** *6* (2), 33–55.

Gryglewski, R. J.; Korbut, R.; Robak, J. On the Mechanism of Antithrombotic Action of Flavonoids. *J. Biochem. Pharmacol.* **1987,** *36*, 317–321.

Hsu, C. Y.; Chan, Y. P.; Chang, J. Antioxidant Activity of Extract From Polygonum cuspidatum. *Biol. Res.* **2007,** *40*, 13–21.

Islam, M.; Ali, E.; Saeed, M. A.; Jamshaid, M.; Khan, M. T. J. Antimicrobial and Irritant Activities of the Extracts of *Malva parviflora* L., *Malvastrum coromandelianum* L. and *Amaranthus viridis* L.—A Preliminary Investigation. *Pak. J. Pharm.* **2010,** *20–23* (1&2), 3–6.

Izhaki, I. Emodin—A Secondary Metabolite With Multiple Ecological Functions in Higher Plants. *New Phytol.* **2002,** *155*, 205–217.

Khan, A.; Hussain, I.; Farooq, S. Phytochemical Screening of Some Pakistanian Medicinal Plants. *Middle-East J. Sci. Res.* **2011,** *8* (3), 575–578.

Khan, R. A.; Khan, M. R.; Sahreen, S.; Ahmed, M. *Evaluation of Phenolic Contents and Antioxidant Activity of Various Solvent Extracts of Sonchusasper (L.) Hill*; *Chem. Cent.* **2012,** *6* (1), 12.

Lone, T. A.; Lone, R. A. Extraction of Cannabinoids From Cannabissativa L Plant and its Potential Antimicrobial Activity. *Univ. J. Med. Dent.* **2012,** *1* (4), 051–055.

Mavi, A.; Yiğit, N.; Yiğit, D.; Kandemir, A. Antioxidant and Antimicrobial Activity of Turkish Endemic *Sonchuserzincanicus* Extracts. *Turk. J. Biol.* **2011,** *35*, 243–250.

Mojab, F.; Kamalinejad, M.; Naysanch, G.; Hamid, R. V. Phytochemical Screening of Some Species of Iranian Plants. *Iran. J. Pharm. Res.* **2003,** *2* (2), 77–82.

Mushatq, S.; Haider, M. S.; Ali, A.; Javed, S.; Khokhar, I.; Mukhtar, I. In Vitro Comparative Screening of Antibacterial and Antifungal Activities of Some Common Weeds Extracts. *Pak. J. Weed Sci. Res.* **2012,** *18* (1), 15–25.

Okwu, D. E. Phytochemicals and Vitamin Content of Indigenous Spices of Southeastern Nigeria. *J. Sustain. Agric. Envir.* **2004,** *6* (1), 30–37.

Panchabhai, T. S.; Kulkarni, U. P.; Rege, N. N. Validation of Therapeutic Claims of Tinosporacardifolia. *Phytother. Res.* **2008,** *22* (4), 425–441.

Sal, M. N.; Hosseinzadeh, H. Review of Pharmacological Effects of Glycyrrhiza sp. and its Bioactive Compounds. *Phytother. Res.* **2008,** *22* (6), 709–724.

Sharanabasappa, G. K. Phytochemical Studies on Bauhinia Racemosa Lam. Bauhinia Purpurea Linn. And Hardwickiabinata Roxb. *E-J. Chem.* **2007,** *4* (1), 21–31.

Singleton, V. L.; Orthofer, R.; Lamuela-Raventos, R. M. Analysis of Total Phenols and Other Oxidation Substrates and Antioxidants by Means of Folin–Ciocalteu Reagent. *Methods Enzymol.* **1999,** *299*, 152–178.

Sittiwet, C.; Jesadanont, S.; Pongpech, P.; Naenna, P.; Pongsamart S. Antibacterial Activity of *Malvastrum coromandelianum* Against Methicillin-sensitive and Methicillin-resistant Strains of *Staphylococcus aureus*. *Curr. Res. Bacteriol.* **2008,** *4* (1), 42–45.

Suman, S. Phytochemical Investigation of Sonchusoleraceus Leaves and Citrullus Colocynth Root. *J. Herbal Med. Toxicol.* **2010,** *4* (2), 159–162.
Wahyuningrum, R.; Utami, P. I.; Dhiani, B. A.; Kumalasari, M.; Kusumawardani, R. S. Screening of Potential Free Radicals Scavenger and Antibacterial Activities of Purwoceng (*Pimpinellaalpina Molk*). *Trop. Life Sci. Res.* **2016,** *27* (Supp. 1), 161–166.
Xia, D.-Z.; Yu, X.-F.; Zhu, Z.-Y.; Zou, Z.-D. Antioxidant and Antibacterial Activity of Six Edible Wild Plants (Sonchus spp.) in China. *Nat. Prod. Res. Form. Nat. Prod. Lett.* **2011,** *25* (20), 1893–1901.

CHAPTER 13

Fungal Biodiversity of Foods and Feeds

FATIMA ABDUL MAJEED and ALI AKBAR*

Department of Microbiology, University of Balochistan, Quetta 87300, Balochistan, Pakistan

Corresponding author. E-mail: aliakbar.uob@gmail.com

ABSTRACT

Fungi is ubiquitous in nature and is capable to grow in harsh conditions. They are important for health, environment, food production and spoilage. They can produce different form of mycotoxins and other related products harmful to human and animal health. The current chapter is focused on the fungal biodiversity of food and feeds and the deteriorating effects of these fungal species on the human and animals.

13.1 INTRODUCTION

Fungi are considered to be the most important organisms owing to their effect on humans and activities related to human beings. It has very important role in the functions of ecosystem (Mueller and Bills, 2004). Fungi are heterotrophic and eukaryotic organisms which require enzymes for breaking down biopolymers and for absorption of nutrients. Fungi show huge diversity which possesses members which have wide and different lifestyle, forms, sizes, and habitats. Fungi are related to animals (Fig. 13.1) which has numerous lineages which have interesting morphologies from mushroom to rusts, molds yeasts, and smuts. Fungi accomplish important role of ecology, likes decomposition and also responsible for different symbiotic associations including parasites (Alexopoulus, 1996).

FIGURE 13.1 (See color insert.) A basic tree of life shows the relations between three eukaryotic groups: fungi and animals are sister groups, with plants as their next closest relative.
Source: Adopted from http://tolweb.org/tree/.

After that lots of discoveries have been made and fungi are a group of organisms which are understudied. Recently it is estimated that about 1.5 million fungal species on the planet have been discovered (Hawksworth, 2001). Out of which only ~7% have been explained. Moreover, fungi live in the communities which are very rich in species, making fungal diversity to be documented a task which is a challenging task. Owing to the indefensible role of fungi in ecosystem documentation is combined with the perspective which are fungi that assists in the comprehension of fungi fast development and advancement in molecular techniques have helped in increase in study of fungal diversity and it also allows the further development in this particular field (Kirk et al., 2008).

13.2 WHAT ARE FUNGI?

Fungi are the microorganisms which are widely spread and they are nonphotosynthetic organisms with very important role in environment, especially, when organic matter are biodegraded in the biodegradation of organic material. Basically, fungi are heterotrophic organisms which are single-celled and have chitin in cell wall. Most fungi grow in multicellular colonies as compared with yeast which is single cell. Approximately, 70,000 species of fungi have been described, but almost 1.5 million species of fungi may exist in Universe (Hawksworth et al., 1995).

Fungi are very important for decomposition, nutrient transport, and nutrient cycling, and also responsible for gaining sustainable development (Palm and Chapela, 1998). Various species of fungi are plant and animal pathogens, which are very important, rest of them form obligate mutualistic obligate mutualistic symbioses with animals, plants, algae, and cyanobacteria. Fungi also possess a great importance economically, bearing both merits and demerits effects on human beings and their activities. Their usages include baking, brewing, pharmaceutical, fermentation industries, and biotechnical industries, on the other hand, some species are cultivated or collected as food uses. It can damage or cause the loss of millions of dollars by destruction of material used by humans, by causing diseases of animals and plants and by spoiling food, due to the abundance and diversity, and the role which it plays both nature and altered ecosystem and human activities (Hawksworth, 1991).

13.3 MORPHOLOGICAL CHARACTERISTICS OF THE FUNGI

Fungal cells are enclosed inside a rigid cell wall typically made up of chitin, chitosan, mannan, glycoprotein, and glucan in several mixtures. These structures of fungal cells are dissimilar from the animal's cell wall while the major component of plants cell wall is cellulose. As all together eukaryotic life forms, fungal cells have a true nucleus with a surrounding membrane, and cell division is accompanied by meiosis or mitosis.

Fungi are heterotrophic. Fungi lack chlorophyll and for their nutrition thus they require preformed organic compounds. Fungi stay alive in a fixed source of medium or food and get their nutrition by secreting enzyme into the exterior substrate and then absorbs the nutrients which are released by fungi cell wall. Fungi are found all through nature playing a vital role in exchanging to the soil nutrients extracted by plants.

Fungi can be multicellular or unicellular. In multicellular living beings, the essential structural unit named a hypha which is a multinucleate chain, tube-shaped, filament alike cells. In most multicellular fungi, the vegetative stage comprises a branching hypha, named a mycelium.

Every specific hypha cell wall is rigid and rises in measurement with mitotic cell division as a result of apical growth. In developed fungi, the hyphae persist lacking cross walls termed aseptate. Henceforth in some of groups, the hyphae are separated into sections of more cross-walls or less common cross walls, called septa. Therefore, such hyphae are referred to as being septate. Fungi which occur in minute multicellular mycelium form are normally termed molds.

Numerous fungi that exist as free single cells proliferate by budding out alike cells from their exterior. The bud becomes separated after the progenitor cell or may be attached and produce additional bud. Fungi that do not create hyphae, however, just comprise a free form of budding cells are named yeasts (Brandt and Warnock, 2011).

13.4 DISTRIBUTION AND ABUNDANCE

Fungi are either terrestrial or aquatic, living in marine or freshwater environments. Species in freshwater do not bear a high degree of salinity and are typically seen in clean, cool water. But certain species are seen in somewhat brackish water and a few flourishes in contaminated streams. On a solid surface of the earth, fungi are seen in all temperate and tropical areas of the biosphere where there is enough moistness to allow them to grow. Soil that is rich in organic matter provides a perfect environment for a variety of species. A small number of species of fungi live in the Arctic and Antarctic areas, while they are infrequent and found as a living symbiosis with algae in the form of lichens. Mycologists evaluate that there might be upwards of 1.5 million of fungal species on Earth out of which about 80,000 fungal species have been distinguished (Publishing and Rogers, 2010).

13.5 FUNGI SPORES

Reproduction by means of spores is a basis in fungi. However, a specific spore may have a small attainment chance of appropriate substrate, in such numbers spores may be formed that separate substrates and can be fragmented by the species. Just a couple of growths manage without spores, living, basically by mycelium. The spores are produced by means of sexual reproduction or asexual reproduction, which are involved in their dispersion and survival.

Fungi replicate by means of spores. From asexual method, conidium spores are formed which involves mitosis process only. But for irregular mutation, asexual conidia are alike to the parent. Several fungi are able by means of sexual reproduction which involves meiosis process. Certain species of fungi are self-fertile that are known as homothallic that form sexual structures among dissimilar cells inside an individual thallous. Furthermost, but some species do not form their sexual structures except two dissimilar isolates come into interaction and are known as heterothallic. In some species, the fungal spores are formed separately on specialized reproductive cells and the structure is microscopic. In some cases, though the fungal spores reproduced in macroscopic "fruiting bodies" like as mushrooms. Sexual reproduction and its accompanying structures from one scheme for classification of the fungi (Brandt and Warnock, 2011).

13.6 TAXONOMY OF FUNGI

Taxonomy is the discipline of classification, that is, the "assigning of objects to described groups" (Kirk et al., 2001). The Classification has three chief roles: it offers a background of familiar structures by which creature under examination can be recognized it is an effort to group organisms together that are related to each other; and it helps in the recovery of information about the recognized organism in the form of a list or catalog. At current, a different, more "natural" arrangement is start to income form, in which sequence data of DNA are compared with minute, ultrastructural characters and biochemical characters. Though, several fungal groups are quiet unwell distinct, and before a comprehensive taxonomic framework and various new trees will fall and that can be predictable to be in place (Brandt and Warnock, 2011).

13.7 CLASSIFICATION OF FUNGI

13.7.1 ASCOMYCOTA

Fungi in the phylum Ascomycota (sac fungi) normally form their generative structures ascospores, inside a sac named the ascus. Most of the nuclei of fungi, usually occur in haploid form. But in life cycle of Ascomycete at one period, by nuclear fusion diploid nuclei are formed. Inside the ascus these nuclei go through process of meiosis which is followed by a process of mitotic division and then difference into eight haploid ascospores.

On the basis of discolored wet form asci may be identified, their shape is spherical, and size of asci is normally 8–15 mm in diameter. Once maturity of asciis nearness of eight ascospores they full inside their walls. Hence, at the stage of maturity, the asci frequently break to discharge the ascospores wall which are thick, very refractile, and trimmed. Further two asci features are that they slowly get mature, afterward incubation for 10 days, and formed inside a larger macroscopic body, named ascocarp. Ascospores are very compressed, refractile, and resistant to chemicals, heat, and pressure. Besides ascospores, Ascomycetes typically produce asexual spores that are formed separately or in a chains afterward mitotic nuclear division. General named for this spore is conidium (plural, conidia). Absent of ascospores, conidial fungi are not typically resistant to heat, though conidia which may be chemical resistant (Pitt and Hocking, 2009).

13.7.2 BASIDIOMYCOTA

Phylum Basidiomycota is a large and diverse group with over 30,000 species. This phylum includes mushrooms, rusts, smuts, jelly, puff-balls, and stinkhorns. The club-shaped organ (basidium) bearing spore commonly form four sexual spores called basidiospores. Basidia are formed on fruiting bodies which is called basidiocarps. Basidiomycetes are most terrestrial, but certain grow in marine habitats or freshwater. Several basidiomycetes are saprotrophic which are involved decay of wood however, some are pathogens of trees. As saprotrophs, recycling of nutrients basidiomycetes plays an essential role (Smith et al., 1992).

13.7.3 CHYTRIDIOMYCOTA

Chytridiomycota is a phylum of fungi, includes above 900 species (Barr, 2001). Fungi that are included in this phylum are termed as "chytrids," members of which are well-known by having a single zoospore (motile cells). Species of Chytridiomycota are microscopic in size, and most are seen in freshwater or in wet soils. Some chytrids are parasites of animals and algae or live as saprobes on organic debris. Certain chytrids are biotrophic parasites of diatoms and filamentous algae which may harshly diminish the phytoplankton of freshwater inhabitants. Chytrids produce motile asexual zoospores in zoosporangia. Thallus of chytrids might be unicellular or filamentous.

A sexual reproduction by means of zygotic meiosis sometimes produces motile sexual spores called zoogametes (Hibbett et al., 2007).

13.7.4 ZYGOMYCOTA

The phylum Zygomycota lacks any motile stage. Zygomycota are distributed by wind, rain, and bugs. Sexual reproduction of Zygomycota is by means of gametangial copulation that results in zygospore formation. Their mycelial body is coenocytic and their cell wall is made up of chitin (Bartnicki-Garcia, 1987). Two classes Zygomycetes and Trichomycetes are compromised in Zygomycota. Zygomycetes including 870 species, 10 orders, whereas, Trichomycetes with 218 species in 3 orders (Kirk et al., 2001). Reproduction by means of sporangiospores, which is an asexually spore that are formed inside a sac. These sporangiospores are formed quickly. From the point view of food spoilage, the properties of Zygomycetes that they grow very fast, particularly in fresh foods with high activity of water, and Zygomycetes lack to heat and chemical resistance. Hence, from food safety point, it have been informed that Zygomycetes produce mycotoxins (Pitt and Hocking, 2009).

13.8 FOOD SPOILAGE BY FUNGI

Food spoilage is defined as "any sensory changes (flavor, visual, physical, or olfactory)" through which user consider to be undesirable. At any period of food chain spoilage of food may occur. Food spoilage can be caused by various factor including enzymatic activity in the plant tissue or animal, damages by insect, by physical damage, or by parasites, bacteria, and fungi that are the key causes of microbial spoilage (Burkepile et al., 2006). Currently, up to one-third food is spoiled or wasted before intake per year which shows about 1.3 billion tons. Results of these losses are problems that occur in the supply chain, from first agricultural making down to the level of consumer (FAO 2011). It is regarded that numerous mycotoxins produced by some fungal species are main risk to human health and animal (Shephard, 2008). In a large range of food yeasts and molds are capable to produce which includes uncooked supplies, for example, cereals, meat, fruits, vegetables, and milk, as well as processed foods (Filtenborg et al., 1996). Severe economic losses are by fungal spoilage for food manufactures. Actually, it has been estimated that worldwide among 5% and 10% production of food production is lost by fungal spoilage (Pitt, 2009).

13.9 FOOD SPOILAGE BY YEAST

Yeasts are usually single-cell microorganisms. Yeasts can grow in the presence or absence of oxygen that are known for their beneficial fermentations. They habitually colonize foods which contain a high sugar or salt concentration. Fruits and juices that are with low pH are also spoiled by yeast, and some grow on the meat and cheese surfaces. Four main groups of yeast that causes food spoilage are: *Zygosaccharomyces* bear high sugar and high salt concentrations and they are normal food spoilage organisms such as jams, honey, dried fruit, and soy sauce. Generally, growth is slowly, forms flavors and bad odors and carbon dioxide which may causes swell and burst to food containers. *Debaryomyces hansenii* grow at high salt concentrations as 24%, accounted for their common isolation from salt which are used to preserve olives, meats, and cheese (Gorski et al., 2006).

Saccharomyces species are well-recognized by their role in formation of fermented food such as bread and wine however, wines and alcoholic beverages are also spoiled by some strains by forming gassiness, turbidity and bad flavors linked with acetic acid and hydrogen sulfide. Some species are heat-resistant some grow on fruits (Elez-Martinez et al., 2004). While in fermented food species of *Dekkera/Brettanomyces* are main cause of spoilage, include some dairy products and alcoholic beverages (Couto et al., 2005).

13.10 FOOD SPOILAGE BY MOLDS

Molds are filamentous fungi. They play a role in recycle of dead plant and animal. Various foods and food materials which are beneficial to human beings are invaded by molds. Molds species are well known for their growth on solid substrates, normally form airborne spores, and for metabolic process they need oxygen. Furthermost molds they grow at pH ranging from 3 to 8 and certain may grow at low water activity. Most spores are sensitive to heat but some can bear severe environmental conditions. Mold species are able to grow in refrigerator different species have different optimal temperatures. Mold species have a various secondary metabolism produces toxic and carcinogenic mycotoxins. Several spoilages by molds are toxigenic whereas others are not (Pitt, 1997).

Molds that are the cause of spoilage can be characterized into four groups: Zygomycetes are widely spread in environment, grow fast on carbon sources in plant and soil debris, and spores of Zygomycetes are normally exist in indoor air. Usually Zygomycetes need high water activity for their growth and producing rots in various fruits and vegetables. Certain are used for formation of enzymes, fermented soy products, and organic chemicals. The common species of Zygomycota are *Mucor* and *Rhizopus*.

Zygomycetes not only produces mycotoxins however, several species produced toxic compounds which have been reported. *Penicillium* and its genera are considered main spoilage organisms and some produce potent mycotoxins. *Penicillium* spp. cause observable decays on apple fruits, pear, and citrus and causes huge damages in these harvests. Certain species also attack on refrigerated foods as well as in treated foods like jams and margarine. *Aspergillus* spp. and associated molds usually grow rapidly and high temperature resistant and low water activity as compared to *Penicillium* species and tend to lead spoilage in warmer weathers. Many *Aspergillus* species produces mycotoxins such as aflatoxins, territrems, and ochratoxin. A wide-ranging of food and nonfood materials like paper, leather, and so on that are spoiled by Aspergilli nevertheless they are maybe well known for spoilage of tree nuts, some spices grains, dried beans, and peanuts.

From spoiled food molds, belonging to numerous genera have been isolated. These normally are not the main reasons of spoilage, however might be problematic for some foods. Species of *Fusarium* are reason for diseases of plant and these species produce mycotoxins, however are not vital spoilage microorganisms. Though, mycotoxins produced by them might be occur in harvested of grains and possess a health threat to humans and animals (Boorand Fromm, 2006).

13.11 FUNGAL SPOILAGE IN CROPS

Usually on growing crops fungi have been divided into "field fungi" and "storage fungi." Plant pathogens are field fungi, before harvest these fungi invade on nut or the growing seed. Spoilage of a specific crop invasion is result from a specific fungus, for the reason that climatic conditions, agricultural practice or plant diversity produces conditions where invasion by these fungal species may arise on a large scales (Pitt and Hocking, 2009).

13.11.1 FUNGAL SPOILAGE IN WHEAT, BARLEY, AND OATS

From a comprehensive study of two year of the Scottish mycoflora of oats, wheat, and barley. Flannigan (1970) determined that cereal crops contamination of field was like. *Alternaria alternata* was the most generally arising fungus that was present on kernels above than 85%. *Alternaria alternata* causes in cereals downgrading because of gray discoloration and due to the mycotoxins production. While in barley and oats species of *Cladosporium* were found very common about 85–95% ingrains, but somewhat less 77% in wheat. Some other generally arising fungi were from species of *Epicoccum nigrum* and *Penicillium*. *Alternaria. alternate* arising more than 75% was dominant species on English barley. Authors have been reported that a large number of species in barley are species of *Penicillium*, mainly P. *hordei, P. verrucosum, P. roqueforti, P. aurantiogriseum, and P. piceum*, whereas with species of *Cladosporium cladosporioides*, *Epicoccum nigrum*, and *Aureobasidium pullulans* were common (Hill and Lacey, 1984).

Alternaria species were considered to be the most prevalent fungi from a two year of study in Slovakia of malting barley grains. Some other species were also included *Cochliobolus sativus* and *Fusarium avenaceum* species. Likewise, in *Argentine angrains* species of *Alternaria alternate* found to be the most common. In Egypt, after recently harvested wheat grains 77 species of fungi, from 26 genera were isolated which include 16 *Aspergillus* species and 21 Penicillium species. However, there was no symptom specified that these species are the reason of spoilage. The main species were *A. Niger* and *P. chrysogenum*. In North American, *Fusarium graminearum* species and *Fusariumpoae* were found most common in wheat. In Japan, red mold disease is termed as an important disease in barley and wheat is caused mainly by *Fusarium graminearum* (Pitt and Hocking, 2009).

13.11.2 SOYBEANS

The most usually isolated fungi in soybeans from Thailand, include species of *Eupenicillium semitectum, Macrophomina phaseolina, Chaetomium brasiliense, C. globosum, and Chaetomium. Funicola,* and *Cladosporium cladosporioides and C. sphaerospermum.* Soya beans were spoiled by weather which were infected with *Alternia alternata, Fusarium graminearum* species. In Ohio, soybean seedlings were spoiled with *F. graminearum.* In Argentinian, survey on soybean was reported that only 5% Fusarium species were present, the species which were common included *F. verticillioides, F. equiseti, and F. semitectum.* In Australian, soybeans common in weather damaged was from recently, harvested soybeans the source was from species of *F. verticillioides, A. ochraceus, A. flavus, F. semitectum,* and *F. semitectum.*

13.11.3 SORGHUM

In Southeast Asia, my coflora of sorghum was detected that was different from other commodities. Species of *Aspergillus flavus* was found at high levels as compare to those that occur in maize. The occurrence of species at high numbers was *Alternaria alternata, Curvulria lunata, A. longissima Dreighton,* C. pallescens, *Fusarium moniliforme, Lasiodiplodia theobromae,* and *Nigrospora oryzae* species which showed that sorghum is consider to be a host to a widespread of field fungi. In Brazil, species of *F. proliferatum, A. flavus,* and *Fusarium verticillioides* were recently isolated from harvested and stored sorghum (Pitt and Hocking, 2009).

13.11.4 CONCLUSION AND FUTURE PERSPECTIVES

Fungal contamination and damage to foods, feeds, and crops are a serious issue in both developed and developing countries. It has the capacity to hard agriculture economy and harm the overall agriculture productivity of the region. Control of fungal growth through safe practices and ecofriendly procedures can help in reduction of fungal contamination in crops, foods, and feeds.

ACKNOWLEDGMENT

The authors acknowledge the funding provided by USAID under the Agriculture Innovation Program through Pakistan Agriculture Research Council.

KEYWORDS

- **fungi**
- **biodiversity**
- **food**
- **feed**
- **spoilage**

REFERENCES

Alexopoulos, C. J.; Mims, C. W.; Blackwell, M. *Introductory Mycology*, 4th ed; John Wiley and Sons, Inc.: New York, USA, 1996.

Barr, D. J. S. Evolution and Kingdoms of Organisms From the Perspective of a Mycologist. *Mycologia* **1992,** *84*, 1–11.

Boor, K.; Fromm, H. *Food Spoilage Microorganisms*; Blackburn, C., Ed.; CRC Press LLC: Boca Raton FL, 2006; pp. 171–193.

Brandt, M. E.; Warnock, D. W. *Taxonomy and Classification of Fungi. Manual of Clinical Microbiology, 10th Ed.*; American Society of Microbiology, United state of America, 2011.

Brasier, C. M. Current Questions in Phytophthora Systematics: The Role of the Population Approach. In *Phytophthora*; Lucas, J. A., Shattock, R. C., Shaw, D. S., Cooke, L. R., Eds.; Cambridge University Press: Cambridge, 1991a; pp 104–128.

Burkepile, D. E.; Parker, J. D.; Woodson, C. B.; Mills, H. J.; Kubanek, J.; Sobecky, P. A.; Hay, H. E. *Ecology* **2006,** *87*, 2821–2831.

Caceres, C. E.; Tessier, A. J. How Long to Rest: The Ecology of Optimal Dormancy and Environmental Constraint. *Ecology* **2003,** *84*, 1189–1198.

Casey, G. D.; Dobson, A.D.W. *J. Appl. Microbiol.* **2003,** *95*, 13–22.

Couto, J. A.; Neves, F.; Campos, F.; Hogg, T. *Int. J. Food Microbiol.* **2005,** *104*, 337–344.

Elez-Martinez, P.; Escola-Hernandez, J.; Soliva-Fortuny, R. C.; Martin-Belloso, O. *J. Food Prot.* **2004,** *67*, 2596–2602.

FAO. Global FOOD Losses and Food Waste—Extent, Causes and Prevention: Rome, 2011. Available online: http://www.fao.org/docrep/014/mb060e/mb060e00.pdf (accessed on July 5, 2017).

Filtenborg, O.; Frisvad, J. C.; Thrame, U. Moulds in Food Spoilage. *Int. J. Food Microbiol.* **1996,** *33*, 85–102.

Gill, M.; Steglich, W. *Pigments of Fungi (Macromycetes)*; Springer-Verlag: Vienna, 1987.

Hawksworth, D. The Magnitude of Fungal Diversity: The 1.5 Million Species Estimate Revisited. *Mycol. Res.* **2001,** *12*, 1422–1432.

Hawksworth, D. L. The Fungal Dimension of Biodiversity: Magnitude, Significance, and Conservation. *Mycol. Res.* **1991,** *95*, 641–655.

Hawksworth, D. L. Fungal Diversity and its Implications for Genetic Resource Collections. *Stud. Mycol.* **2004,** *50*, 9–18.

Hawksworth, D. L.; Kirk, P.; Sutton, B. C.; Pegler, D. N. Ainsworth and Bisby's Dictionary of the Fungi. *Rev. Inst. Med. Trop. Sao Paulo* **1995,** *38*, 272.

Jones, S.; Lennon, J. Dormancy Contributes to the Maintenance of Microbial Diversity, 2009, www.pnas.org/cgi/doi/10.1073/pnas.0912765107.

Kirk, P. M.; Cannon, P. F.; Minter, D. W.; Stalpers, J. A. *Dictionary of the Fungi*, 10th ed.; CABI Publishing: Wallingford, UK, 2008.

Kirk, P. M.; Cannon, P. F.; David, J. C.; Stalpers, J. A., Eds. *Dictionary of the Fungi*, 9th ed.; CABI Publishing: Wallingford, UK, 2001.

Lang, Gregory I.; Murray, Andrew W. Estimating the Per-Base-Pair Mutation Rate in the Yeast *Saccharomyces cerevisiae*. *Genetics* **2007,** *178*, 67–82.

Mandrell, R. E.; Gorski, L.; Brandl, M. T. In *Microbiology of Fresh Fruits and Vegetables*; Sapers, G. M., Gorney, J. R., Yousef, A. E., Eds.; Taylor and Francis Group: New York, 2006; pp 33–73.

Mueller, G. M.; Bills, G. F. *Introduction*. In *Biodiversity of Fungi: Inventory and Monitoring Methods*; Mueller, G. M., Bills, G. F., Foster, M. S., Eds.; Elsevier Academic Press: San Diego, 2004; pp. 1–4.

Palm, M. E.; Chapela, I. H. *Mycology in Sustainable Development: Expanding Concepts, Vanishing Borders*; Parkway: Boone, North Carolina, 1998.

Pitt, J. I.; Hocking, A. D. *Fungi and Food Spoilage*; Springer: US, 2009.

Pitt, J. I.; Hocking, A. D. *Fungi and Food Spoilage*; Blackie Academic and Professional: New York, 1977.

Publishing, B. E.; Rogers, K. *Fungi, Algae, and Protists*; Rosen Publishing Group, Incorporated, 2010.

Rast, D. M.; Pfyffer, G. E. Acyclic Polyols and Higher Taxa of Fungi. *Botan. J. Linnean Soc.* **1989,** *99*, 39–57.

Russo, G.; Libkind, D.; Ulloa, R. J.; de García, V.; Sampaio, J. P.; van Broock, M. R. *Cryptococcus agrioensis* sp. nov., A Basidiomycetous Yeast of the Acidic Rock Drainage Ecoclade, Isolated From Acidic Aquatic Environments of Volcanic Origin (River Agrio, Argentina). *Int. J. Syst. Evol. Microbiol.* **2010,** *60*, 996–1000.

Shephard, G. S. Impact of Mycotoxins on Human Health in Developing Countries. *Food Addit. Contam. A* **2008,** *25*, 146–151.

Webster, J. *Introduction to Fungi*, 2nd ed.; Cambridge University Press: Cambridge, 1980.

CHAPTER 14

Biodiversity and Distribution of Algae in Egypt

HEBA S. ABBAS*

Microbiology Department, National Organization for Drug Control and Research (NODCAR), Giza, Egypt

**Corresponding author. E-mail: hebasalah18@gmail.com*

ABSTRACT

Over the last century, the research of phycology in Egypt is concerned with water resources especially the River Nile and its two main branches as the main source of freshwater and also some lakes, streams, and ponds. Nowadays the research is not restricted to water sources of Egypt but also arid and desert lands that are considered rich algae biodiversity habitat. Also, limited studies on the biodiversity of microalgae in the saltwater environment are reported. Thus, this chapter summarizes distribution and biodiversity of algae a in River Nile, Springs of Bahariya Oasis (Western Desert), Ain Helwan (near Cairo), Ain El-Sokhna (near Suez town), El-Temsah Lake (Ismailia town), Hammam Musa Faroun (South Sinai), WadiEl-Natrun Lakes (Western Desert), and Farafra Oasis. Besides, Mediterranean lakes, like Mariut, Edku, Manzala, Burullus, and Bardawil, are natural resources for fish production in Egypt and rich with phytoplankton. Even rice fields at El-Kharga Oasis are suitable places for algal growth. Also, it was found that there is a relationship between salinity or ions concentrations (sodium, chloride, sulfate, magnesium, and calcium) and algae biodiversity.

14.1 INTRODUCTION

Algae like plants are photosynthetic organisms that have the ability to absorb and store sunlight. The broad-based distribution of the algae at diverse geographic latitudes, on all continents, covering air, water, and land, is the consequence of wide adaptive variations. Algae play essential rule as a primary producer in the marine and freshwater food chain. Also, it can be used as an indicator organism for controlling ecosystems and for biotechnology.

The classification of algae depends on their size, habit and habitat, motility, pigment color (red, blue, green, etc.), food saving, and reproduction type. They can be prokaryotic unicellular/colonial forms; or Protista –eukaryotic unicellular/colonial forms or plant-eukaryotic multicellular forms (Fritsch, 1935; Whittaker, 1969; Sieburth, 1978; Van den Hoek et al., 1995) and phylogenetic tree of microalgae classes showed (Lee, 2008) in Figure 14.1. There are around 26,000 species of algae all over the world (Bold and Wynne, 1985; Stevenson, 1996). However, Norton et al. (1996) documented 37,300 species. In other studies, the number of diatoms is more than 1000 of 200 with planktonic, periphytonic, and benthic forms (Mann and Droop, 1996). On the other hand, there are many studies on a large group of seaweeds that live in the marine environment (macroalgae) belonging to the algal classes Chlorophyta, Phaeophyta, and Rhodophyta; although they have a lesser varied in comparison with the microscopic microalgae.

Microalgae are widely distributed and take many forms it can be unicellular, colonial, or filamentous, with different colors and size, with or without cellulose in their cell wall and several other modifications. They are important for the earth because they give the half percentage of atmospheric oxygen and use the greenhouse carbon dioxide because they are photoautotrophic (Lee, 2008). The biodiversity of algae is huge it needs more researches because the total number of species belonging to taxa may be varied due to varied habitat. Microalgae may be suspended in the water and it called planktonic or live in association with the substrate (benthic). They can be motile or nonmotile.

Planktonic may be macroplanktonic (>1 mm), microplankton (<1 mm), nanoplankton (5–50 μm), ultraplankton (5 μm), and picoplankton (>1 μm) (Gopinathan, 2004). Differences in species composition and diversity may cause changes in community-level parameters, like growth, change of plankton and the photosynthetic response to solar irradiance or others

(Duarte et al., 2006). Nutrient changes can be the indicator for marine biodiversity and competition between species. Planktonic microalgae have many modifications, like gas vacuole for cyanophyte, flagellation for movement, and so on (Tomaselli, 2004).

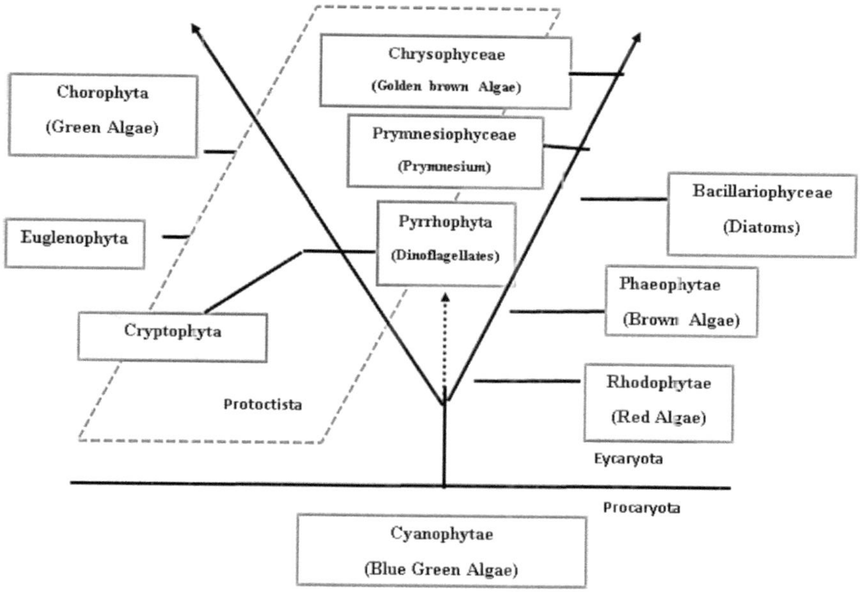

FIGURE 14.1 Phylogenetic tree of microalgae classes.
Source: Lee (2008).

On the other hand, benthic algae may be microscopic or macroscopic seaweeds that attach to rocks, corals, and sand particles. They are very small with an average size of about 62 μ (Underwood and Kromkamp, 1999).

The high increased in the numbers of algae may cause discoloration of water or red tides (algal blooms). Changing the color of the water into white, green, olive, brown, red, yellow, or even black, is an indicator of the existence of algal blooms (Richardson, 1997). Blooming algae is considered an energy producer at the base of food web but some are harmful algal blooms by toxin production. Two percent from 5000 species are known to be harmful or toxic (Landsberg, 2002).

Several factors influenced the existence of phytoplankton such as dissolved oxygen, salinity, pH, total hardness and nutrient level and other physical factors like temperature and light intensity (Reynolds, 1984). The increase of water temperature and nutrient in spring are both cause increasing of phytoplankton diversity while the increase of nutrients in autumn as a result of plankton disintegration in late summer and entrance of sewages causes a sharp increase in algae biodiversity (Nowrouzi and Valavi, 2010).

14.2 COLLECTION OF ALGAE FROM DIFFERENT LOCATIONS IN EGYPT

Effect of water quality in river Nile (Cairo, Egypt) on algae distribution was studied since 1976–1982 where three major phytoplankton groups (green algae, blue-green algae, and diatoms) were found. Green algae group represent the less dominant group. On other hand, diatoms group represent 42–96% from phytoplankton community while the percentage of blue-green algae group ranged from 0.7% to 48% of total cell number (Salwa et al., 1985).

Limited Studies on biodiversity of microalgae in the saltwater environment is reported compared with that of terrestrial flora. El-Naggar (1995) collected 32 species of Egyptian Marine algae of Chlorophycophyta, Phaeopycophyta, and Rhodophycophytae from the Mediterranean and red sea.

Other studies was focused on five important Mediterranean lakes in Egypt (Mariut, Edku, Burullus, and Manzala, and Bardawil) that are important sites for fish production (Shaltout and Khalil, 2005) and their algae biodiversity. The location map of the five northern lakes of Egypt is shown in Figure 14.2.

There are 867 species of algae flora are collected from five northern lakes of Egypt. Bacillarohyta (537 species) was the dominant group followed by Chlorophyta (141species) then Cyanphyta (104 species), Dinophyta (57 species), and Euglenophyta (19 species) as in Table 14.1.

It is clear that Bacillarophyta is the most diversified in Mariut Lake. Also, Manzala lake is the richest lake with phytoplankton (Bacillarophyta, Chlorophyta, and Cyanophyta) compared with other four lakes. Lake Baradawil comes after Mariut and Manzala lakes in the species diversity of phytoplankton. On the other hand, Edku lake showed the least diverse of phytoplankton species (Table 14.2) (Khairy et al., 2015).

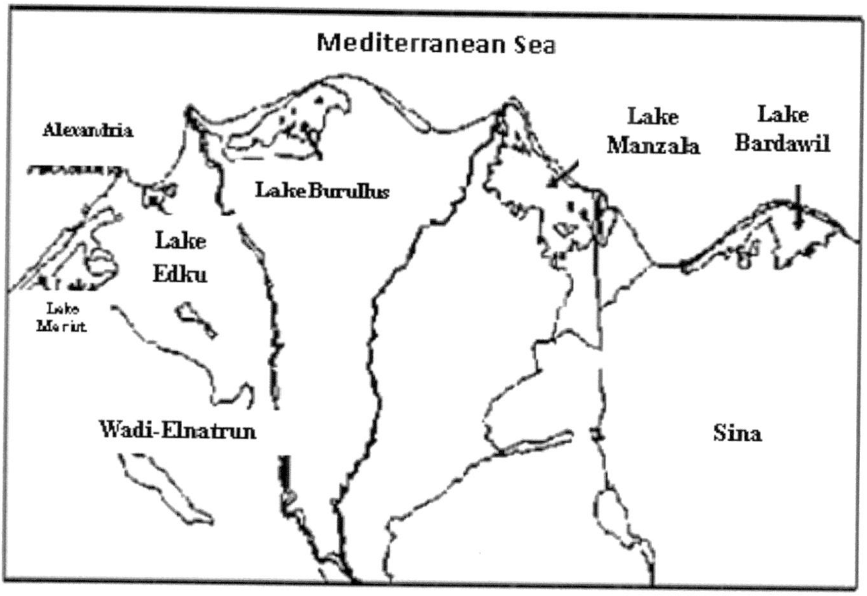

FIGURE 14.2 Location map of the five Egyptian northern lakes (after Shaltout and Galal, 2006).

TABLE 14.1 Taxa Diversity of Major Algal Divisions in the Five Egyptian Northern Lakes.

Divisions	Family (F)	Species (S)	Genus (G)	S/G	G/F
Bacillarophhyta	45	537	94	5.7	2.1
Chlorophyta	28	141	47	3	1.7
Cyanophyta	12	104	34	3	2.8
Euglenophyta	1	19	4	4.7	4
Dinophtyta	10	57	17	3.3	1.7
Chrysophyta	3	6	4	1.5	1.3
Crypophyta	1	1	1	0.0	0.0
Rhodophyta	1	1	1	0.0	0.0
Phaeophyta	1	1	1	0.0	0.0

Hamed (2008) collected 353 planktonic and benthic algae from natural water resources in Egypt including River Nile, Springs of Bahariya Oasis (Western Desert) Ain Helwan (near Cairo), Ain El-Sokhna (near Suez town) El-Temsah Lake (Ismailia town), Hammam Musa Faroun (South Sinai), and WadiEl-Natrun Lakes (Western Desert) as shown in Figure 14.3.

TABLE 14.2 Algal Species Richness in the Five Mediterranean Lakes of Egypt.

Divisions	Mariut	Edku	Burullus	Manzala	Bardawil	Total species
Bacillarophyta	255	87	126	253	238	537
Chlorophta	65	48	66	70	14	141
Cyanophyta	43	33	36	49	22	104
Dinophyta	1	2	7	4	53	57
Euglenophyta	10	10	11	7	1	19
Chrysophyta	1	10	0	0	5	6
Cryotophyta	0	0	1	0	0	1
Rhodophyta	0	1	0	0	0	1
Phaeophyta	1	1				
Total species	376	183	247	383	333	867

The maximum values of the algal division in relation to the five lakes are underlined.

Desert and arid habitats are poor in nutrients specifically C, N, and P (Housman et al., 2006; Mager and Thomas, 2011) and also rich by algae biodiversity habitat (Bohunická et al., 2015; Mareš et al., 2015). Algal flora of ambient and hot springs; in addition to other comparable habitats drilled wells" (e.g., Uher et al., 2001; Mohamed and Al-Shehri, 2008) was aim of several researchers (Cantonati et al., 2006, 2012a, 2012b,c, 2014, 2015; Spitale et al., 2012; Kaštovský et al., 2014), also in the arid lands (Aboal et al., 1998; Amarouche-Yala et al., 2014).

During the summer of 2011 and winter 2012, 82 algae were collected from 23 stations of Farafra Oasis wells and springs (West Desert, Egypt) as in Figure 14.4. Algae taxa lie under five algae divisions (cyanoprokaryotes, diatoms, xanthophytes, euglenoids, chlorophytes, and charophytes) from both benthic and water cultivated samples (Table 14.3). A total of 55.88% of these identified taxa were Cyanoprokaryotes, followed by chlorophytes (27.94%), with a little percentage of euglenoids, xanthophytes, diatoms, and only unique expression in winter 2012 for charophytes. These hyper-arid-desert dwelling water systems are excellent potable water resources according to the Egyptian Higher Committee for Water and World Health Organization (Shaaban et al., 2015).

Forty-seven soil algae species identified from the desert soils (Egypt) of El-Farafra OasisThese algae taxa belonging to four algal divisions: Cyanophyta followed by Chlorophyta (five taxa), Xanthophyta (five taxa) and Diatoms (only three species). The most dominant group was Cyanophyta (33 different species) belonging to eighteenth genera (Shaaban et al., 2017).

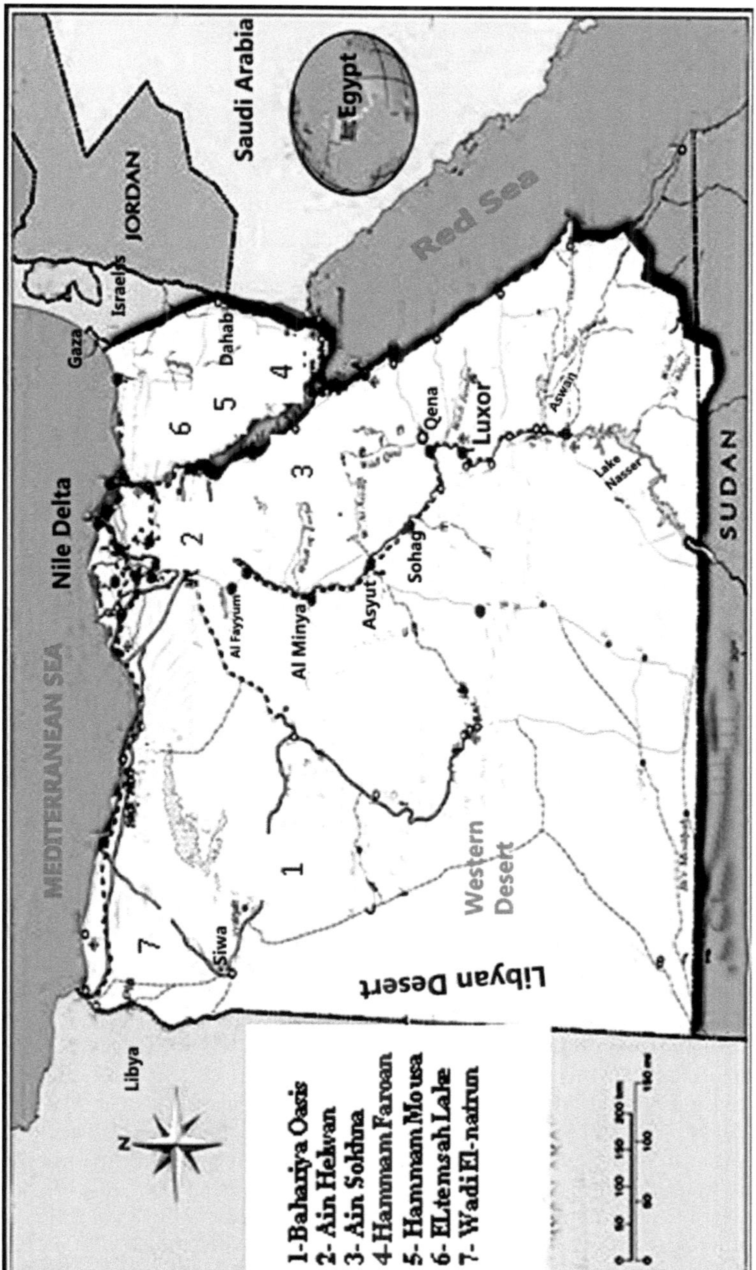

FIGURE 14.3 (See color insert.) Investigation locations for cyanobacteria and diatoms in Egypt.

Source: Hamed (2008).

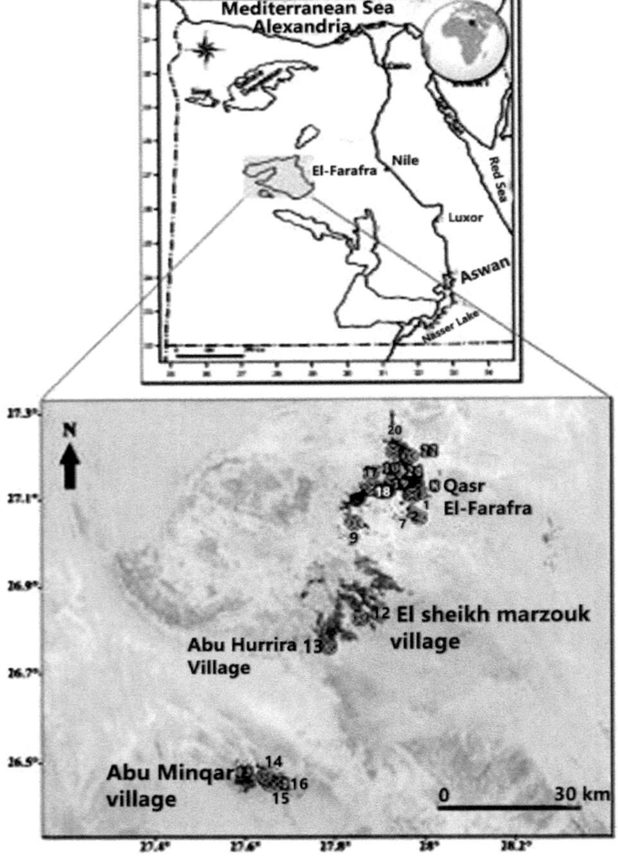

FIGURE 14.4 Stations of materials collection from El-Farafra Oasis, Western Desert (Egypt) (Shaaban et al., 2015). **1** = Ain El-Balad (Qasr El-Farafra), **2** = Agricultural ditch in close to Ain El-Balad (Qasr El-Farafra), **3** = Bir 1(Qasr El-Farafra), **4** = well of drinking water treatment plant, Qasr El-Farafra (before watertreatment), **5** = well of drinking water treatment plant, Qasr El-Farafra (after water treatment), **6** =Ain El-Hateyya (Qasr El-Farafra), **7** = Ain Bishwa (Qasr El-Farafra), **8** = Bir Gelaw (Qasr El-Farafra), **9** = Ain Goshna (Qasr El-Farafra), **10** = Bir El-Hurra (Qasr El-Farafra), **11**= Bir Sitta (QasrEl-Farafra), **12** = Bir El-Sheikh Marzouk (El-Sheikh Marzouk village), **13** = Bir 51 (Abu Hurriravillage), **14** = Bir 1A (Abu Minqar village), **15** = Bir 1B (Abu Minqar village), **16** = Bir 9 (AbuMinqar village), **17** = well of drinking water treatment plant, Abu Minqar village (before water treatment), **18** = well of drinking water treatment plant, Abu Minqar village (after water treatment), **19** = Bir 7(Lewa Soubah village), **20** = agricultural drainage (Lewa Soubah village), **21** = Lewa Soubah lake (Lewa Soubah village), **22** = Bir 4 "Bir El-Gazwarin" (El-Nahda village), **23** =Abu Nuss lake (El-Nahda village), **24** = Bir 3A (Esha Abd El-Rahman village).

TABLE 14.3 Naturally propagated Algal Assemblages of Different Watercourses (Drilled Wells, Ambient-to-Thermal and Slightly Hot Springs, DWTPs, Agricultural Ditch and Drainage, and Lakes).

	Seasons	1	2	3	4	5	6	7	8	9	10	11	12	13	14	15	16	17	18	19	20	21	22	23	24
Cyanoprokaryota	Summer 2011	14	15	5	-	-	7	1	3	1	4	7	4	2	6	8	2	-	-	5	3	4	3	4	4
	Winter 2012	9	3	6	2	1	6	1	3	1	5	15	5	9	5	6	5	2	2	2	2	4	6	6	4
Bacillarophyta	Summer 2011	1	3	-	-	-	-	-	-	-	-	-	-	-	-	-	-	-	-	-	-	-	-	-	-
	Winter 2012	3	3	-	-	-	-	-	-	-	-	-	-	-	-	-	-	-	-	-	-	-	-	-	-
Xanthophyta	Summer 2011	-	-	-	-	-	-	-	-	-	-	-	-	-	-	-	-	-	-	-	-	-	-	-	-
	Winter 2012	-	3	-	-	-	-	-	-	-	-	-	-	-	-	-	-	-	-	-	-	-	-	-	-
Euglenophyta	Summer 2011	-	-	-	-	-	-	-	-	-	-	-	-	-	-	-	-	-	-	-	4	2	-	-	-
	Winter 2012	-	-	-	-	-	-	-	-	-	-	-	-	-	-	-	-	-	-	-	-	1	-	2	-
Chloropyhta	Summer 2011	-	10	-	-	-	-	-	-	-	-	-	-	-	-	-	-	-	-	-	1	1	-	2	1
	Winter 2012		4	-	-	-	-	-	-	-	-	5	-	-	1	-	-	-	-	-	2	2	1	1	1
Charophyta	Summer 2011	-	-	-	-	-	-	-	-	-	-	-	-	-	-	-	-	-	-	-	-	-	-	-	-
	Winter 2012	-	1	-	-	-	-	-	-	-	-	-	-	-	-	-	-	-	-	-	-	-	-	-	-
Total algal taxa													68												

Several studies on the algal flora showed the importance of cyanophytes for nitrogen fixation (Brammer, 1983; Hung and Chow, 1988). Cyanophytes or blue–green algae have a crucial role in soil fertility in rice fields especially when these fields are placed in nitrogen-poor desert soils (McGregor and Johnson, 1971).

Rice fields of El-Kharga Oasis (three sites) were investigated for algae flora distribution as in Figure 14.5. The algae samples were collected from good illuminated places during the period from 1998 to 1999 (cultivation season). A total of 52 bacillarophytes, 23 cyanophytes, 19 chlorophytes, and 5 euglenophytes were recorded. In this study, blue-green algae were subdominant compared while the study of Pandy, 1965 showed that the blue-green algae were the most dominant because in this study both Planktonic and sessile algae here included. Blue-green algae present in numbers, including heterocystous and nitrogen fixers like *Anabaena variabilis* and *Cylindrospermum alatosporum*. Also, the appearance of green algae in the form of desmids was noticed in spite of hypothesis the acidophilic nature of these organisms. The water pH was alkaline (above 8) throughout the period of study (El-Naghy et al., 2004).

Also, benthic macroalgae were reported in Port Said coast which is one of the Egyptian coasts in the Mediterranean Sea (Aleem, 1993). Four sites form Portsaid were investigated for algae diversity where Site I was at the east corner near El-Tafriaa (absent from any effluent), Site II is near the entrance of the Suez Canal (affected by different human activities), Site III is located at El-Gameel, and Site IV present at the entrance of Lake Manzala and exposed to the brackish water discharges from Lake Manzala. A total of 55 species of benthic macroalgae from Port Said sites were collected during four successive seasons from summer 2004 to spring 2005 as in Figure 14.6. The dominant group was Rodophyta that represent 56.4% from algae diversity. Beside that Chlorophyta represents 38.2% with 21 species while Phaeophyta with their three species can be neglected.

The persistence of macroalgae species was noticed due to man-made modifications in Port Said coast during the last decades or Lessepsian immigration. These modifications include rocks pavements and the new eastern branch in the northern Suez Canal (Madkour and El-Shoubaky, 2007).

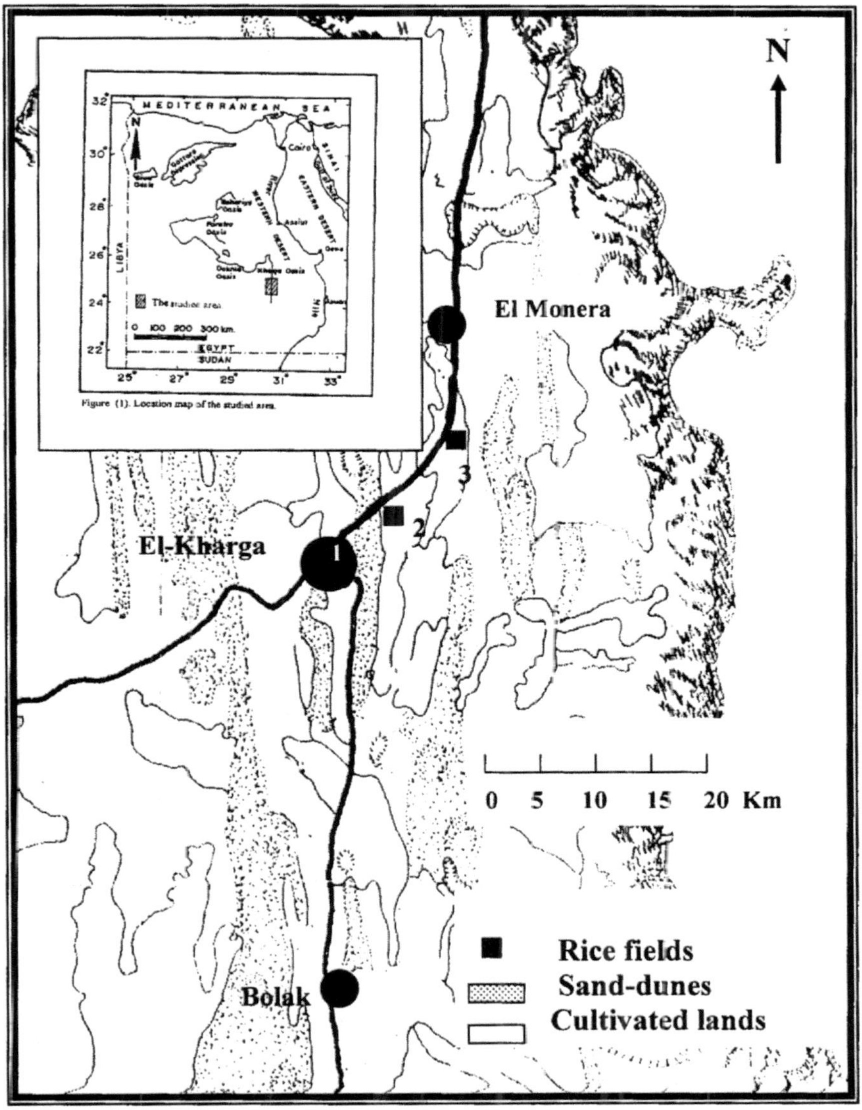

FIGURE 14.5 Map showing the sampling sites of rice fields in El-Kharga oasis.

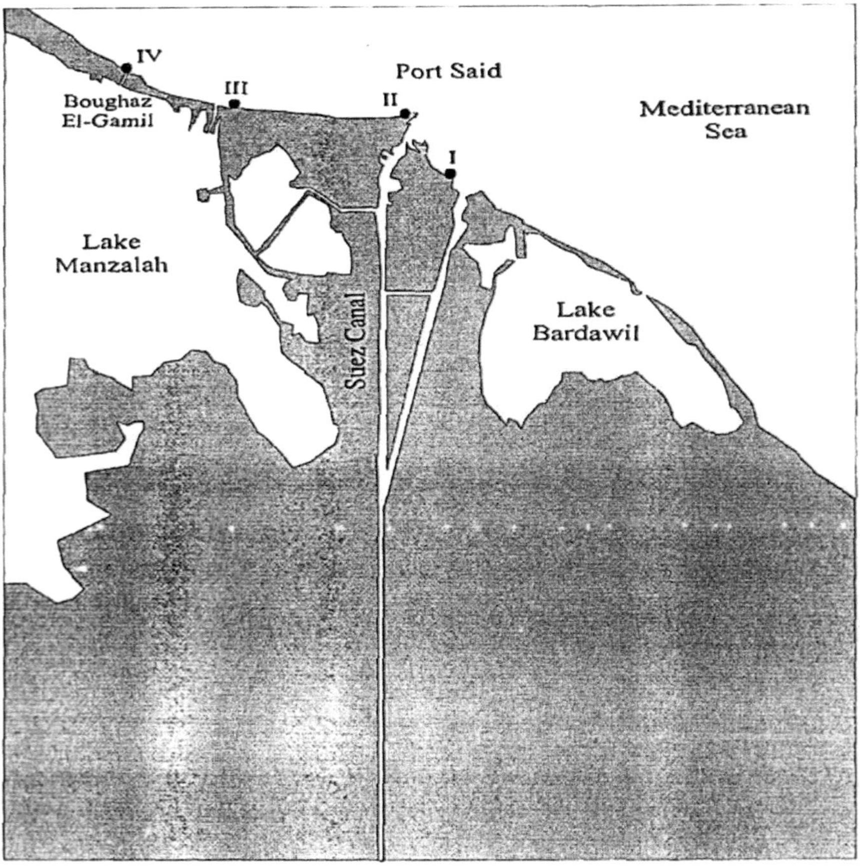

FIGURE 14.6 Map of the selected site in Port Said coast.

14.3 INFLUENCE OF WATER CONDUCTIVITY AND ION CONCENTRATIONS ON ALGAE DISTRIBUTION

It is well known that conductivity, due to salinity, and ions have a direct relationship on distributions of individual cyanoprokaryotes (Sigee, 2005) and diatom taxa (Cholonky, 1968).

Water conductivity ranges from 60 to 1,213,300 μScm^{-1} and different ions, like chloride, sulfate, and sodium, have a strong influence on the distribution of Cyanophyta and diatoms. Sodium was the dominant cation in all investigated sites except some localities, such as River Nile, Ain El-Mahabes (Bahariya Oasis), and Ain El-Hobga (Bahariya Oasis) where Mg^{+2}/Ca^{+2} were the most dominant cations. The highest value of water conductivity due to increasing the concentration of of sodium and chloride. The water conductivity of different regions influenced by marine water, geological and climate conditions. The study showed that the most productive habitat was saline water followed by freshwater then hypersaline and brackish waters as in Tables 14.4 and 14.5 (Hamed, 2008).

TABLE 14.4 Water Conductivity of Different Locations in Egypt

Locality	Conductivity μScm^4	Water habitat type
River Nile	60	Freshwater habitats
Ain Halfa (Bahariya Oasis)	592	
Ain El-Goag (Bahariya Oasis)	756	
Ain El-Ramla (Bahariya Oasis)	826	
Ain El-Mahabes (Bahariya Oasis)	904	
Ain El-Nibika (Bahariya Oasis)	982	
Ain El-Hobga (Bahariya Oasis)	1028	
Ain Helwan	8091	Brackish water habitats
Hammam Musa (South Sinai)	10,164	
Ain El-Sokhna	15,064	Saline water habitats
El-sabkha Lake (Wadi El-Natrun)	29,734	
Hammam Faroun (South Sunai)	32,332	
El-Fasda Lake (Wadi El-Natrun)	43,006	
El-Tamsah Lake (Suez Canal)	43,500	
El-Khadra Lake (Wadi El-Natrun)	62,204	
El-Hamra Lake (Wadi El-Natrun)	447,933	Hypersaline habitats
El-zaagig Lake (Wadi El-Natrun)	529,852	
El-Gaar Lake (Wadi El-Natrun)	121,330	

Source: Reprinted from Hamed (2008).

The highest value of iodine was recorded in the Mediterranean in brown species (Phaeopycophyta) compared with red and green species (Rhodophycophytae and Chlorophycophyta) as in Table 14.6 (El-Naggar, 1995).

TABLE 14.5 Concentration of Major Ions in the Investigated Localities in ppm.

Parameters	1	2	3	4	5	6	7	8	9	10	11	12	13	14	15	16	17	18
TDS	36	296	378	413	452	464	514	4045	5082	7532	14,867	16,166	21,503	21,705	3112	22,3966	26,3926	60,6650
K^+	–	14.8	22.6	19.5	13.2	25.7	12.5	12.5	46.4	4921	345	148.4	342	469.8	463	559	1545	2288
Na^+	–	48.7	54.9	49.6	30.5	47.7	28.5	1011.4	2486.8	1748.9	4703	4443.6	6506	13,800	10057	84,662	93,264	93,264
Mg^{1+}	8	36.8	38.6	21.1	55.8	36.4	45.8	150.6	512.8	321.1	34	526.3	283.8	1392	243.3	1873	431.8	104.6
Ca^{3+}	6	40	40	51.1	51.1	328	67.6	303	310.5	418.8	45.7	6443	309	340	50.6	24.8	32	48
Cl	14	95.7	79.4	111.7	99.2	79.7	63.8	1388.2	3620.5	3545.4	2269	7659.5	5177	24,140	6693	60,992	72,694	95,071
SO_4^-	–	57.6	153.8	57.6	173.5	111.5	245.6	1214.9	3081.7	988.9	64,118	25,242	8247	2680	11,568	84,968	88,348	95,662
HCO_4^-	–	36.5	91.41	121	18.2	73.1	365	134.1	91.4	1835	829	60.9	488	244	1493	2927	2406	2024
CO_4^-	–	72	36	72	57	42	36	156.1	43.8	0.0	22	808.1	150	150	6201	5418	6201	2781
PO_4^-	0.21	4.7	4.7	4.76	4.76	4.7	4.76	9	0.95	2.38	3.4	95	1.88	0.153	4.6	29	1.66	
NO_4^+	0.28	0.6	1.2	`0.0	0.6	12	0.6	3.7	62	18.6	19.6	18	5.6	113	30.5	13	20.4	21

Source: Reprinted from Hamed (2008)

TABLE 14.6 The Mean Value of Iodine (%) in the Algal Divisions Occurring in the Mediterranean and the Red Sea.

Location	Algal division		
	Chlorophycophyta	Phaeophycophyta	Rhodophycophyta
Mediterranean Sea	0.0471±0.009	0.1433±0.034	0.0770±0.025
Red Sea	0.017±0.004	0.1063±0.021	0.0545±0.025

The iodine concentration of the brown species ranged from 0.069% to 0.203%. *Dictyota divaricata* and *Sargassum subrepandum* were characterized by having a high percentage of iodine (El-Naggar, 1995) as in Table 14.7.

TABLE 14.7 Concentrations of Iodine (% of dry wt) in Marine Algae of the Red Sea.

Algal species	Site of collection	Concentration (%)
Chlorophycophyta		
Entermorpha clathrate	Ghardaka	0.0045
Ulva lactuca L.	Ghardaka	0.0057
Caulerpa serrulate (Forsk) Borg.	Abushaar	0.0230
Halimeda opuntia (L.) Lamour	Safaga	0 0191
Cladophora heteronema (C.Ag) Kutz	Safaga	0 0060
Phaeophycophyta		
Dictyota divaricate Lamour	Ghardaka	0.203
Padina boryana Thivy	Safaga	0.079
Hydroclathrus clathratus (C. Ag) Howe	Ghardaka	0.069
Hormophysa triquetra (C. Ag) Kutz.	Ghardaka	0.076
Turbinaria triquetra (J. Ag) Ag.	Ghardaka	0.084
Sargassum subrepandum	Safaga	0.127
Rhodophycophyta		
Actanthophora Najadifomis (Delile) Papenfuss	Sharam El Naga	0.053
Digenea simplex (Wulfen) C. Ag.	Ghardaka	0.116
Luarencia Obtusa (Huds.) Lamour.	Safaga	0.017
Hypnea musciformis (Wulfen) Lamour.	Safaga	0.032

Manzala Lake is high in chloride concentration with low nutrient concentration during winter and spring that makes it rich with diatoms while Chlorophytes appear in autumn. Euglenophytes were present in

January depending on inorganic sources of nitrogen and phosphorus (El-Sherif and Gharib, 2001). Besides, Cyanophytes reached its maximum in July linked with high concentrations of ammonia (Tilman et al., 1986). The hypersaline nature of Bardawil Lake, a lower level of water pollution and the lower human impact was the reason for a high number of unique species (Khairy et al., 2015). On the other hand, the persistence of Euglenophyte in Edku Lake during autumn is considered an indicator of organic pollution (Munawar, 1972; Khairy et al., 2015). Besides, the persistence of bacillarophyta was recorded due to high pH and low values of nutrients (El-Shinawy, 2002). Chlorophyta, especially *Scendesmus* genus, developed in Burulus Lake near high level of organic matter and phosphate (Radwan, 2002) also the diatoms especially Cyclotella genus persist due to high load of organic matter discharges with water drainage (Abdallah et al., 1991) while cyanophtes are rarely recorded in this lake where only Oscillatoria and Gleocapsa may be present due to high load of domestic sewage and nutrient salts discharged water (Radwan, 2002).

Also increasing the contents of minerals, cations, and anions specifically nitrate, nitrite, and soluble reactive phosphorous as an indicator for contamination was discussed (Macioszczyk and Dobrzyński, 2002). It was shown the presence of algae species with some taxa that indicates water pollution and eutrophication (Shaaban et al., 2015). As *Oscillatoria limosa, O. princeps,* and *O. tenuis* from Cyanophyta associated with high accessibility of nitrates (Cantonati et al., 1996; Nowicka-Krawczyk and Żelazna-Wieczorek, 2013). Oligosaprobous shapes of algal flora in El-Farafra Oasis (e.g., *Vaucheria geminata* and *V. sessilis*) with presence of some species tolarate β-mesosaprobic, like *Oscillatoria limosa, O. princeps, Euglena acus,* and *Coelastrum microporum*, and a little presence of α-mesosaprobic biomonitors (e.g., *Oscillatoria tenuis* and *Euglena oxyuris*) with rarely polysaprobic characters of *Anabaena constricta* was detected (Shaaban et al., 2015).

14.4 IDENTIFICATION OF ALGAE FROM DIFFERENT LOCATIONS IN EGYPT

A total of 164 Cyanobacteria taxa were identified from different water habitat in Egypt, including River Nile, Springs of Bahariya Oasis (Western Desert), Ain Helwan (near Cairo), Ain El-Sokhna (near Suez town),

El-Temsah Lake (Ismailia town), Hammam Musa Faroun (South Sinai), and WadiEl-Natrun Lakes (Western Desert) that take different forms as coccoid species as *Synechococcus aeruginosus*, *Synechocystis sallensis*, *Synechocystis salina*, and *Synechocystis pevalekii* (plate I.1, 2, 3, and 4) or colonies arranged in rows as *Gloeocapsa atrata*, *Gloeocapsa decorticans*, *Gloeocapsa lacustris*, *Gloeocapsa minor*, and *Gloeocapsa turgida* (plate I.5, I.6, I.7, I.8, and I.9), or permanently spiral coiled *Spirulina gigantea*, *Spirulina labyrinthiformis*, *Spirulina platensis*, and *Spirulina subtilissima* as in plate II, or straight form as *Oscillatoria annae, O. curviceps,* and *O.limosa* (plate III.1). Some filamentous species are characterized by true cell differentiation and form heterocystes with unlike vegetative cells as *Anabaena affinis* (plate III.2.), *Anabaena bergif. minor, Anabaena inaequalis, Anabaena iyengarii, Anabaena laxa* and *Nodularia spumigena.* Other heterocystous Cyanobacteria also form a second type, an akinate, which can germinate for growth in favorable conditions as *Nostoc sp.* as plate III.3. 189 diatom taxa were identified. Eighteenth species belonged to class Centrophyceae, while the rest of pennatophycean forms. The frustules of distinct centric diatom species showed three circular undulations as *Terpsinoe americana* (Plate IV.1 and 2). Valves of the pennate diatoms are generally elongated (linear) with their markings arranged in transverse rows along each margin. Frustule may be linear, solitary with striae forming pseudoraphe on both valves as *Synedra ulna* and *Synedra tabulate* (Plate IV.3 and 4). In *Navicula cryptocephala*, *Navicula cryptocephala var. veneta*, *Naviculacuspidate*, *Navicula heufleri var. leptocephali,* and *Navicula halophila f. tenuirostris* the valves may be symmetrical about the transapical plane and asymmetrical about the apical plane (Plate V.1, 2, 3, 4, and 5). Other pennate diatoms, their valves are symmetrical about both the transapical and apical plane, while some are represented by symmetrical valves along the apical axis and a symmetrical along the transapical axis as *Gomphonema parvulum*, *Surirella ovalis,* and *Surirella elegans* (Plate V.6, 7, 8, and 9). Also Valves with longitudinal shaddow lines present in *Neidium bisculatum var. subundulatum* and *Anomoeonies sphaerophora* as in plate VI. Other pennatophycean species, like *Nitzschia amphibian*, *Nitzschia elegantula*, *Nitzschia microcephala, Nitzschia apiculata*, *Nitzschia Constricta,* and *Nitzschia filiformis*, raphe may be enclosed in a keel on one margin of the valve (Plate VII). The keel may extend along both margins of valves, like *Surirella ovalis* and *Surirella elegans* (Plate V.7, 8, and 9) (Hamed, 2008).

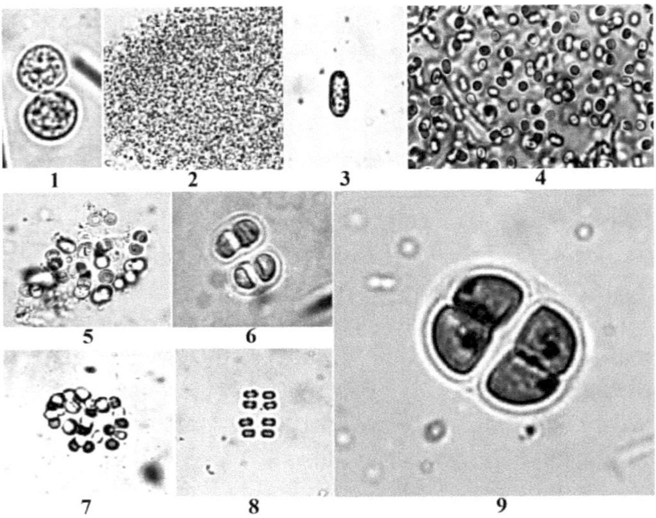

PLATE I 1, *Synechococcus aeruginosus* (1000X); 2, *Synechocystis sallensis*(600X); 3, *Synechocystis salina* (400X); 4, *Synechocystis pevalekii* (1500X); 5, *Gloeocapsa atrata* (600X); 6, *Gloeocapsa decorticans* (600X); 7, *Gloeocapsa lacustris* (600X); 8, *Gloeocapsa minor* (600X); 9, *Gloeocapsa turgida* (600 X).

Source: Reprinted from Hamed, 2008.

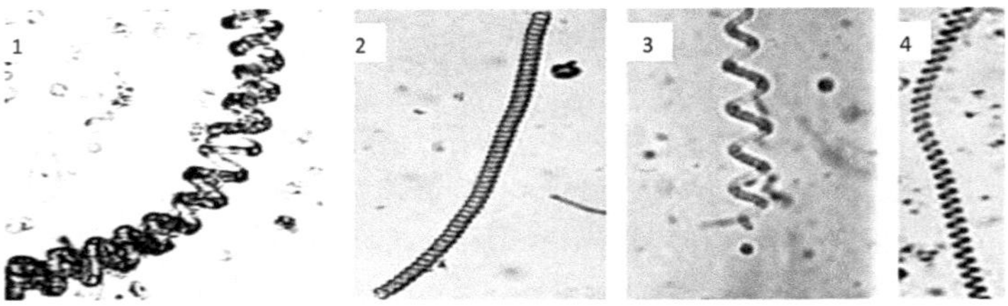

PLATE II 1, *Spirulina gigantea* (400X); 2, *Spirulina labyrinthiformis* (450X); 3, *Spirulina platensis* (450X); and 4, *Spirulina subtilissima* (600X).

Source: Reprinted from Hamed, 2008.

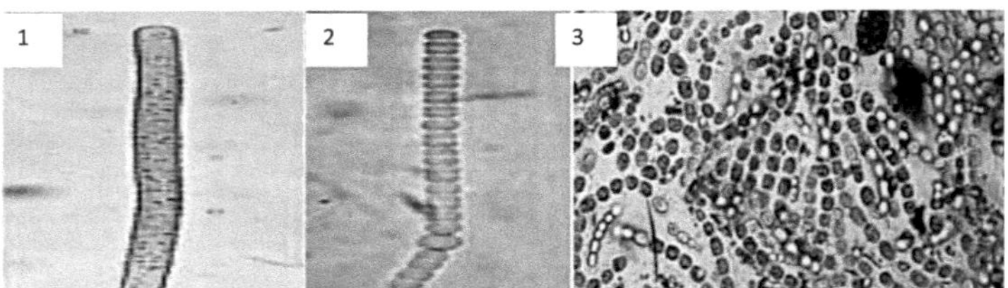

PLATE III 1, *Oscillatoria Vannae*; 2, *Anabaena affinis*; and 3, *Nostoc sp.*

Source: Reprinted from Hamed, 2008.

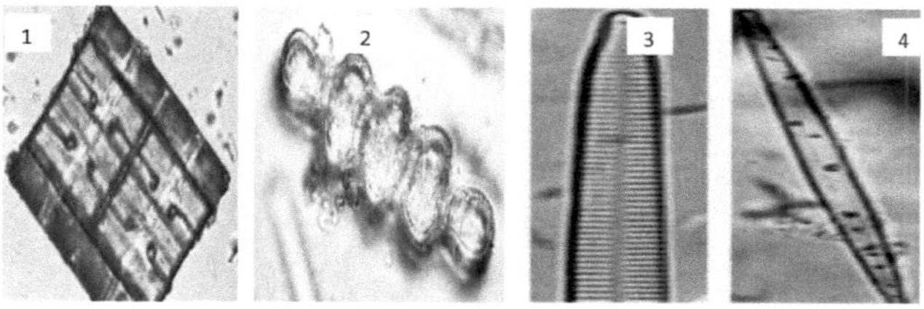

PLATE IV 1, 2, *Terpsinoe americana*; 3, *Synedra ulna*; and 4, *Synedra tabulate*

Source: Reprinted from Hamed, 2008.

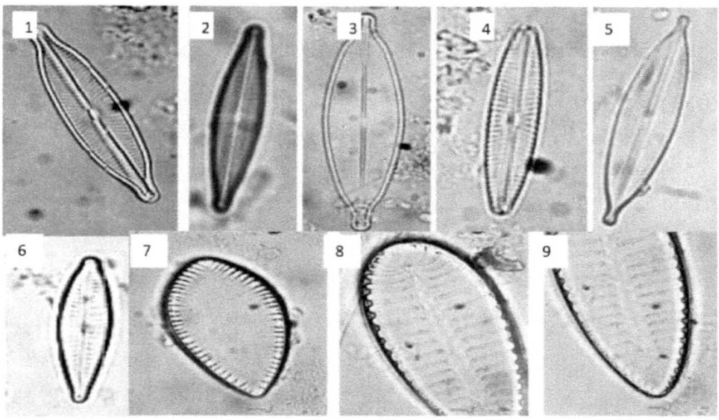

PLATE V 1, *Navicula cryptocephala* (1200X); 2, *Navicula cryptocephala var. veneta* (1200X); 3, *Naviculacuspidate* (1200X); 4, *Navicula heufleri var. leptocephala* (1200X); 5, *Navicula halophila f. tenuirostris* (1200X); 6, *Gomphonema parvulum* (1100X); and 7, 8, *Surirella ovalis*; and 9, *Surirella elegans* (1100X).

Source: Reprinted from Hamed, 2008.

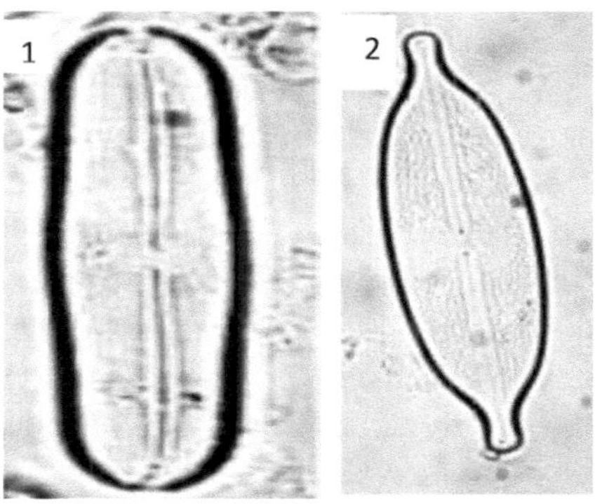

PLATE VI 1, *Neidium bisculatum var. subundulatum*; and 2, *Anomoeonies sphaerophora*.

Source: Reprinted from Hamed, 2008.

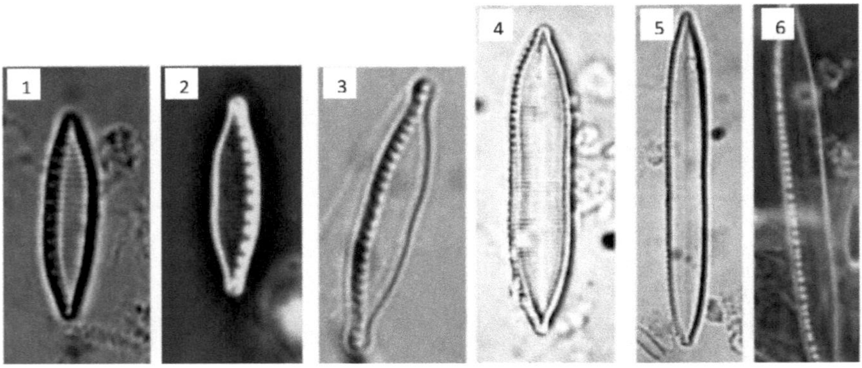

PLATE VII 1, *Nitzschia amphibian* (1500X); 2, *Nitzschia elegantula* (1500X); 3, *Nitzschia microcephala* (1500X); 4, *Nitzschia apiculata* (1500X); 5, *Nitzschia constricta* (1500X); 6, *Nitzschia filiformis* (1500X).

Source: Reprinted from Hamed, 2008.

In Farafra oasis, many blue green algae were identified as *Amorphonostoc punctiforme* f. *populorum*, *Aphanothece microscopica*, *Cyacnosarcina chroococcoides*, *Phormidiumterebriforme*, *Synechococcus aeruginosus*, and *Synechocystis pevalekii*. Besides, *Chloridella simplex*, *Heterothrix ulotrichoides*, *Monodus dactylococcoides*, and *Pleurochloris pyrenoidosa* representing xanthophytes and *Chlorella vulgaris*, *Scenedesmus bijugatus*, *S. obliquus*, and *S. quadricauda* var. *eualternans* belong to Chlorophytas in plate VIII.

Most of the identified Cyanophytes were related to order Nostocales and order Oscillatoriale in Egyptian soil. On previous studies, variations of nitrogen fixation rate of blue-green algae were recorded (Moisander et al., 1996; Osman et al., 2010; Gheda and Ahmed, 2015; Khanna et al., 2016). The isolated blue-green algae taxa from El-Farafra Oasis were specifically *Scytonema ocellatum* and *Westiellopsis prolifica* (Fig. 14.6) exhibited the highest large Nitrogen fixation rate with mean value, 30.52 and 33.01 μmole/mL^{-1}h^{-1}. Thus, they could be widely used as excellent biofertilizers and in the reclamation of the low nutrient desert soils of El-Farafra Oasis (Shaaban et al., 2017).

González (2003) reported that *Scytonema. ocellatum* isolated from the tropical soil in Mexico not only could be used as an excellent soil bio-fertilizer. Also, other studies showed *Westiellopsis prolifica*as soil bio-fertilizer (Jha et al., 1987; Jha and Kaushik, 1988; Prabu and Udayasoorian, 2007; Singh and Dhar, 2010; Paudel et al., 2012).

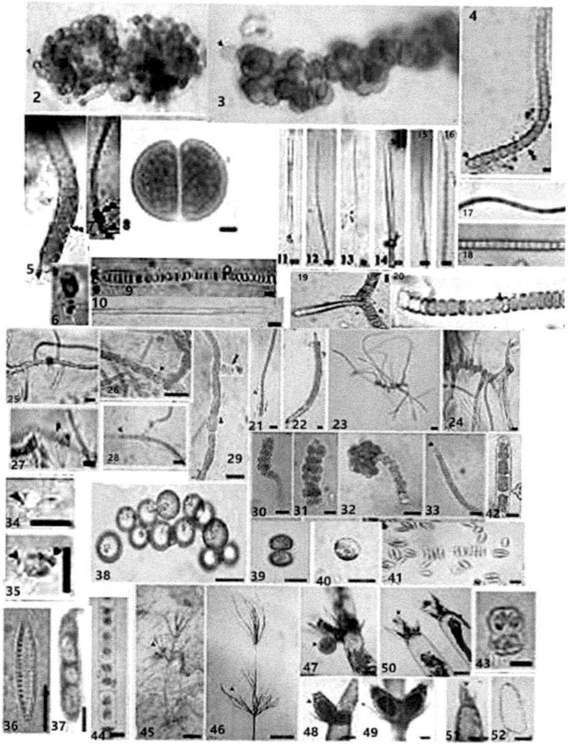

PLATE VIII Some common naturally propagated and isolated algal strains in El-Farafra Oasis during summer 2011 and winter 2012. (**2-3**) *Amorphonostoc punctiforme, f. populorum* with spherical or subspherical heterocysts (arrowheads). (**4-5**) *Calothrix elenkinii* with immature basal heterocyst (arrowhead, **4**) and bent at the base (double arrowheads, **4**); with mature basal heterocyst (arrowhead, **5**) and thin, nonlamellated sheath close to the trichome (double arrowheads, **5**). (**6-7**) *Cylindrospermum alatosporum* with ellipsoidal basal heterocyst (arrowhead) and akinete (double arrowheads). (**8**) *Chroococcus turgidus* (**9**) *Johannesbaptistia pellucida* (**10-11**) *Planktolyngbya limnetica* showing liberated cell (arrowhead, **10**). (**12-13**) *Geitlerinema jasorvense*. (**14**) *Phormidium acuminatum*. (**15-16**) *Geitlerinemasplendidum*. (**17**) *Jaaginema subtilissimum*. (**18**) *Oscillatoria tenuis*. (**19-20**) *Scytonemaocellatum* showing false branching (**19**) and subquadrate heterocyst (arrowhead, **20**). (**21-22**) *Symploca dubia* showing outer yellowish firm uneven sheath (arrowhead, **21**). (**23–35**) Complete lifestages of *Westiellopsis prolifica* showing true T-shaped branching (**23–25**), heterocysts (arrowheads, **26–29**). Double arrowheads in (**29**) depicting the early stage of lateral branch formation), pseudohormocysts (**30–32**), escaping of gonidia (arrowhead, **33**), free gonidial cells with refractive bodies (arrowhead, **34**), the early stage of germinating gonidial cell (arrowheads, **35**). (**36**) *Nitzschia amphibia*. (**37**) *Euglena oxyuris*. (**38-39**) *Chloridella simplex*. (**40**) *Pleurochloris pyrenoidosa*. (**41**) *Scenedesmusobliquus*. (**42**) *Penium margaritaceum*. (**43**) *Euastrum insulare*(**44**) *Zygnemopsis* sp. (**45–52**) habitus of *Nitellopsis obtusa* showing branchlets (arrowheads, **45-46**); antheridium (**47**); Oogonium(**48-49**); terminal pointed conical cells of branchlets (**50-51**); eccorticated axial cell (**52**).

The diagnosis of many algae flora of rice field in El-Kharga Oasis was illustrated by El-Naghy et al., 2004 as in plate IX where cells with rounded ends and flattened apex (73–75 µm long and 65 µm broad) identified as *Cosmarium pachydermum* var. *pachydermum, Cosmarium pseudopyramidatum* described as longer cells without granules, pyrenoids one in each semicell and with rounded apex, besides *Pediastrum tetra* (8–16 polygonal cells arranged in a stellate plates), *Scenedesmus alternans* (eight cells arranged in two alternating rows and *Scenedesmus magnus* (four cells arranged in one row).

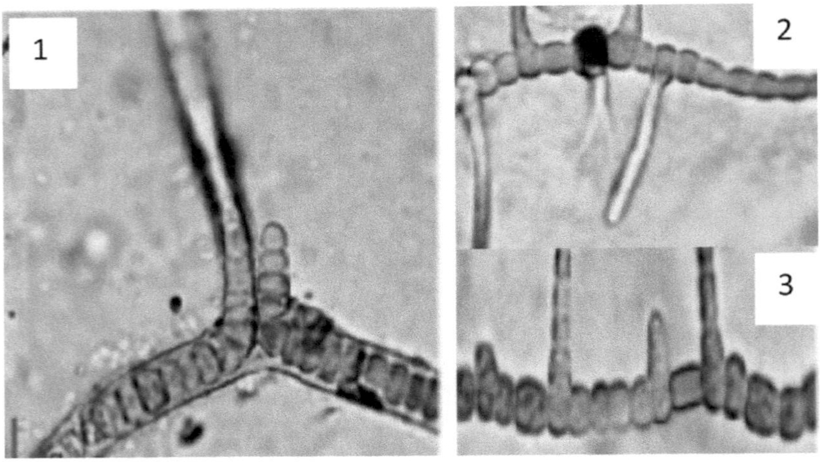

FIGURE 14.6 1, *Scytonemaocellatum Lyngb*; 2, 3, *Westiellopsis prolifica Janet.*

A great number of epiphytic marine algae can grow on other macroalgae or seaweeds (Rindi and Guiry, 2004). Several epiphytic algal species were recorded in Port Said shores as mostly from Rhodophyta, such as *Achrochaetum unifilum, Ceramium* spp., *Fosliella farinose,* and *Erythrotrichia carnea* as in Figure 14.7.

Also, *Hypnea cornuta, H. musciformis,* and *Sarcomma filiformis* were noticed at the east corner near to El-Tafriaa (PortSaid Coast). This region characterized by high salinity and low nutrient content as Figure 14.8.

Epiphytes may use other algae as a substrate for physical support, in spite of the biological contact between them and benefits from the host's buoyancy stimulating it to be near to the sunlight (Harlin, 1980; Pearson and Evans, 1990). Moreover, *Pilayella litoralis,* the brown alga, was observed in high abundance at the entrance of Lake Manzala. At a site near the entrance of the Suez Canal, prosperous growth of *Ulva* spp. especially *U. fasciata* was observed as in Figure 14.9.

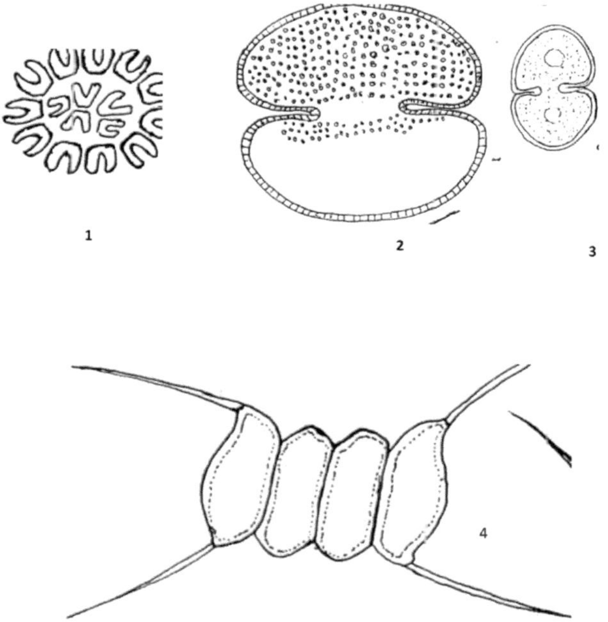

PLATE IX 1, *Pediastrum tetras*; 2, *C. pachydermum*; 3, *C. pseudopyramidatum*; and 4, *Scenedesmus magnus*.

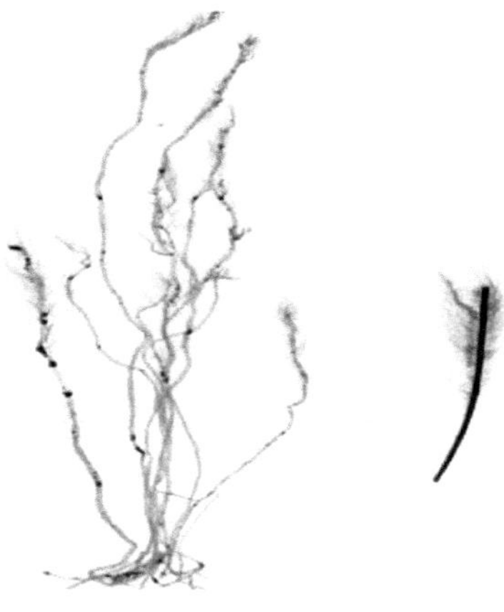

FIGURE 14.7 *Erythrotrichia carnea*.

FIGURE 14.8 1, *Hypnea cornuta*; 2, *Sarconema*; and 3, *H. musciformis*.

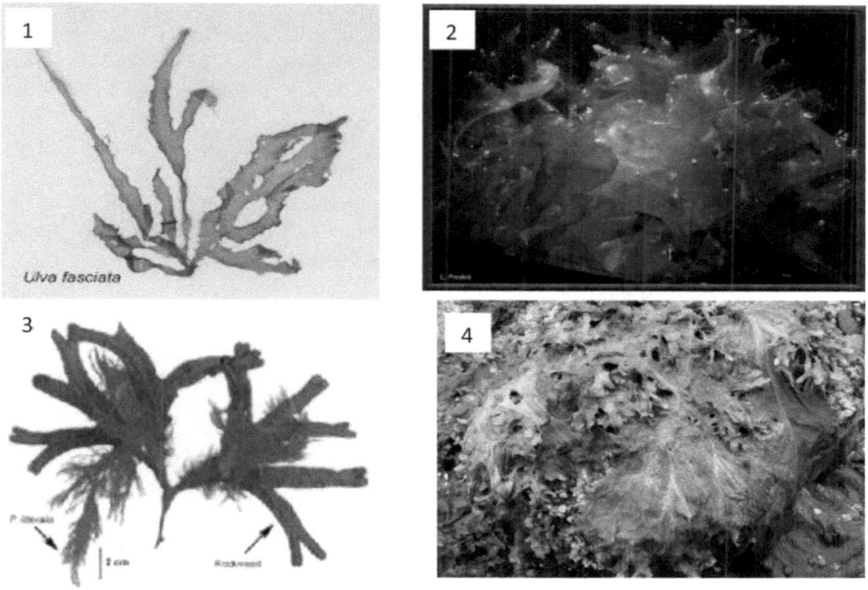

FIGURE 14.9 1,2, *Ulva fasciata*; and 3,4, *Pilayella litoralis.*

KEYWORDS

- **algae biodiversity**
- **algae classification**
- **algae in Egypt**
- **microalgae**
- **chlorophyta**

REFERENCES

Abdalla, R. R.; Samaan, A. A.; Ghobrial, M. G. Eutrophication in Lakr Mariut. *Bull. Natio. Inst. Oceanog. Fish. ARE.* **1991,** *17* (1), 157–166.

Aleem, A. A. *The Marine Algae of Alexandria;* Alexandria: Egypt, 1993; pp 154.

Brammer, H. Agriculture and Food Production in Poders Areas. A Case Study in Bangaladesh. *Water Int.* **1983,** 8, 74–81.

Bohunická, M.; Pietrasiak N.; Johansen, J. R.; Gómez, E. B.; Hauer, T.; Gaysina, L. A. and Lukešová, A. *Roholtiella*, gen. nov. (Nostocales, Cyanobacteria)-a Tapering and Branching Cyanobacteria of the Family Nostocaceae. *Phytotaxa* **2015,** *197* (2), 84–103.

Bold, H. C.; M. J. Wynne, *Introduction to the Algae, Structure & Reproduction;* Prentice Hall: Eaglewood Cliffs, New Jersey, 1985; pp 1–720.

Cholonky, B. J. *Die Ökologie der Diatomeen in Binnengewässern.* J. Cramer, Lehre. 1968, 4.

Cantonati, M.; Rott, E.; Pipp, E. Ecology of Cyanophythes Inmountain Springs of the River Sarca Catchment (Adamello-Brenta RegionalPark, Trentino, Northern Italy). *Algological Stud.* **1996,** *83*, 145–162.

Duarte, P.; Macedo, M. F.; Fonseca, L. C. The Relationship Between Phytoplankton Diversity and Community Function in a Coastallagoon. *Hydrobiology* **2006,** *555*, 3–18.

El-Shinawy, I. El-Burullus Wetland's Hydrological Study. Med Wet Coast, Global Environmental Facility (GEF) & Egyptian Environmental Affaris Agency (EEAA), Cairo, 2002.

El-Sherif, Z. M.; Gharib, M.; Samiha, Spatial and temporal patterns of phytoplankton communities in Manzalah lagoon. *Bull. Nat. Inst. Oceanog. Fish. ARE* **2001,** *227*, 239.

El-Naggar, M. E. E. Distribution of Iodine in Egyptian Marine Algae. JKAU: *Mar. Sci.* **1995,** *6*, 151–100.

El-Naghy, M. A.; El-Shahed, A. M.; Fathi, A. A.; Ahmed, G. G. Algal Flora of Rice Fields at El-Kharga oasis, Egypt. *Egyptian J. Phycol.* **2004,** *5*, 51–69.

Fritsch, F. E.; *Structure and Reproduction of the Algae;* Cambridge University Press, London, 1935; Vol. I.

Gheda, S. F. Ahmed, D. A. Improved Soil Characteristics and Wheat Germination as Influenced by Inoculation of *Nostoc kihlmani* and *Anabaena cylindrical. Rend. Fis. Acc. Lincei* **2015,** *26,* 21–131.

Gopinathan, C. P. In *Marine Microalgae*. Proceedings of Ocean Life Food and Medicine Expo, CMFRI, Cochin, 2004, pp 51–61.

Hamed, A. F. Biodiversity and Distribution of Blue-Green Algae/Cyanobacteria and Diatoms in Some of the Egyptian Water Habitats in Relation to Conductivity.A. *J. Basic Appl. Sci.* **2008,** *2* (1), 1–21.

Hung, T. C.; Chow, T. J. Comparative Studies of Some Nitrogen-fixing Unicellular Cyanobacteria Isolated from Rice Fields. *J. gen. Microbiol.* **1988,** *134*, 3089–3097.

Housman, D. C.; Powers, H. H.; Collins, A. D.; Belnap, J. Carbon and nitrogen fixation differ between successional stages of biological soil crusts in the Colorado Plateau and ChihuahuanDesert. *J. Arid Environ.* **2006,** *66*, 620–634.

Jha, M. N.; Kaushik, B. D. Response of *Westiellopsis prolifica* and *Anabaena* sp. to Salt Stress II. Uptake of Na+ in the Presence of K+ as Chloride, Nitrate, and Phosphate. *Curr. Sci.* **1988,** *57*, 667–668.

Jha, M. N.; Venkataraman, G. S.; Kaushik, B. D. Response of *Westiellopsis prolifica* and *Anabaena* sp. to Salt Stress. *Mircen J.* 1987, *3*, 99–103.

Khanna, N.; Raleiras, P.; Lindblad, P. Fundamentals and Recent Advances in Hydrogen Production and Nitrogen Fixation in Cyanobacteria. In *The Physiology of Microalgae, Developments in Applied Phycology;* Borowitzka, M. A., Ed.; pp 101–127. Springer: Switzerland, 2016.

Khairy, H. A.; Shaltout, K. H.; EL-sheekh, M. M.; Eassa, D. I. Algae Diversity of Mediterranean Lakes in Egypt. International Conference on Advances in Agricultural, Biological & Environmental Sciences (AABES-2015), London (UK), pp 127–135.

Landsberg J. H. The Effects of Harmful Algal Blooms on Aquatic Organisms. *Rev. Fisheries Sci.* **2002**,*10*, pp 113–390.

Lee, R. E. *Phycology;* 4th ed.; Cambridge University Press: Cambridge, 2008; pp 1–560.

Madkour, F. F. El-Shoubaky, G. A. Seasonal Distribution and Community Structure of Macroalgae along Port Said coast, Mediterranean Sea, Egypt. *Egypt J. Aquat. Biol. Fish* **2007**, *11*, 221–236.

Mager, D. M.; Thomas, A. D. Extracellular polysaccharides from cyanobacterial soil crusts: a review of their role in dryland soil processes. *J. Arid Environ.* **2011,** *75*, 91–97.

Mann, D. G.; Droop, S. J. M. *Biodiversity, Biogeography, and Conservation of Diatoms. Hydrobiology* **1996,** *336*, pp 19–32.

Mareš, J.; Lara, Y.; Dadáková, I.; Hauer, T.; Uher, B.; Wilmotte, A.; Kaštovský, J. Phylogenetic Analysis of Cultivation-resistant Terrestrial Cyanobacteria with Massive Sheaths (*Stigonema* spp. and *Petalonema alatum*, Nostocales, Cyanobacteria) Using Single-cell and Filament Sequencing of Environmental Samples. *J. Phycol.* **2015,** *51* (2), 288–297.

Macioszczyk, A.; Dobrzyński, D. Hydrogeochemia, strefy aktywnejwymiany wód podziemnych. Wydawnictwo Naukowe PWN, Warszawa, 2002, pp 448.

McGregor, A. W.; Johnson, D. E Capacity of Desert Algal Crusts to Fix Atmospheric Nitrogen. *Soil Sci. Soc. Amer. Proc.* **1971,** *35* 843–844.

Munawar M. Ecological studies of Eugleninae in Certain Polluted and Unpolluted Environments. *Hydrobiologia* **1972,** *39*, 307–320. http://dx.doi.org/10.1007/BF00046647

Moisander, P.; Lehtimaki, J.; Sivonen, K.; Kononen, K. Comparison of 15N2 and Acetylene Reduction Methods for the Measurement of Nitrogen Fixation by Baltic Sea Cyanobacteria. *Phycologia* **1996,** *35* (6),140–146.

Nowicka-Krawczyk, P. B. and Żelazna-Wieczorek, J. (2013). Cyanobacteria micro flora in a Limestone Spring (Troniny spring, Central Poland). *Acta Societatis Botanicorum Poloniae*, **2013,** *8* (3), 219–224.

Nowrouzi, S.; Valavi, H. Effects of Environmental Factors on Phytoplankton Abundance and Diversity in Kaftar Lake. *Biology* **2010,** *46*, 63–74.

Norton, T. A.; Malkonian, M.; Anderson, R. A. Algal biodiversity. *Phycologia***1996,** *35*, 308–326.

Osman, M. E. H.; El-Sheekh, M. M.; El-Naggar, A. H.; Gheda, S. F. Effect of Two Species of Cyanobacteria as Biofertilizers on Some Metabolic Activities, Growth, and Yield of Pea Plant. *Biol. Fert. Soils* **2010,** *46*, 861–875.

Paudel, Y. P.; Pradhan, S.; Pant, B.; Prasad, B. N. Role of Blue Green Algae in Rice Productivity. *Agricul. Biol. J. North Am.* **2012,** *3* (8), 332–335.

Prabu, P. C.; Udayasoorian, C. Native Cyanobacteria *Westiellopsis* (TL-2) sp. for Reclaiming Paper Mill Effluent Polluted Saline Sodic Soil Habitat of India. *Electron. J. Environ. Agri. Food Chem.* **2007,** *6* (2),1775–1786.

Radwan, A. M. Comparative Study on the Phytoplankton Standing Crop in the Different Sectors of Burullus Lake during 1997-1998. *Bull. Nat. Inst. Oceanog. Fish. ARE* **2002,** *28*, 189–305.

Reynolds, C. S. Phytoplankton Periodicity: The Interactions of Form, Function and Environmental Variablility. *Freshwater Biol.* **1984,** *14*, 111–142. http://dx.doi.org/10.1111/j.1365-2427.1984.tb00027.x

Richardson, K. Harmful or Exceptional Phytoplankton Blooms in the Marine Ecosystems. *Advan. Mar. Biol.* **1997,** *36*, 301–385.

Shaaban, A. M.; Mansour, H. A.; Saber A. A. Unveiling algal Biodiversity of El-farafra Oasis (Western Desert, Egypt) and Potential Relevance of its use in Water Bio-assessment: Special Interest on Springs and Drilled Wells. *Egyptian J. Phycol.* **2015,** *16*, 47–75.

Shaaban, A. M.; Mansour, H. A.; Saber A. A. Soil Algae of El-Farafra Oasis (The Western Desert, Egypt), and N2-fixation Efficiency of Five Heterocytous Cyanophytes. *Egypt. J. Bot.* **2017,** *57* (3), 517–524.

Shaltout, K. H.; Khalil, M. T. Lake Burullus (Burullus protected area). Publication of National Biodiversity Unit. No. 13, 2005.

Shaltout, K. H.; T. M. Galal Report on Ecosystem of Lake Manzala, 2006.

Sieburth, J. M. Bacterioplankton: Nature, biomass, Activity, and Relationships to the Protist Plankton. *Arch. Microbiol.* **1978,** *100*, pp 419–436.

Sigee, D.C.; *Freshwater Microbiology: Biodiversity and Dynamic Interactions of Micro-organisms in the Aquatic Environments;* John Wiley and Sons Ltd, Chichester, West Sussex England, 2005, pp 544.

Singh, N. K.; Dhar, D. W. Cyanobacterial Reclamation of Salt-affected Soil. In *Genetic Engineering, Biofertilisation, Soil Quality, and Organic Farming. Sustainable Agriculture Reviews;* 4. Lichtfouse, E. Ed.; Springer Science + Business Media B. V: Dordrecht, 2010; pp 243–275.

Stevenson, R. J. In *Algal Ecology -Freshwater Benthic Ecosystems;* Stevenson, R. J., Bothwell, M. L., Lowe, R. L., Eds.; Academic Press: San Diego, 1996, pp 3–30.

Tilman, D.; Kieling, R.; Sterner, R.; Kilham, S. S.; Johnson, F. A. Green-blue Algae and Diatoms; Taxonomic Differences in Competitive Ability for Phosphorus, Silicon, and Nitrogen. *Arch. Hydrobiologia 106*, 473–485.

Tomaselli, L. The Microalgal Cell. In *Handbook of Microalgal Culture: Biotechnology and Applied Phycology.* Richmond A, Ed.; Blackwell Publishing Ltd: Oxford, 2004, pp 3–19.

Underwood, G. J. C.; Kromkamp, J. Primary Production by Phytoplankton and Microphytobenthos in Estuaries. *Advan. Ecol. Resr.* pp 93–153.

Van den Hoek, C.; Mann, D. G.; Jans, H. M. Algae. *An Introductionto Phycology*, Cambridge University Press: Cambridge,1995; pp 623.

Whittaker, R. H. New concepts of Kingdoms of Organisms. *Science* **1969,** *163,* pp 150–159.

CHAPTER 15

Role of Diverse Medicinal Algae in a Therapeutic World

HEBA S. ABBAS*

Microbiology Department, National Organization for Drug Control and Research (NODCAR), Giza, Egypt

**Corresponding author. E-mail: hebasalah18@gmail.com*

ABSTRACT

Algae are a diverse group that are used in medicine from long times. They produce secondary metabolites, such as proteins, vitamins, fatty acids, pigments, and polysaccharides, which have a pharmaceutical value and play an important role in medicine. Such biologically active products (bromophenols, sulfated polysaccharides, c-phycocyanin, phycobiliprotein, ß-carotene, peptides, etc.) can be used as antioxidants, anticancer, antimicrobial (antibacterial, antifungal), anti-inflammatory, anticoagulant, immune enhancer and modulator, antidiabetic, reduction of fats, cosmetic, and thalassotherapy. And this not restricted on marine algae or macroalgae but even microalgae and cyanobacteria can produce pharmaceutical and nutraceutical value products. This study concludes that algae are natural future drugs.

15.1 INTRODUCTION

Algae are autotrophic producers that can produce their own food by using sunlight and photosynthesis. They divided into macroscopic and microscopic algae (Singh et al., 2013). Algae are a rich source of vitamin, minerals, and protein (Pulz and Gross, 2004). Algae species have various applications in food, pharmaceuticals, cosmetic, and industry. It can be used as antibiotic, antioxidant and also it can stop viral propagation. Food products set from algae may have a positive or negative effect on mammalians where poisonous elements in algae, like cadmium or fucotoxins, should be avoided. So, bioactive compounds in algae play an essential role in the assessment of food quality (Mabeau and Fleurence, 1993). The bioactive compounds involve alkaloids, polyketides, cyclic peptide, polysaccharide, phlorotannin, diterpenoids, sterols, quinones, lipids, and glycerol (Cabrita et al., 2010).

15.2 ALGAE AND MEDICINE

Marine algae have been utilized as food and medicine from centuries. Marine macroscopic algae produce chemical active compounds that have several biological activities, such as antiviral, antibacterial, antifungal, antiallergic, anticoagulant, antioxidant, anticancer activities (Bouhlal et al., 2011; Kim and Karadeniz, 2011; De Felício et al., 2010; Na et al., 2005; Dayong et al., 2008; Kim et al., 2011) and could be used in pharmaceutical industries.

Bromophenol compounds in red algae exhibited a wide variety of biological products like enzyme inhibition, cytotoxic, antioxidant, feeding deterrent, antiflammatory, and antimicrobial activities (Tapiero et al., 2002; Matanjunand et al., 2008; Jaganathan and Mandal, 2009; Williamson and Carughi, 2010).

Microalgae, like Cyanobacteria, are rich with secondary metabolites, such as proteins, vitamin, fatty acids, or pigments that have biological activities (Patterson et al., 1994; Falch, 1996) like antibacterial diterpenoids in *Nostoc commune* (Jaki et al., 1999, 2000) or antifungal peptides in *Tolypotrixbyssoidea* (Jaki et al., 2001).

15.3 ANTICANCER ACTIVITIES OF ALGAE

Preliminary studies of ß-carotene showed that it has a beneficial effect in the treatment of precancerous, like oral leukoplakia, the cause of oral cancer (Boopathy and Kathiresan, 2010). *Spirulina platensis* is a rich source of sulfated polysaccharide, selano compounds, phenolic compounds, c-phycocyanin, and phycobiliproteins (Dilon et al.,1995). It has various biological effects including the prevention of diabetes (Padyana et al., 2002). C-phycocyanin contains phycocyanobilin (PCB) that is tetrapyrrolic compounds (Fig. 15.1). PCB can be degraded by biliverdin reductase to phycocyanorubin, like biliverdin, which is reduced into bilirubin in the body of a human (Konickova et al., 2014). Also, chlorophyll resembles bilirubin. Bilirubin is an antioxidant that has relation to oxidative stress-mediated diseases, like cardiovascular, cancer, and autoimmune disease (Novotony and Vitek, 2003; Jiraskova et al., 2012; Vitek et al., 2010).

S. platensis and associated tetrapyrroles have antiproliferative activities on human pancreatic cancer cells (PA-TU-8902) (Konickova et al., 2014) as in Figure 15.2. The viability of PA-TU-8902 was decreased after treatment with 0.16 g/L *S. platensi*. Also, 10 μM of bilirubin decreased the viability of PA-TU-8902 cell. On the other hand, high doses of PCB (60 μM) and chlorophyllin (125 μM) give similar effects. The antiproliferative activities of *S. platensis* are due to their antioxidant activity and inhibition the production of reactive oxygen species (ROS) by mitochondria then changes of intracellular redox status (Konickova et al., 2014). Thus, *S. platensis* has a chemopreventive role in cancer treatment and this is supported in many researche that use it as anticancer for liver and breast cancer (Nazarewicz et al., 2013; Ouhtt et al., 2013).

15.4 ALGAE AS ANTIOXIDANTS

Polyphenols, phycobiliprotein, and vitamin are important water-soluble antioxidants in algae, antioxidants able to cause degeneration premalignant lesions and inhibit their progress into cancer. Algae prohibited oxygen damage by hunting free radicals and active oxygen and, therefore, the incidence of cancer will be avoided (Richardson, 1993). These antioxidants can be used against various diseases, like cardiovascular disorder, chronic inflammation, and atherosclerosis (Kohen and Nyska, 2002).

C-Phycocyanin is a pigment of blue-green algae that comprise an open chain of tetrapyrrole, has the ability to scavenge alkoxy and hydroxyl radicals. Romay et al. 1997 evaluated hydroxy radicals' inhibitions c-phycocyanin and they observed the inhibition at high concentrations. Besides, the inhibition was found on a microsomal lipid peroxidation because of the ability of phycocyanin to bind metals that lead to constant

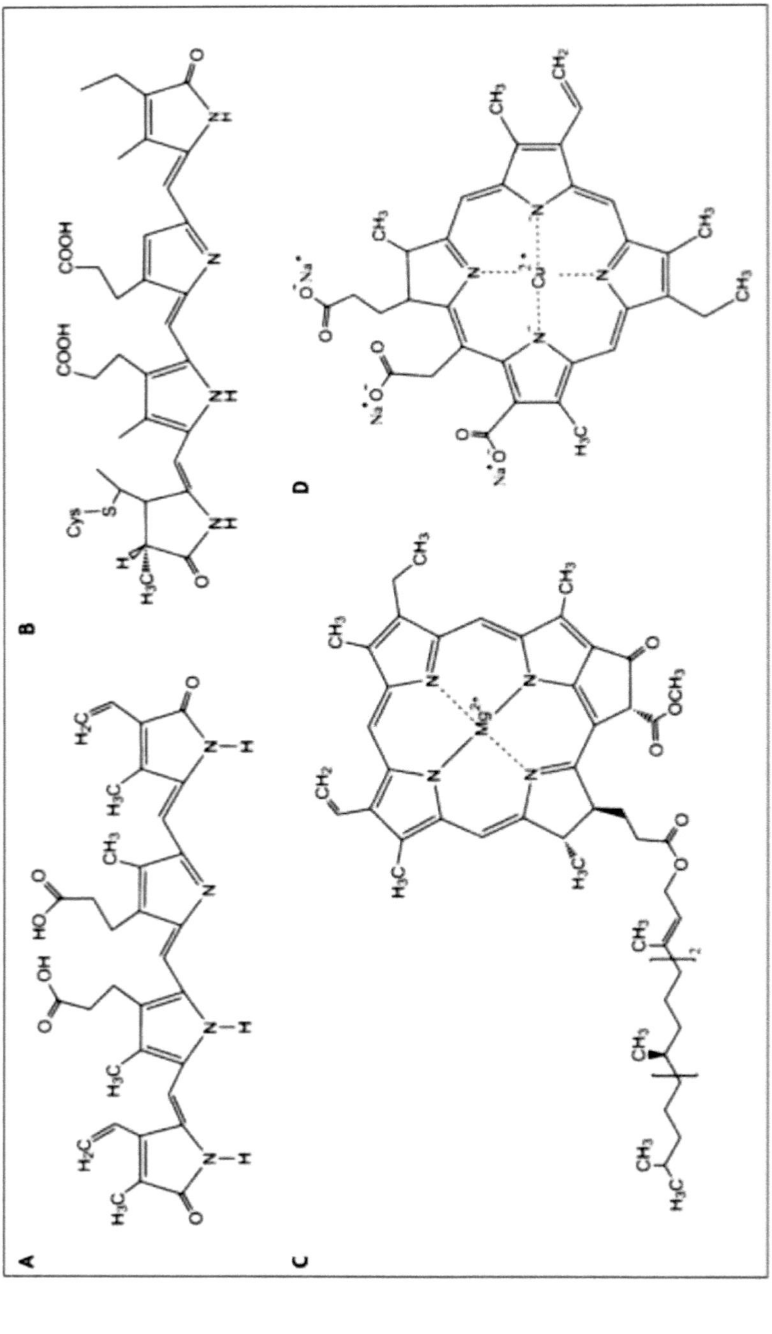

FIGURE 15.1 Tetrapyloric structures of human and algal source. (A) Beliviridin, (B) phycocyanobilin, (C) chlorophyll a, (D) chlorophyllin.

inhibition during the reaction. It is also known that c- phycocyanin inhibits deoxyribose damage in absence of ethyl diamine tetraacetic acid. Phycocyanin may have an influence on enzymes involved in the generation of ROS by phagocytes activation. NADPH oxidase and myeloperoxidase. The action of phycocyanin against hydroxyl radicals is related to its anti-inflammatory effect (Halliwell, 1990).

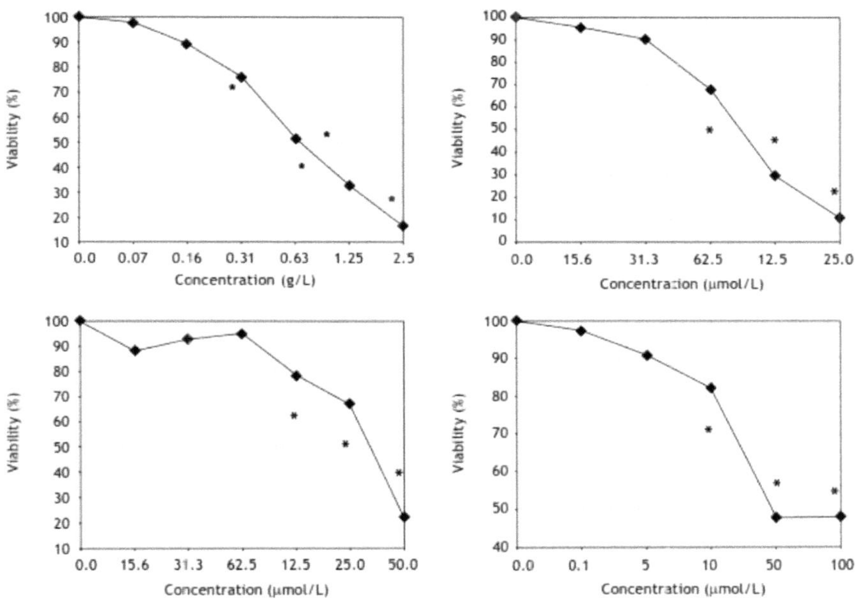

FIGURE 15.2 Effect of *S. platensis* extract and related tetrapyrrolic compounds on PA-TU-8902 Pancreatic cancer cell viability. (A) *S. platensis* extract, (B) phycocyanobilin, (C) chlorophyllin, and (D) bilirubin. Viability was measured using MTT assay after 24-h exposure of each compound.

Seaweeds are rich with pigments, such as fucoxanthin, astaxanthin, carotenoids, and polyphenols that is considered a source for antioxidants. They act by inhibiting lipid peroxidation where the indicator for oxidation in erythrocytes (TBRAS) was lower after seaweeds supplementation compared with control; however, the catalase and GSH-px were higher compared with control (Table 15.1). Thus, it reduces the oxidative stress in vivo (Kim et al., 2008).

TABLE 15.1 Lipid Peroxidase Level and Antioxidant Activities After Seaweeds Supplements.

Group[a]	CG	SG
TBARS μg/mg Hb)	25.1 ±3.6[b]	16.9 ±3.78*
SOD (Unit/mg Hb)	0.22±0.03	0.20 ±0.02
Catalase (Unit/mg Hb)	0.09 ±0.02	0.67±0.17*
GSH-Px (Unit/mg Hb)	0.03±0.01	0.32 ±0.1*

[a]CG: control group, 3G : seavieed group
[b]Data was expressed as mean ± S,E,
*Significant dfference between groups by independent t-test: $p < 0.05$

Source: Kim et al. (2008).

Therefore, antioxidant activities reduce the hazard of cardiovascular and diabetes type 2 diseases. Many studies are required for that (Kim et al., 2008).

15.5 ANTIMICROBIAL ACTIVITIES OF ALGAE

Michrochaetetenera (blue-green algae), *Sphareopleaannulina,* and *Nitellatenuissima* (green algae) collected from Bheema river adjacent to Gulbarga, India. The methanol extract of these algae was screened against bacteria like *Proteus vulgaris, Bacillus cereus, Escherichia coli, and Pseudomonas aeruginosae* and fungi like *Aspergillus niger*, *Aspergillus flavus,* and *Rhizopus nigricans* (Prashantkumar et al., 2006) as shown in Table 15.2.

All algae are active against *Pseudomonas aeruginosae. Michrochaetetenera* showed strong antimicrobial activity against *A. niger* and *Proteus vulgaris.* It is assumed that phytoconstituents are responsible for the antimicrobial activity.

Two cyanobacteria *Nodulariaharveyana* and *Nostoc insulare* were studied for the production of exometabolites (norharmane (9H pyrido (3,4-b) indole) and 4,4-dihydroxy biphenyl). The structure of the compound showed in Figure 15.3.

Table 15.3 shows the minimum inhibitory concentrations for Norharmane against *P. aeruginosa* and *E. coli* was 32 μg/mL, for *B.cereus, B. subtilus* is 160 and 128 μg/mL, respectively, 16 μg/mL for *Staphylococcus aureus* and 40 µg/mL for *Candida albicans*. However, 128 µg/mL or more for Gram-negative bacteria and gram-positive bacteria except for *B.cereus* and *C. albicans*was 32 µg/mL (Volk and furkert,2006). Also, 2-hydroxybiphenyl shows that for inhibition of Gram-negative bacteria higher concentrations are needed than for Gram-positive bacteria like 4,4-dihydroxy biphenyl (Wallha¨ußer, 1995).

TABLE 15.2 Antimicrobial Activity of *Michrochaete tenera, Sphaeroplea annulina,* and *Nitella tenuissantma.*

Organisms	Zone of inhibition (mm)*					
	1	2	3	4	5	6
Proteus vulgaris	14±0.0	13±0.0	12.5±0.01	17.5±0.01	ND	NI
Bacillus cereus	12±0.05	13±0.11	13±0.05	16±0.20	ND	NI
Escherichia coli	12.5±0.01	11.5±0.2	12±0.05	16.5±0.50	ND	NI
Pseudomonas aerusinosa	19.5±0.06	19±0.50	19.5±0.20	18±0.03	ND	NI
Aspergillus niger	17±0.02	13.5±0.1	14±0.40	ND	19±0.5	NI
Aspergillus ftavus	12±0.01	10±0.03	11±0.05	ND	18.5±0.3	NI
Rhizopus nigricans	12.5±0.06	11±0.5	12.5±0.03	ND	16±0.05	NI

*All the values are mean ± standard deviation of three determinations. 1. *Miaochaete tenera,* 2. *Nitello tenuissima,* 3, *Sphaeroplea annulina,* 4. Streptomycin sulphate (4 mg/mL of distilled water), 5. Bavistin (4 mg/mL of distilled water) and 6. Control (dimethylformamide). ND-Not done; NI-No Inhibition.

Source: Prashantkumar et al. (2006).

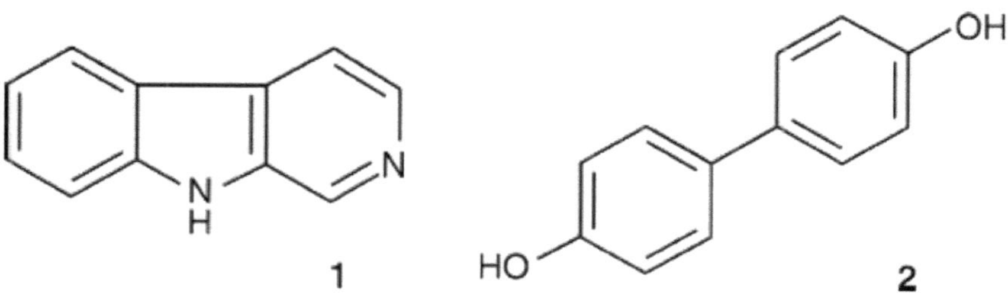

FIGURE 15.3 Exometbolites of cyanobacteria. 1. Norharmane (9H pyrido(3,4-b) indole). 2. 4,4-dihydroxy biphenyl.

TABLE 15.3 Minimum Inhibitory Concentrations of Norhamane and 4,4-Dihydroxybiphenyl Against Selected Bacteria and Yeast Fungi.

Test organisms	Norharmane-HCl (9H-pyrido(3,4-b) indole-HCl)	4,4'-dihydroxybiphenyl
Gram-negative bacteria		
Escherichia coli	32	>128
Pseudomonas aeruginosa	32	>128
Gram-positive bacteria		
Bacillus cereus	160	32
Bacillus subtilis	128	128
Staphylococcus aureus	16	128
Yeast		
Candida albicans	40	32

Concentrations are given in $\mu g\ mL^{-1}$

Also, algae extracts were tested on different viruses included herpes virus (Serkedjieva, 2004). Brown algae showed antiviral activities. Antiviral compounds must have a wide variety of activity, stall the virus finally, to have positive pharmacodynamic characteristics and not be immunosuppressive, which is necessary, as there should be no repression of the usual immune processes (Murray et al., 2004).

15.6 ANTI-INFLAMMATORY DRUGS FROM ALGAE

Inflammation is the main characteristic of many diseases and it considered an essential host defense mechanism (Wakefield and Kumar, 2001). It may cause stomach ulcers or disorders of the digestive tract when it happens for the mucosal of the gastrointestinal tract. Stomach ulcers are caused due to infection of *Helicobacter pylori* (Graham, 1997) or using of nonsteroidal anti-inflammatory drugs (NSAIDS), like aspirin and ibuprofen, for a long period of time (Al-Mofleh and Al-Rashed, 2007).

Bioactive compounds in red algae can aid in treatment the inflammation of gastrointestinal tract (Kang et al., 2008) by preventing stomach ulcers or cancers triggered by oxidative stress (Gonzalez et al., 1999) or reducing the production of inflammatory mediators (Khan et al., 2007, 2008; Lee, 2009) and encourage cancer apoptosis in stomach (Kown and Nam, 2007). Bioactive compounds from edible algae are benign when it used in medicine.

Gracilariachangii (red algae) is extensively used in folk medicine. Ultra-fine liquid chromatography and mass spectrometry analysis of methanol extracts of *G. changii* showed the mass in the positive ion spectrum of m/z 457.405 and had fragments of 398.327 (base peak), 380.316, and 158.082. The neutral loss of 59.077 mass revealing the loss of trimethylamine thus, it proposes that compound contains trimethyl ammonium but the compounds not known definitely by the *Dictionary of Natural Products*. On the other hand, it was observed that association with two identified chlorophylls methyl 10-hydroxyphaeophorbide a and 10-hydroxypheophytin a (Chu et al., 2013) (Table 15.4; Fig. 15.4).

TABLE 15.4 Identification of the Most Significant Mass Found in Methanolic Extract of *G. changii*.

Mass (ion)	Found in batch	Possible ID
417.298	1,2	Most likely contaminant
457.405*	1,2	Unidentified
461.326	1,2	Most likely contaminant
505.353	1,2	Most likely contaminant
573.256	1,2	Unidentified
645.280	1,2	Methyl 10-hydroxyphaeophorhide *a*
743.636	1,2	Unidentified
887.579	1,2	10-Hydrcwypheophytin *a*

Dictionary of Natural Products, CRC Press with high resolution MS spectra were used to analyse the total positive ion spectra for the identification of possible chemical constituents.

*MS/MS (relative intensities %) 457.405 → 398.327 (100%), 380.316 (50%), 158.082 (40%).

Source: Chu et al. (2013).

It was studied the gastric histological lesions of rat's stomach treated with methanolic extract of *G. changii* subsequent induction of gastric mucosal induction. The reduction in gastric ulcers area was recorded in pretreated rats with 2, 2.5, 5, 250, and 500 mg/kg methanolic extract of *G. changii* (Fig. 15.5).

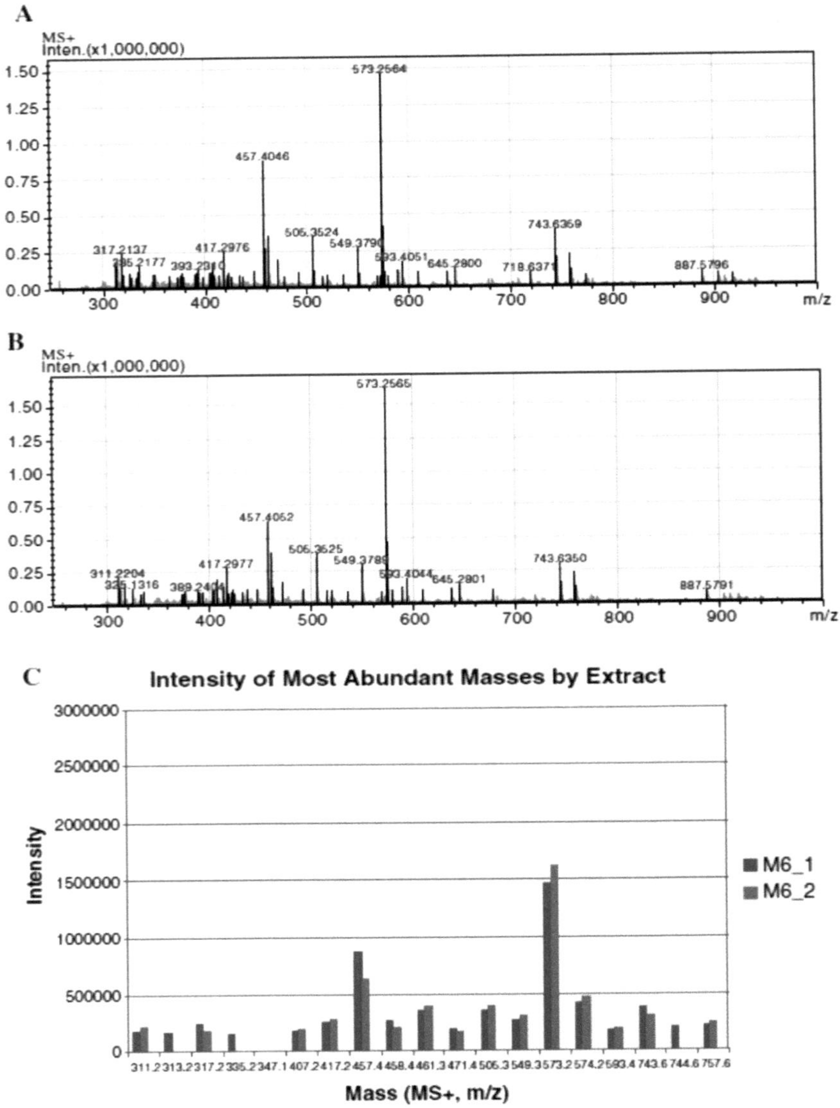

FIGURE 15.4 (See color insert.) Profiling of different batches of methanol extract *G. changii* (Chu et al., 2013). (A) batch 1; (B) batch 2; (C) comparison of the intensity of the most abundant masses of the two methanol extracts *G. changii.*

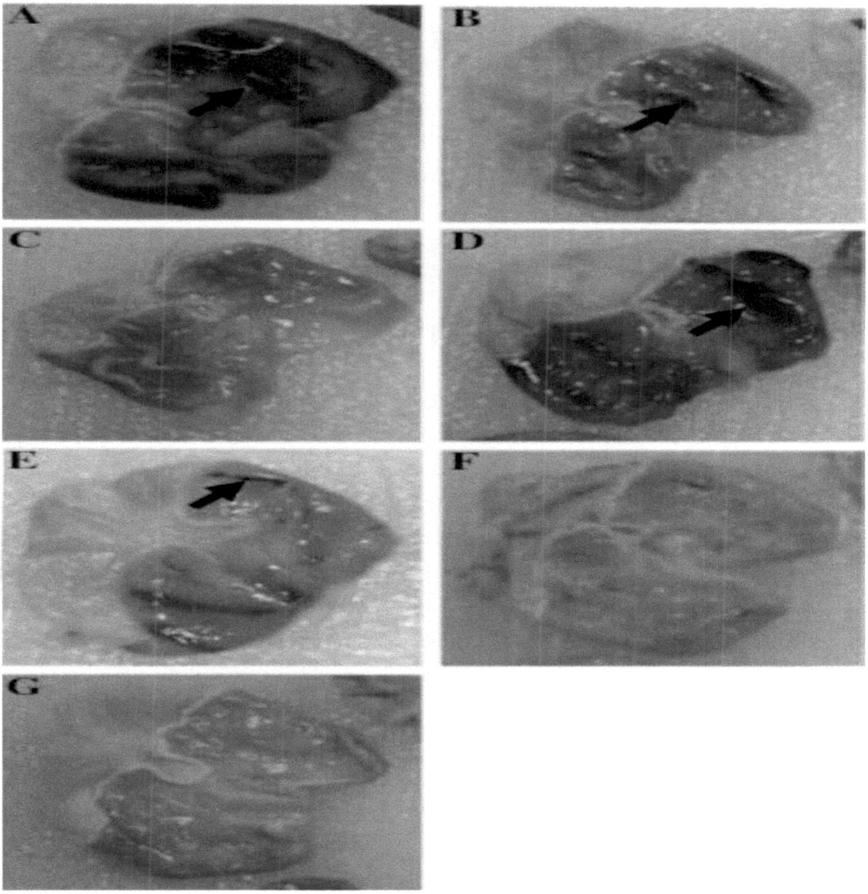

FIGURE 15.5 (See color insert.) Influence of methanol extract *G.changii*on gastric lesions following induction of gastric mucosal injury (Chu et al., 2013). (A) Diluent. (B) Non-pecific plant extract (NSPE). (C) Positive control from chemical company of Malaysia (OMP). (D) 2.5 mg/kg b.w. MeOHGCM6. (E) 5 mg/kg b.w. 6. (F) 250 mg/kg b.w. (G) 500 mg/kg b.w.

The nonspecific plant extract did not gave the same gastric responses and considerable inhibition of the EtOH-induced severe gastric mucosal injury.

Thus, the methanol extract of *G. changii* has gastroprotective and antiulcerogenic activities. The mechanism is explained by reduction gastric juice and gastric mucous excretion and lowered the stomach juice pH to 4.74±0.38 at 250 mg/kg b.w. However, the positive control (OMP) decreased the gastric mucous excretions and counteracted stomach juice pH to 6.97±0.37 that causes changing in the normal flora of the stomach and leading to diarrhea disease (Chu et al., 2013). Also, algae have antioxidant activities which help in preventing stomach ulcers (Okuzumi et al., 1993; Sreenivasan et al., 2007; Ranković et al., 2012).

15.7 ANTICOAGULANT ACTIVITY OF ALGAE

Marine algae contain high concentrations of sulfated polysaccharides that have various biological activities (Leite et al., 1998; Farias et al., 2001; Mazumder et al., 2002; Qi et al., 2005; Lins et al., 2009; Ananthi et al., 2010; Rodrigues et al., 2011). Many studies on the anticoagulant and antithrombotic properties of marine algae have been performed (Farias et al., 2000; Pereira et al., 2002; Mourão, 2004; Pereira et al., 2005; Athukorala et al., 2006; Fonseca et al., 2008; Pushpamali et al., 2008; Rodrigues et al., 2009). Seaweeds rich with phycocolloids (sulfated polysaccharides), with stabilizing properties, are used in biotechnological applications (Compo et al., 2009). Sulfated polysaccharide galacto arabinoglucan in green algae *Codium pugniformis* and sulfated galactan in green algae *C. cylindricum* with anticoagulant activity were recorded (Matsubara et al., 2000, 2001). Also, sulfated polysaccharide extracted from marine green algae *Monostroma* genus by Mao et al. (2008) and Zhang et al. (2008). In *C. cupressoides,* high sulfated polysaccharide and sugars were obtained in Fraction II compared with fraction I and III as in Figure 15.6. These sulfated polysaccharides similar to that produced by *Gracillaria cornea* (Melo et al., 2002), *Monostroma latissimum* (Zhang et al., 2008) and *M. nitidum* (Karnjanapratum and You, 2011)

15.8 ALGAE AS ANTIDIABETIC

Marine algae can be helpful in the detection of novel antidiabetic compounds with high biomedical rate. Brown and red algae have antidiabetic activity. It controls the blood glucose levels through the hanging up of carbohydrate, hydrolyzing enzymes, and protein tyrosine phosphatase 1B enzymes, insulin sensitization (Unnikrishnan and Jayasri, 2018). Edible seaweeds have high contents of nonstarch polysaccharides, proteins, and minerals while the lipid contents are very low. (Jurkovic et al., 1995; Urbano and Goni, 2002).

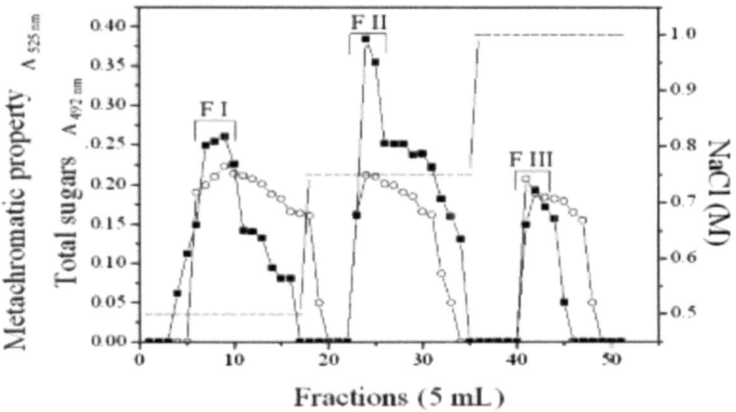

FIGURE 15.6 Separation of Sulfated Polysaccharides from *C. cupressoides* by DEAE-cellulose. Fractions were collected and checked by metachromasia using 1,9-dimethymethylene blue (O–O) and phenol-H_2SO (■–■). Vertical steps represent the NaCl concentration (–).

A study was done on the effect of seaweeds supplements on blood glucose levels, lipids, and antioxidant activities in patients with diabetes type 2. The increase of the concentrations of fast blood glucose level and postprandial-2h blood glucose were decreased significantly compared with baseline by consistent increasing of seaweed supplements (30.1 g/day) due to delay of glucose absorption by seaweeds fibers (Kim et al., 2008) as shown in Table 15.5.

Diabetic patients are disposed to have atherogenic blood lipid profile with increased triglyceride (TG), low high-density lipoproteins (HDL) cholesterol and thick low-density lipoprotein (LDL) cholesterol (Saudek and Eder, 1979; Kilo, 1987). Seaweeds feeding caused lower levels of TG and higher levels of HDL-cholesterol. Small changes were observed in the levels of T-cholesterol, and LDL-cholesterol, however, these changes were nonsignificant statistically as in shown Table 15.6. The mechanism of seaweed fiber to decrease cholesterol is still unknown (Kim et al., 2008).

Schroder (2007) and Schafrir (1996) propose that during the construction intraluminal of micelles, the seaweeds fibers bind with cholesterol or bile acids others, proposed the production of short-chain fatty acids, like acetate, as a fermentation product that could inhibit fatty liver and the alteration of intestine movement reduce macronutrient absorption, increase the sensitivity for insulin that leads to the reduction the total energy uptake (Schroder, 2007).

TABLE 15.5 Blood Glucose Levels Before and After Seaweeds Supplements.

	Baseline	4 wks	Paired t-tast[a]
FBG (mg/dL)			
CG[b]	177.0±14.2[c]	172.9±13.6	NS
SG	181.7±8.2	151.8±8.7	p<0.01
PP-2hrBG (mg/dL)			
CG	266.7±25.8	254.4±22.8[*]	NS
SG	263.2±13.0	203.1±12.3	p<0.05
HbA_1c (%)			
CG	7.9±0.3	7.1±0.4	NS
SG	8.4±0.3	7.7±0.3	NS

[a] Paired t-test was used to compare means by experimental period (baseline vs. 4wk) in each group.

[b] CG: control group, SG: seaweed group.

[c] Data was expressed as meat ± S.E.

[*] Significant difference between groups by independent t-text: $p<0.05$, FBG: Fasting blood glucose level, PP-2hr. BG: Postprandial blood glucose level, HbA_1c: Glycosylated hemoglobin A_1C level

Source: Kim et al. (2008).

15.9 ALGAE AS AN IMMUNE ENHANCER

Bacteria, fungi, and virus are present naturally in the human body as normal flora but when it is exposed to the high infectious micro-organism, the disease occurs (acute infectious, chronic inflammation, and autoimmune diseases) (Jensen and Drapeau, 2001).

The effect of blue-green algae *Aphanizomenonflos aquae* was studied in healthy human volunteers as shown in Table 15.7. They found out it slightly reduces the phagocytic activity of polymorph nucleated cells in vitro that is a sign for the anti-inflammatory relative to antiphagocytic effect on human neutrophils (Jensen et al., 2000).

TABLE 15.6 Blood Lipids Levels Before and After Seaweeds Supplements.

	Baseline	4 wks	Paired t-test[a]
T-C (mg/dL)			
CG[b]	194.4±11.0[c]	187.1±8.3	NS
SG	191.6±6.5	174.4±12.9	NS
TG (mg/dL)			
CG	182.6±24.0	178.3±20.7	NS
SG	171.2±36.7	111.8±17.6*	$p<0.05$
LDL-C (mg/dL)			
CG	121.7±12.4	123.4±13.3	NS
SG	126.1±10.0	95.0±9.0*	NS
HDL-C (mg/dL)			
OG	39.6±2.0	42.0±1.5	NS
SG	37.1±3.2	44.6±2.9	$p<0.05$

[a] Paired t-test was used to compare means by experimental period (baseline vs. 4wk) in each group.
[b] OG: control group, SG : seaweed group.
[c] Data was expressed as mean ± S.E.
* Significant difference between groups (CG vs. SG) by independent t-test; $p<0.05$
T-C: total cholesterol, TG: triglyceride, LDL-C: low density lipoprotein cholesterol HDL-C-high density lipoprotein cholesterol.

Source: Kim et al. (2008).

The effect of algae on building specific immune response was studied by Hayashi et al. (1994) as in Table 15.8. They showed that feeding the mice with the algae-supplemented diet causes an increase the number of splenic IgM antibody-producing cells compared with control and this happened only for a primary immune response while was IgG scarcely affected.

TABLE 15.7 Research on Blue-Green Algae as Biomodulators.

Studies on	Route of administration	Compounds investigated
Human	Oral consumption*	Whole algae
Chicken	Oral consumption	Whole algae
Rodents	Oral consumption**	Whole algae
	Injection	Isolated fractions
	Injection	Purified compounds
In vitro	Added to media.	Isolated fractions
	Added to media	Purified compounds

* Humans: oral dose was 1.5–2.8 grams per day for adult subjects.
** Mice: oral dose varied up to 800 mg/kg.

15.10 THE USE OF ALGAE IN COSMETIC AND THALASSOTHERAPY

A few studies have occurred on the use of microalgae and cyanobacteria on cosmetics that are listed in Table 15.9.

Also, active compounds from algae inhibit blemishes, repair damaged skin, as anti-inflammatory agents. Besides, microalgae extracts have many bioactive substances that increase the skin healing and preserve the moisture of the skin. The antioxidizing activities of colorings produced by microalgae and cyanobacteria could use in the cosmetic industry as natural pigments, such as ß-carotenes from *Dunaliellasalina*; astaxanthin from *Haematococcuspluvialis* (red color) and PCBs (blue color) and phycoerythrobilins from *Spirulina* and *Porphyridium* that could be used in eyeliners and lipsticks (Hamed,2016). Astaxanthin could be used as protective agents against photo-oxidation induction by UV (Koller et al., 2014).

More than 120 examples of cosmetic products present in cosmetic ingredients Database (COSIng) https://ec.europa.eu/growth/sectors/cosmetics/cosing_as shown in Table 15.10.

Thalassotherapy comprises the using of seawater, algae, plankton, sea mud, and sand for therapeutic and preventive applications. Thus, thalassotherapy centers use hydrothermal methods, antistress programs combined with marine cosmetics and message methods to eliminate the negative influence of modern life (Lucchetta et al., 2013).

TABLE 15.8 Immunomodulatory and Anti-inflammatory Effects of Whole Blue-Green Algae.

Algae Species	Introduced as:	Test Species	Effetes	Reference
Spiyitlrna sp.	Food	Human	Reversal of tobacco-induced oral cancer	Mathew et al. 1995
	Food	Mouse	Proportional reduction of IgE, increase of IgA	Hayashi et al. 1998
	Food	Mouse	Increased phagocytic activity Increased spleen cell proliferation Increased antibody production	Hayashi et al, 1994
	Food	Chicken	Increased phagocytic activity Increased NK cell-mediated anti-tumor activity Increased antibody production	Qureshi et al, 1996
	Extract	In vitro, cat	Increased phagocytic activity	Qureshi & Ali, 1996
	IP injection	Rat	Inhibition or must cells Decrease in local allergic reaction Decrease in serum histamine levels Reduced allergy-induced mortality	Kim et al. 1998 Yang et al, 1997
Aphanizomenon flos-aquae	Food	Human	Increased transient recruitment of NK cells into tissue Increased mobilization of T and E cells into blood Mild modulation of PMN-mediated phagocytic response	Jensen et a], 2000
	Food	Rat	Decreased serum levels of arachidonic acid	Kushak et al, 2000
	Food	Rat	Source of linolenic acid (omega-3) Increased serum levels of EPA and DHA	Kushak et al, 2000
	Extract	In vitro rat	Activation of macrophages (NF-kappaB. cytokines)	Pasco, in press

TABLE 15.9 The Use of Bioactive Compounds from Cyanobacteria and Microalgae in Cosmetics.

Bioactive compounds	Microalgae/Cyanobacteria	Potential activities and uses in cosmetics	References
Polysaccharides	Chlorella	Moisturizing and thickener agent	Jain et al. 2005
Methanolic extracts of exopolysaccharides	*Arthrospira platensis*	Antioxidant	Raposo et al. 2015
Chrysolaminarin	*Odontella aurita*	Antioxidant	Xia et al. 2014
Sulfated polysacharides	*Porphyridium and Rhodella reticulata*	Antioxidant	Raposo et al. 2015
β-1,3-Glucan	*Chlorella* *Skeletonema* *Porphyridium* *Nostoc flegelliforme*	Free-radical collector Immune system booster Anti-inflammatory	Spolaore et al. 2006 Roller et al. 2014 Bin et al. 2013 Hamed, 2016
β-carotenes	*Dunaliella salina*	Antioxidant	Hamed, 2016
Asthaxanthin	*Haematococcus pluvialis*	Antioxidant Sunscreen protection	Hamed, 2016 Koller et al. 2014
Phycocyanobilin phycoerythrobilin	*Spirulina Porphxjridium*	Antioxidant pigment for eye-liner and lipsticks	Hamed, 2016
β-Cryptoxanthin	*Dunaliella salina*	Anti-inflammatory Promote Hialuronan synthesis	Tang and Suter, 2011
Chlorophyll	*Chlorella* sp.	To mask odors in dentifrices and deodorants	Hosikian et al. 2010
Canthaxanthin	*Nannochloropsis salina* *Nannochloropsis oculata* *Nannochloropsis gaditana*	Tanning cosmetics and cosmeceutics	Keller et al. 2014

TABLE 15.9 *(Continued)*

Bioactive compounds	Microalgae/Cyanobacteria	Potential activities and uses in cosmetics	References
Phycocyanin	*Porphyridium cruentum* *Spirulina platensis*	Eye-shadows	Bermejo et al. 2003 Arad and Yaron, 1992
Lycopene	*Anabaena vaginicola*	Antioxidant anti-ageing sunscreen	Singh et al. 2012 Hashtroudi et al. 2013
Scytonemin	Marine cyanobacteria	Sunscreen	Takamatsu et al. 2003
Vitamin E (α-Tocopherol)	*Dunaliella tertiolecta* *Tetraselmis suecica*	Antioxidant	Hashtroudi et al.1999
Biopterin glucose	Marine planktonic cyanobacterium	Sunscreen	Matsunaga et al. 1993
Ectoine	*Halomonas elongata* *Halomonas boliviensis* *Brevibacterium epidermis* *Chromohalobacter israelensis* *Chromohalobacter salexigens*	Immune protection UV protection Stress protection Moisturizing agent	Kim et al. 2008 Shivanand and Mugeraya, 2011
Phytohormones (auxin, absrisic acid, cytokinin ethylene, gibberellins)	Broad spectrum of microalgal lineages *Nannochloropsis oceanica*	Anti-ageing	Lu and Xu, 2015 Michelet et al. 2012
Micosporine-like amino acids	Cyanobacteria	Sunscreen	Llewellyn and Airs, 2010 Singh, 2017

Source: Mourelle et al. (2017).

TABLE 15.10 INCI Names (International Nomenclature of Cosmetic Ingredients) and CAS Number (Chemical Abstracts Service) of Cosmetic Substances and Ingredients on CosIng (Cosmetic Ingredient Database).

INCI name	Description
ALGAE EXOPOLYSACCHARIDES	Algae Exopolysaccharides are exopolysaccharides released by the fermentation of various species of microalgae of the divisions, *Rhodophyta* and *Chlorophyta.*
PARACHLORELLA BEIJERINCKII EXOPOLYSACCHARIDES	Parachlorella Beijerinckii Exopolysaccharides are exopolysaccharides produced through the fermentation of the microalgae *Parachlorella beijerinckii.*
APHANOTHECE SACRUM POWER	Anacardoyl Tripeptide-35 is the product obtained by the reaction of anacardic acid and Tripeptide-35, Cyanobacteriaceae.
ARTHROSPIRA EXTRACT	Arthrospira Extract is an extract of the Cyanobacterium, *Arthrospira maximum.*
ARTHROSPIRA PLATENSIS CULTURE CONDITIONED MEDIA	Arthrospira Platensis Culture Conditioned Media is the growth media removed from cultures of *Arthrospira platensis* after several days of growth, Phormidiaceae.
CYANOBACTERIUM APONINUM FERMENT	Cyanobacterium Aponinum Ferment is the product obtained by the fermentation of *Cyanobacterium aponinum.*
ODONTELLA AURITA EXTRACT	Odontella Aurita Extract is an extract of the phytoplankton, *Odontella aurita,* Bacillariophyceae.
ODONTELLA AURITA OIL	Odontella Aurita Oil is the oil obtained from the phytoplankton, *Odontella aurita,* Bacillariophyceae.
PARALLELOSTROMBIDIUM SICULUM EXTRACT	Parallelostrombidium Siculum Extract is the extract of the plankton, *Parallelostrombidium siculum,* Strombidiidae
PLANKTON EXTRACT (CAS NO. 91079–57–1)	Plankton Extract is an extract obtained from marine plankton.
TETRASELMIS SUECICA EXTRACT	Tetraselmis Suecica Extract is an extract of the plankton *Tetraselmis suecica.*
CHLORELLA ELLIPSOIDEA EXTRACT	Chlorella Ellipsoidea Extract is the extract of the alga, *Chlorella ellipsoidea,* Chlorellaceae.

TABLE 15.10 *(Continued)*

INCI name	Description
CHLORELLA EMERSONII EXTRACT (CAS NO. 223749–78–8)	Chlorella Emersonii Extract is an extract of the Alga, *Chlorella emersonii,* Oocystaceae.
CHLORELLA VULGARIS CALLUS CULTURE EXTRACT	Chlorella Vulgaris Callus Culture Extract is the extract of a culture of the callus of the alga, *Chlorella vulgaris,* Chlorellaceae.
PHORMIDIUM UNCINATUM EXTRACT	Phormidium Uncinatum Extract is the extract of the Alga, *Phormidium uncinatum.*
PHORMIDIUM FERMENT	Phormidium Ferment is the product obtained through fermentation by the microorganism, *Phormidium.*
APHANIZOMENON FLOS-AQUAE EXTRACT	Aphanizomenon Flos-Aquae Extract is the extract of the alga, *Aphanizomenon flos-aquae,* Nostocaceae.
ALPHA-GLUCAN OLIGOSACCHARIDE	Alpha-Glucan Oligosaccharide is a glucose oligomer exhibiting a degree of polymerization ranging from 2–10. It is prepared by the action of a Leuconostoc mesenteroides glucosyl transferase on sucrose.
NOSTOC FLAGELLIFORME EXTRACT	Nostoc Flagelliforme Extract is an extract of the Alga, *Nostoc flagelliforme,* Nostocaceae.
DUNALILLA SALINA/HAEMATOCOCCUS PLUVIALIS EXTRACT	Dunaliella Salina/Haematococcus Pluvialis Extract is the extract of the alga, *Dunaliella salina* and *Haematococcus pluvialis.*
PORPHYRIDIUM CRUENTUM EXTRACT (CAS No. 223751-77-7)	Porphyridium Cruentum Extract is an extract of the Alga, *Porphyridium cruentum,* Porphyridaceae.
THALASSIOSIRA PSEUDONANA EXTRACT	Thalassiosira Pseudonana Extract is the extract of the alga, *Thalassiosira pseudonana.*
NANNOCHLOROPSIS OCULATA POWDER	Nannochloropsis Oculata Powder is the powder obtained from the dried, ground alga, *Nannochloropsis oculata.*

TABLE 15.10 *(Continued)*

INCI name	Description
SKELETONEMA COSTATUM EXTRACT	Skeletonema Costatum Extract is an extract of *Skeletonema costatum*.
DUNALIELLA SALINA/ISOCHRYSIS GALBANA/NANNOCHLOROPSIS GADITANA/ PHAEODACTYLUM TRICORNUTUM/ TETRASELMIS CHUII EXTRACT	Dunaliella Salina/Isochrysis Galbana/Nannochloropsis Gaditana/Phaeodactylum Tricornutum/Tetraselmis Chuii Extract is the extract of the whole plants of *Dunaliella salina, Isochrysis galbana, nannochloropsis gaditana, Phaeodactylum Tricornutum* and *Tetraselmis chuii*.
PSEUDANABAENACEAE FERMENT FILTRATE	Pseudanabaenaceae Ferment Filtrate is a filtrate of the product obtained by the fermentation of a growth media by the microorganism, Pseudanabaenaceae.
RHODELLA VIOLACEA EXTRACT	Rhodella Violacea Extract is the extract of the alga, *Rhodella violacea,* Rhodellaceae.
HALOMONAS ANTICARIENSIS FERMENT EXTRACT FILTRATE	Halomonas Anticariensis Ferment Extract Filtrate is a filtrate of an extract of the product obtained by the fermentation of the bacteria, *Halomonas anticariensis*.

Source: Mourelle et al. (2017).

KEYWORDS

- **algae biodiversity and medicine**
- **algae as antimicrobial**
- **anti-inflammatory**
- **antidiabetic**
- **cosmetic**

REFERENCES

Al-Mofleh, I. A.; Al-Rashed, R. S.; Nonsteroidal, Anti-inflammatory Drug-induced Gastrointestinal Injuries, and Related Adverse Reactions: Epidemiology, Pathogenesis, and Management (2007). *Saudi J. Gastroenterol.* **2007,** *13* (3), 107–113.

Athukorala, Y.; Jung, W. K.; Vasanthan, T.; Jeon, Y. J. An Anticoagulative Polysaccharide from an Enzymatic Hydrolysate of Ecklonia cava. *Carbohydr. Polym.* **2006,** *66*, 184–191.

Bhat, V. B.; Madyastha, K. M. C-phycocyanin: A Potent Peroxyl Radical Scavenger In Vivo and In Vitro. *Biochem. Biophys. Res. Commun.* **2000,** *275* (1), 20–25.

Bin, S.; Wang, Z. P.; Liu, X. Y.; Yu, J. X.; Lan, J. J.; Wang, J. M.; Ma, L. F.; Chen, Z. Y. Breeding of a Chlorella Strain with High Yield of Polysaccharide and its Effect on Growth and Immunoregulation of Litopenaeus vannamei, *J.Nucl. Agric. Sci.* **2013,** *27*, 168–172.

Bouhlal, R.; C. Haslin, J. C.; Chermann, S.; Colliec-jouault, C.; Sinquin, G.; Simon, S.; Cerantola, H.; Riadi Bourgougnon, N. Antiviral Activities of Sulfated Polysaccharides Isolated from *Sphaerococcus coronopifolius* (*Rhodophytha, Gigartinales*) and *Boergeseniellathuyoides* (Rhodophyta, Ceramiales). *Marine Drug.* **2011,** *9*, 1187–1209.

Boopathy, N. S.; Kathiresan, K. Anticancer Drugs from Marine Flora: An Overview. *J. Oncol.* **2011**, 1–18 https://www.hindawi.com/journals/jo/2010/214186/abs/.

Boyd M. Protein Isolated from Blue-green Algae Inactivates HIV. Antimicrob. Agents Chemother. **1997,** *41*, 1521–1530.

Cabrita, M. T.; Vale, C.; Rauter, A. P. Halogenated Compounds from Marine Algae. *Mar. Drugs* **2010,** *8*, 2301–2317.

Campo, V. L.; Kawano, D. F.; Silva, D. B.; Carvalho, I. Carrageenans: Biological Properties, Chemical Modifications and Structural Analysis–A Review. *Carbohydr. Polym.* **2009,** *77*, 167–180.

Dayong, S.; Jing, L.; Shuju, G.;Lijun, H. Antithrombotic Effect of Romophenol, The Alga-derived Thrombin Inhibitor. *J. Biotechnol.* **2008,** *136*, 577–588.

Dillon, J. C.; Phuc, A. P.; Dubacq, J. P. Nutritional Value of the Algae Spirulina. *World Rev. Nutr. Diet.* **1995,** *77*, 32–46.

De Felício, R.; De Albuquerque, S.; Young, M. C. M.; Yokoya, N. S.; Debonsi, H. M. Trypanocidal, Leishmanicidal and Antifungal Potential from Marine Red Alga *Bostrychia tenella* J. Agardh (Rhodomelaceae, Ceramiales). *J. Pharma. Biomed. Ana.* **2010,** *52*, 763–769.

Falch, B.; What Remains of Cyanobacteria? *Pharm. i. u. Zeit.* **1996,** *25*, 311–319.

Farias, W. R. L.; Valente, A. P.; Pereira, M. S.; Mourão, P. A. S. Structure and Anticoagulant Activity of Galactans: Isolation of a Unique Sulfated Galactan from the Red Algae *Botryocladia occidentalis* and Comparison of its Anticoagulant Action with that of Sulfated Galactans from Invertebrates. *J. Biol.Chem.* **2000,** *275*, 29299–29307.

Fonseca, R. J. C.; Oliveira, S. N. M. C. G.; Melo, F. R.; Pereira, M.G.; Benevides, N. M. B.; Mourão, P. A. S. Slight Differences in Sulfatation Ofalgal Galactanas Account for Differences in their Anticoagulant and Venous Antithrombotic Activities. *Thromb. Haemost.* **2008,** *99*, 539–545.

Graham, D. Y.; Helicobacter Pylori Infection in the Pathogenesis of Duodenal Ulcer and Gastric Cancer: A Model. *Gastroenterol* **1997,** *113* (6), 1983–1991.

Gonzalez, R.; Rodriguez, S.; Romay, C.; Ancheta, O.; Gonzalez, A.; Armesta, J.; Remirez, D.; Merino, N. Anti-inflammatory Activity of Phycocyanin Extract in Acetic Acid-induced Colitis in Rats. *Pharmacol. Res.* **1999,** *39* (1), 55–59.

Gustafson, K.; Cardellina II, J.; et al. AIDS-antiviral Sulfolipids from Cyanobacteria (blue-green algae). *J. National Cancer Institute* **1989,** *81*, 1254–1258.

Hayashi, O.; Katoh, T.; Enhancement of Antibody Production in Mice by Dietary *Spirulina platensis*. *J. Nutr. Sci. Vitaminol.* **1994,** *40*, 431–441.

Hayashi, K.; Hayashi, T. A Natural Sulfated Polysaccharide, Calcium Spirulan, Isolated from *Spirulina platensis*: In Vitro and Ex Vivo Evaluation of Anti-herpes Simplex Virus and Antihuman Immunodeficiency Virus Activities. AIDS Res. Hum. Retrovir. **1996,** *12* (15), 1463–1471.

Halliwell, B. How to Characterize a Biological Antioxidant *Free Rad. Res. Comm.* **1990,** *9*, 1–32.

Hamed, I. The Evolution and Versatility of Microalgal Biotechnology: A Review. *Compr. Rev. Food Sci. Food Saf.* **2016,** *15*, 1104–1123.

Hashtroudi, M. S.; Shariatmadari, Z.; Riahi, H.; Ghassempour, A. Analysis of *Anabaena vaginicola* and *Nostoc calcicola* from Northern Iran, as Rich Sources of Major Carotenoids. *Food Chem.* **2013,** *136*, 1148–1153.

Jensen, G. S.; Ginsberg, D. I. Consumption of Aphanizomenon flos-aquae has Rapid *JANA*, Winter on the Circulation and Function of Immune Cells in Humans. *JANA* **2000,** *2* (3), 50–58.

Jensen, G. S.; Drapeau, C. Blue Green Algae as an Immune Enhancer and Biomodulator. *Winter* **2001,** *3* (4), 24–30.

Jaganathan, S. K.; Mandal, M.; Antiproliferative Effects of Honey and of Its Polyphenols. A Review. *J. Biomed. Biotechnol.* **2009,** *830616*, 1–13.

Jaki, B.; Orjala, J.; Sticher, O. A Novel Extracellular Diterpenoid with Antibacterial Activity from the Cyanobacterium *Nostoc commune*. *J. Nat. Prod.* **1999,** *62*, 502–503.

Jaki, B.; Orjala, J.; Heilmann, J.; Linden, A.; Vogler, B.; Sticher, O. Novel Extracellular Diterpenoid with Biological Activity from the Cyanobacterium *Nostoc commune*. *J. Nat. Prod.* **2000,** *63*, 339–343.

Jaki, B.; Zerbe, O.; Heilmann, J.; Sticher, O. Two Novel Cyclic Peptides with Antifungal Activity from the Cyanobacterium Tolypothrix byssoidea (EAWAG 195). *J. Nat. Prod.* **2001,** *64*, 154–158.

Jain, R.; Raghukumar, S.; Tharanathan, R.; Bhosle, N. B. Extracellular Polysaccharide Production by Thraustochytrid Protists. *Mar. Biotechnol.* **2005,** *7*, 184–192.

Jirascova, A.; Novotny, J.; Novotny, L.; Vodicka, P; Pardini, B.; et al. Association of Serum Bilirubin and Promotor Variations in HMOX1 and UGT1A1 Genes with sporadic Colorectal Cancer. *Int. J. Cancer* **2012,** *131*, 1549–1555.

Kang J. Y.; Khan, M. N. A.; Park N. H.; Cho, J. Y.; Lee, M. C.; Fujii, H.; Hong, Y. K. Antipyretic, Analgesic and Anti-inflammatory Activities of the Seaweed *Sargassum fulvellum* and *Sargassum thunbergii* in Mice *J Ethnopharmacol* **2008,** *116*, 187–190.

Karnjanapratum, S.; You, S. Molecular Characteristics of Sulfated Polysaccharides from *Monostroma nitidum* and Their In Vitro Anticancer and Immunomodulatory Activities. *Inter. J. Biol. Macromol.* **2011,** *48*, 311–318.

Khan, M. N. A.; Choi, J. S.; Lee, M. C.; Kim, E.; Nam, T. J.; Fujii, H.; Hong, Y. K. Anti-inflammatory Activities of Methanol Extracts from Various Seaweed Species. *J. Environ. Biol.* **2008,** *29* (4), 465–469.

Kilo, C. Vascular Complications of Diabetes. *Cardiovasc. Rev. Rep.* **1987,** *8*, 18–24.

Kim, S. K.; Karadeniz, F. Anti- HIV Activity of Extracts and Compounds from Marine Algae. *Adv. Food. Nutri. Res.* **2011,** *64*, 255–265.

Kim, S. K.; Thomas, N. V.; Li, X. Anticancer Compounds from Marine Macroalgae and Their Application as Medicinal Foods. *Adv. Food. Nutri. Res.* **2011,** *64*, 213–224.

Kim, M. S.; Kim, J. Y.; Choi, W. H.; Lee, S. S. Effects of Seaweed Supplementation on Blood Glucose Concentration, Lipid Profile, and Antioxidant Enzyme Activities in Patients with type 2 Diabetes Mellitus. *Nutr. Res. Prac.* **2008,** *2* (2), 62–67.

Kim, S. K.; Ravichandran, Y. D.; Khan, S. B.; Kim, Y. T. Prospective of the Cosmeceuticals Derived from Marine Organisms. *Biotechnol. Bioprocess Eng.* **2008,** *13*, 511–523.

Kohen, R.; Nyska, A. Oxidationof Biological Systems: Oxidative Stress Phenomena, Antioxidants, Redox Reactions, and Method for their Quantification. *Toxicol. Pathol.* **2002,** *30* (6), 620–650.

Koller, M.; Muhr, A.; Braunegg, G. Microalgae as Versatile Cellular Factories for Valued Products. *Algal Res.* **2014,** *6*, 52–63.

Konickova, R.; Vankova, K.; Vankova, J.; Vanova, K., et al. Anticancer Effects of Blue Green Algae *S. patensis*, is a Natural Source of Bilirubin like Tetrpyrrolic Compounds. *A. Hepatol.* **2014,** *13* (2), 271–283.

Kwon, M. J.; Nam T. J. A Polysaccharide of the Marine Alga *Capsosiphon fulvescens* Induces Apoptosis in AGS Gastric Cancer Cells Via an IGF-IR -mediated PI3K/Akt Pathway. *Cell Biol. Int.* **2007,** *31* (8), 768–775.

Lee, H. J.; Dang, H. T.; Kang, G. J.; Yang, E. J.; Park, S. S.; Yoonm, W. J.; Jung, J. H.; Kang, H. K.; Yoo, E. S. Two Enone Fatty Acids Isolated from *Gracilaria verrucosa* Suppress the Production of Inflammatory Mediators by Down-regulating NF-κB and STAT1 Activity in Lipopolysaccharide-stimulated Raw 264.7 Cells. *Arch. Pharmacal. Res.* **2009,** *32* (3), 453–462.

Llewellyn, C.A.; Airs, R.L. Distribution and Abundance of MAAs in 33 Species of Microalgae Across 13 Classes. *Mar. Drugs* **2010,** *8*, 1273–1291.

Lu, Y.; Xu, J. Phytohormones in Microalgae a New Opportunity for Microalgal Biotechnology? *Trends Plant Sci.* **2015,** *20*, 273–282.

Lucchetta, M. C.; Monaco, G.; Valenzi, V. I.; Fraioli, A. The Historical-scientific Bases of Thalassotherapy: State of the Art. *Clin. Ter.* **2013,** *158*, 533–541.

Matanjun, P.; Mohamed, S.; Mustapha, N.; Muhammad, K.; Ming, C. Antioxidant Activities and Phenolics Content of Eight Species of Seaweeds from North Borneo. *J. Appl. Phycol.* **2008,** *20*, 367–373.

Mabeau S., Fleurence J. Seaweed in Food Products: Biochemical and Nutritional Aspects. *Trends Food Sci. Technol.* **1993,** *4*, 103–107.

Mao, W. J.; Fang, F.; Li, H. Y.; Qi, X. H.; Sun, H. H.; Chen, Y.; Guo, S. D. Heparinoid-Active Two Sulfated Polysaccharides Isolated from Marine Green Algae *Monostroma nitidum. Carbohydr. Polym.* **2008,** *74*, 834–839.

Matsubara, K.; Matsubara, Y.; Hori, K.; Miyazawa K. An Anticoagulant Proteoglycan from Thmarine Green Alga, Codium pugniformis. *J. App.Phycol.* **2000,** *12*, 9–14.

Matsunaga, T.; Burgess, J. G.; Yamada, N.; Komatsu, K.; Yoshida, S.; Wachi, Y. An Ultra-violet (UV-A) Absorbing Biopterin Glucoside from the Marine Planktonic Cyanobacterium Oscillatoria sp. *Appl. Microbiol. Biotechnol.* **1993,** *39*, 250–253.

Melo, M. R. S.; Feitosa, J. P. A.; Freitas, A. L. P.; Paula, R. C. M. Isolation and Characterization of Soluble Sulfated Polysaccharide from the Red Seaweed *Gracilaria cornea. Carbohydr. Polym.* **2002,** *49*, 491–498.

Michelet, J. F.; Olive, C.; Rieux, E.; Fagot, D.; Simonetti, L.; Galey, J. B.; Dalko-Csiba, M.; Bernard, B. A.; Pereira, R. The Anti-ageing Potential of a New Jasmonic Acid Derivative (LR2412): In Vitro Evaluation Using Reconstructed Epidermis EpiSkin™. *Exp. Dermatol.* **2012,** *21*, 390–400.

Mishima, T.; Murata, J.; et al. Inhibition of Tumor Invasion and Metastasis by Calcium Spirulan (Ca-SP), a Novel Sulfated Polysaccharide Derived from a Blue-green algae, *Spirulina platensis. Clin. Exp. Metastasis* **1998,** *16* (6), 541–550.

Mourão, P. A. S. Use of Sulfated fucans as Anticoagulant and Antithrombotic Agents: Future Perspectives. *Curr. Pharm. Des.* **2004,** *10*, 967–981.

Mourella, M. L. Gomaz, C. P. Legido, J. L. The Potential Use of Microalgae and Cyanobacteria in Cosmetics and Thalassotherapy. *Cosmetics* **2017,** *4* (46), 1–14.

Murray, H. G.; Munro 1; Richard, T.; Luibrand 2; John W. The Search for Antiviral and Anticancer Compounds from Marine Organisms, Bioorganic marine chemistry. Springer: Berlin, Heidelberg, 1987; pp 93–176. 2004.

Na, H. J.; Moon, P.D.; Lee, H. J.; Kim, H. R.; Chae, H. J.; Shin, T.; Seo, Y.; Hong, S. H.; Kim, H. M. Regulatory Effect of Atopic Allergic Reaction by Carpopeltisaffinis. *J. Ethnopharmacol.* **2005,** 101, 43–48.

Novotony, L.; Vitek, L. Inverse Relationship Between Serum Bilirubin and Atherosclerosis in Men: A Meta-analysis of Published Studies. *EXP. Biol. Med.* **2003,** *228*, 568–571.

Nazarewicz, R. R.; Dikalova, A.; Bikineyeva, A.; et al. Does Scavenging of Mitochondrial Superoxide Attenuate Cancer Prosurvive Signaling Pathway? *Antiox. Redox Signal.* **2013,** *19*, 344–349.

Okuzumi, J.; Takahashi, T.; Yamane, T.; Kitao, Y.; Inagake, M.; Ohya, K.; Nishino, H.; Tanaka, Y.; Inhibitory Effects of Fucoxanthin, a Natural Carotenoid, on N-ethyl-N'-nitro-N-nitrosoguanidine Induced Mouse Duodenal Carcinogenesis. *Cancer Lett.* **1993,** *68*, 159–168.

Ouhtit, A.; Ismail, M. F.; Osman, A.; Fernando, A.; et al. Chemopreventive of Rat Mammary Carcinogensis by Spirulina. *Am. J. Pathol.* **2013,** {EPub ahead of print}.

Padyana, A. K.; Bhat, V. B.; Madyastha, K. M.; Rajashankar, K. R.; Ramakumar S. Crystal Structure of a Light Harvesting Protein C-phycocyanin from *Spirulina platensis. Biochembiophys. Res commun.* **2001,** *282*, 568–71.

Parada, J. L.; Zulpa de Caire, G.; et al. Lactic Acid Bacteria Growth Promoters from *Spirulina platensis. Int. J. Food Microbiol.* **1998,** *45* (3), 225–228.

Patterson, G. M. L.; Larsen, L. K.; Moore, R. E.; Bioactive Natural Products from Blue–green Algae. *J. Appl. Phycol.* **1994,** *6*, 151–157.

Pereira, M. G.; Benevides, N. M. B.; Melo, M. R. S.; Valente, A. P.; Melo, F. R.; Mourão, P. A. S. Structure and Anticoagulant Activity of a Sulfated Galactan from the Red Alga, Gelidiumcrinale. Is there a Specific Structural Requirement for the Anticoagulant Action? *Carbohydr. Res.* **2005,** *340*, 2015–2023.

Pereira, M. S.; Melo, F. R. and Mourão, P. A. S. Is There a Correlation Between Structure and Anticoagulant Action of Sulfated Galactans and Sulfated Fucans? *Glycobiology* **2002,** *12*, 573–580.

Pushpamali, W. A.; Nikapitiya, C.; Zoysa, M. D.; Whang, I.; Kim, S. J.; Lee, J. Isolation and Purification of an Anticoagulant from Fermented Red seaweed *Lomentaria catenata. Carbohydr. Polym.* **2008,** *73*, 274–279.

Rodrigues, J. A. G.; Torres, V. M.; Alencar, D. B.; Sampaio, A. H. and Farias, W. R. L. Extraçãoeatividadeanticoagulante dos polissacarídeossulfatados da alga marinhavermelha *Halymenia pseudofloresia. Rev. Ciên. Agron.* **2009,** *40*, 224–231.

Prashantkumar, P.; Angadi, S. B.; Vidyasagar, G. Antimicrobial Activity of Blue Green Algae and Green Algae. *Indian J. Pharm. Sci.* **2006,** *68* (5), 647–648.

Pulz, O.; Gross, W. Valuable Products from Biotechnology of Microalgae. *Appl. Microbiol. Biotechnol.* **2004,** *65* (6), 635–648.

Rankovic, B. R.; Kosanic, M. M.; Stanojkovic, T. P. Antioxidant, Antimicrobial and Anticancer Activity of the Lichens *Cladoniafurcata, Lecanoraatra,* and *Lecanoramuralis. BMC Complement Alter. Med.* **2012,** *11*, 97.

Richardson, J. S. Free Radicals in the Genesis of Alzheimer's Disease. *Annals New York Acad. Sci.* **1993,** *695*, 73–76.

Romay, C.; Armesto, J.; Remirez, D.; Gonza, R.; Ledon, N.; Garci, I. Antioxidant and Anti-inflammatory Properties of C-phycocyanin from Blue-green algae. *Inflamm. Res.* **1998,** *47*, 36–41.

Romay, C.; Ledon, N.; Gonzalez, R.; Phycocyanin Extract Reduces Leukotriene B4 Levels in Arachidonic Acid-induced Mouse-ear inflammation Test. *J Pharm Pharmacol.* **1999,** *51* (5), 641–642.

Raposo, M. F. J.; Morais, A. M. M. B.; Morais, R. M. Marine Polysaccharides from Algae with Potential Biomedical Applications. *Mar. Drugs* **2015,** *13*, 2967–3028.

Reddy, C. M.; Bhat, V. B.; Selective Inhibition of Cyclooxygenase-2 by C-phycocyanin, a Biliprotein from *Spirulina platensis Biochem. Biophys. Res. Commun.* **2000,** *277* (3), 599–603.

Saudek, C. D. Eder, H. A. Lipid Metabolism in Diabetes Mellitus. *Am. J. Med.* **1979,** *66*, 843–852.

Schroder, H. Protective Mechanisms of the Mediterranean Diet in Obesity and Type 2 Diabetes. *J. Nutr. Biochem.* **2007,** *18*, 149–160.

Serkedjieva, J. Antiviral Activity of the Red Marine Algae *Ceramium rubrum*. *Phytother. Res.* **2004,** *18*, 480–483.
Shafrir, E. Development and Consequences of Insulin Resistance: Lessons from Animals with Hyperinsulinaemia. *Diabetes Metab.* **1996,** *22*, 122–131.
Shivanand, P.; Mugeraya, G. Halophilic Bacteria and Their Compatible Solutes—Osmoregulation and Potential Applications. *Curr. Sci.* **2011,** *100*, 1516–1521.
Shu, M.; Appleton, D.; Zandi, K.; Abubakar, S. (2013) Anti-inflammatory, Gastroprotective, and Antiulcerogenic Effects of Red Algae *Gracilaria changii* (Gracilariales, Rhodophyta) Extract. *Complementary and Alternative Medicine BMD* **2013,** *13* (61), 1–13.
Singh, P.; Rani, B.; Chauhan, A.K.; Maheshwari, R. Lycopene's Antioxidant Activity in Cosmetics Meadow. *IRJP* **2013,** *3*, 46–4.
Singh, R.; Parihar, P.; Singh, M.; Bajguz, A.; Kumar, J.; Singh, S.; Singh, V. P.; Prasad, S. M. Uncovering Potential Applications of Cyanobacteria and Algal Metabolites in Biology, Agriculture and Medicine: Current Status and Future Prospects. *Front. Microbiol.* **2017,** *8*, 1–37.
Spolaore, P.; Joannis-Cassan, C.; Duran, E.; Isambert, A. Commercial applications of microalgae. *J. Biosci. Bioeng.* **2006,** *101*, 87–96.
Sreenivasan, S.; Ibrahim, D.; Mohd-Kassim, M. J. N. Free Radical Scavenging Activity and Total Phenolic Compounds of *Gracilariachangii*. *Int. J. Nat. Eng. Sci.* **2007,** *1* (3), 115–117.
Takamatsu, S.; Hodges, T. W.; Rajbhandari, I.; Gerwick, W. H.; Hamann, M. T.; Nagle, D. G. Marine Natural Products as Novel Antioxidant Prototypes. *J. Nat. Prod.* **2003,** *66*, 605–608.
Tapiero, H., Tew, K. D., Naguyen, G., Mathe, G. Polyphenods: Do They Play a Role in the Prevention of Human Pathologies? *Review. Biomed. Pharmacother*. **2002,** *56*, 200–207.
Tang, G.; Suter, P. M. Vitamin A, Nutrition, and Health Values of Algae: Spirulina, Chlorella, and Dunaliella. *J. Pharm. Nutr. Sci.* **2011,** *1*, 111–118.
Tokuda, H.; Nishino, H.; et al. Inhibition of 12-O-tetradecanoylphorbol-13-acetate Promoted Mouse Skin Papilloma by Digalactosyl Diacylglycerols from the Fresh Water Cyanobacterium *Phormidiumtenue*. *Cancer Lett.* **1996,** *104* (1), 91–95.
Unnikrishnan, P. S.; Jayasri, M. A. Marine Algae as Prospective Source for Antidiabetic Compounds. *Curr. Diabetes Rev.* **2018,** *14* (3), 237–245.
Vadiraja B. B.; Gaikwad, N. W.; Madyastha, K. M. Hepatoprotective Effect of C-phycocyanin: Protection for Carbon Tetrachloride and R-(+)-pulegone-mediated Hepatotoxicity in Rats. *Biochem. Biophys. Res. Commun.* **1998,** *249* (2), 428–431
Vitek, l.; Muchova, L.; Jancova, E.; Pesickova, S. Association of Systemic Lupus Erythematosus with Low Serum Bilirubin Levels. *Scand J. Heumatol.* **2010,** *39*, 480–484.
Volk, R.-B.; Furker, F.H. Antialgal, Antibacterial and Antifungal Activity of Two Metabolites Produced and Excreted by Cyanobacteria During Growth. *Microbiol. Res.* **2006,** *161*, 180–186.
Wakefield, D.; Kumar R. K. Inflammation: Chronic. In *Encyclopedia of Life Sciences* (ELS); Nature Publishing Group: London, 2001.
Williamson, G.; Carughi, A. Polyphenol Content and Health Benefits of Raisins. *Nutr. Res.* **2010,** *30* (8), 511–519.

Wallhaüßer, K. H.; **1995**. Antimikrobielle Wirkstoffe. In *Praxis der Sterilisation Desinfektion-Konservierung;* Wallhaüßer, K. H., Ed.; Georg Thieme Verlag: Stuttgart, 1995; pp 550–552.

Xia, S.; Gao, B.; Li, A.; Xiong, J.; Ao, Z.; Zhang, C. W. Preliminary Characterization, Antioxidant Propertiesand Production of Chrysolaminarin from Marine Diatom *Odontella aurita Mar. Drugs* **2014,** *12*, 4883.

Zhang, H. J.; Mao, W. J.; Fang, F.; Li, H. Y.; Sun, H. H.; Chen, Y.; Qi, X. H. Chemical Characteristics and Anticoagulant Activities of a Sulfated Polysaccharide and its Fragments from *Monostromalatissimum. Carbohydr. Polym.* **2008,** *71*, 428–434.

Index

B

C

D

F

G

H

I

R

S

T

W